Foundations in Sound Design for Linear Media

This volume provides a comprehensive introduction to foundational topics in sound design for linear media, such as listening and recording; audio postproduction; key musical concepts and forms such as harmony, conceptual sound design, electronica, soundscape and electroacoustic composition; the audio commons; and sound's ontology and phenomenology.

The reader will gain a broad understanding of the key concepts and practices that define sound design for its use with moving images as well as important forms of composed sound. The chapters are written by international authors from diverse backgrounds who provide multidisciplinary perspectives on sound in its linear forms.

The volume is designed as a textbook for students and teachers, as a handbook for researchers in sound, media and experience, and as a survey of key trends and ideas for practitioners interested in exploring the boundaries of their profession.

Michael Filimowicz, PhD, is Senior Lecturer in the School of Interactive Arts and Technology (SIAT) at Simon Fraser University and coeditor of *The Soundtrack* journal. He develops new forms of general-purpose multimodal and audiovisual display technology, exploring novel product lines across different application contexts including gaming, immersive exhibitions, control rooms, telepresence and simulation-based training. He has published across disciplines in journals such as *Organised Sound, Arts and Humanities in Higher Education, Leonardo, Sound Effects, Parsons Journal for Information Mapping* and *Semiotica*. His art has been exhibited internationally at venues such as SIGGRAPH, Re-New, Design Shanghai, ARTECH, Les Instants Vidéo, IDEAS, Kinsey Institute and Art Currents, and published in monographs such as *Spotlight: 20 Years of the Biel/Bienne Festival of Photography*, *Reframing Photography* and *Infinite Instances*. His personal website is http://filimowi.cz.

Sound Design

The Sound Design series takes a comprehensive and multidisciplinary view of the field of sound design across linear, interactive and embedded media and design contexts. Today's sound designers might work in film and video, installation and performance, auditory displays and interface design, electroacoustic composition and software applications and beyond. These forms and practices continuously cross-pollinate and produce an ever changing array of technologies and techniques for audiences and users, which the series aims to represent and foster.

Series Editor
Michael Filimowicz

Titles in the Series

Foundations in Sound Design for Linear Media

A Multidisciplinary Approach

Foundations in Sound Design for Interactive Media

A Multidisciplinary Approach

Foundations in Sound Design for Embedded Media

A Multidisciplinary Approach

For more information about this series, please visit: www.routledge.com/Sound-Design/book-series/SDS

Foundations in Sound Design for Linear Media

A Multidisciplinary Approach

Edited by Michael Filimowicz

Routledge
Taylor & Francis Group

NEW YORK AND LONDON

First published 2020
by Routledge
605 Third Avenue, New York, NY 10017

and by Routledge
2 Park Square, Milton Park, Abingdon, Oxon, OX14 4RN

Routledge is an imprint of the Taylor & Francis Group, an informa business

© 2020 Taylor & Francis

The right of Michael Filimowicz to be identified as the author of the editorial material, and of the authors for their individual chapters, has been asserted in accordance with sections 77 and 78 of the Copyright, Designs and Patents Act 1988.

Library of Congress Cataloging-in-Publication Data
Names: Filimowicz, Michael, author.
Title: Foundations in sound design for linear media : a multidisciplinary
    approach / Michael Filimowicz.
Description: New York : Routledge, 2019. | Series: Sound design series;
    volume 1 | Includes bibliographical references and index.
Identifiers: LCCN 2019004240| ISBN 9781138093959 (hbk) |
    ISBN 9781138093966 (pbk) | ISBN 9781351603812 (mobi) |
    ISBN 9781351603829 (epub3) | ISBN 9781315106335 (ebk)
Subjects: LCSH: Sound—Recording and reproducing. | Audio-visual
    materials—Design. | Sounds. | Music.
Classification: LCC TK7881.4 .F55 2019 | DDC 621.389/3—dc23
LC record available at https://lccn.loc.gov/2019004240

ISBN: 978-1-138-09395-9 (hbk)
ISBN: 978-1-138-09396-6 (pbk)
ISBN: 978-1-315-10633-5 (ebk)

Typeset in Times New Roman
by Apex CoVantage, LLC

DOI: 10.4324/9781315106335

Contents

List of Contributors   vii
Series Preface   xiii
    MICHAEL FILIMOWICZ

Volume Introduction   xv
    MICHAEL FILIMOWICZ

1   The Nature of Sound and Recording   1
    ANDREW KNIGHT-HILL

2   Invisible Seams: The Role of Foley and Voice
    Postproduction Recordings in the Design
    of Cinematic Performances   61
    SANDRA PAULETTO

3   Media Management, Sound Editing and Mixing   82
    GEORGE KALLIRIS, CHARALAMPOS A. DIMOULAS,
    AND MARIA MATSIOLA

4   Audio Effects in Sound Design   113
    BRECHT DE MAN

5   The Mix Stems: Voice, Effects, Music, Buses   129
    NEIL HILLMAN

6   Mixing and Mastering   152
    PAUL GELUSO

7   Designing Sound for 3D Films   191
    DAMIAN CANDUSSO

  8   Compositional Techniques for Sound Design   211
      SARAH PICKETT

  9   Music Theory for Sound Designers   227
      ADAM MELVIN AND BRIAN BRIDGES

 10   Leveraging Online Audio Commons Content
      for Media Production   248
      ANNA XAMBÓ, FREDERIC FONT, GYÖRGY FAZEKAS,
      AND MATHIEU BARTHET

 11   Sound Ontologies: Methods and Approaches for the
      Description of Sound   283
      DAVIDE ANDREA MAURO AND ANDREA VALLE

 12   Electroacoustic Music: An Art of Sound   303
      ANDREW KNIGHT-HILL

 13   Electronic Dance Music in Narrative Film   327
      ROBERTO FILOSETA

 14   Soundscape Composition: Listening to Context
      and Contingency   358
      JOHN L. DREVER

 15   From Feeling Vibrations to Building Audiovisual Scenes:
      The Perceptual Practice of Storytelling With Sound   380
      ISABELLE DELMOTTE

Index   400

Contributors

Mathieu Barthet is a Lecturer in Digital Media at the Centre for Digital Music at Queen Mary University of London and Technical Director of the qMedia Studios. He was awarded a PhD in Acoustics, Signal Processing and Computer Science applied to Music from Aix-Marseille University and CNRS-LMA in 2008. His research lies at the confluence of several disciplines including Music Information Retrieval, Interaction and Perception. He is Principal Investigator of the EU-funded project "Towards the Internet of Musical Things" and Co-Investigator of the EU-funded project "Audio Commons." He was chair of several audio conferences (CMMR 2012, AM'17) and serves as guest editor for the *Journal of the Audio Engineering Society*.

Brian Bridges is a composer and lecturer based in Derry~Londonderry, Northern Ireland, where he lectures in music technology at Ulster University and is Research Director for music and associated subjects. He is the current president of ISSTA (Irish Sound, Science and Technology Association) and serves on the editorial board of *Interference: a Journal of Audio Culture*. Much of his work is inspired by connections between perceptual processes, creative practices and technologies, and his creative output includes sound-based installations, audiovisual pieces and electroacoustic and acoustic composition, including microtonal and spatial music. He is a member of the Spatial Music Collective and is represented by the Contemporary Music Centre.

Damian Candusso is an Associate Professor at Charles Sturt University, Australia and an internationally awarded Sound Designer, Supervising Sound Editor and Re-recording mixer. Some of his notable screen credits include: *Better Watch Out* (2017), *The LEGO Movie* (2014), *The Great Gatsby* (2013), *Legend of The Guardians: The Owls of Ga'Hoole* (2010), *Australia* (2008) and *Happy Feet* (2006). This unique nexus between academia, research and professional practice experience provides areas of expertise in sound design for film, spatial sound, immersive media and virtual reality.

Isabelle Delmotte (PhD) explores professional sound design practices for the screen and their relationships to perceptual agency and acoustic ecologies. She also investigates representations of neurological diseases in media and fiction films. Her interests include the role of creative practices as research tools and as vehicles for interdisciplinary collaborations. Isabelle is a lecturer in Screen and Media Studies at the University of Waikato (New Zealand).

Brecht De Man is a sound engineer, researcher and founder of Semantic Audio Labs Ltd. He holds a PhD on music mixing practices from the Centre for Digital Music at Queen Mary University of London and has published, presented and patented research on intelligent audio effects the perception of recording and mix engineering, and the analysis of music production practices. Prior to that, he received a BSc and an MSc in Electronic Engineering from the University of Ghent, Belgium.

Charalampos A. Dimoulas holds a diploma degree (1997) and a PhD degree (2006), both received from the School of Electrical & Computer Engineering, AUTh. In 2008, he received scholarship on postdoctoral research at the Laboratory of Electronic Media, School of Journalism and Mass Communications, AUTh. Both his doctoral dissertation and his post-doc research deal with advanced audiovisual processing and content management techniques for intelligent analysis of prolonged multichannel recordings. He was elected Lecturer (November 2009), Assistant Professor (June 2014) and Associate Professor (October 2018) of Electronic Media Technology in the School of Journalism and Mass Communications, AUTh, where he is currently serving. He leads the audiovisual team for the PANDORA robotic vehicle (http://pandora.ee.auth.gr/) project, mainly focusing on audio detection—localization and machine vision tasks. Dr. Dimoulas is member of IEEE, EURASIP and AES. He is also coeditor of the 2016 published *JAES* special issue(s) (2) on "Intelligent Audio Processing, Semantics, and Interaction." His current scientific interests include media technologies, signal processing, machine learning, media authentication, audiovisual content description and management automation, multimedia semantics and more.

John L. Drever is Professor of Acoustic Ecology and Sound Art at Goldsmiths, University of London, where he leads the Unit for Sound Practice Research and is Deputy Dean of the Graduate School. He was a cofounder (1998) of the UK and Ireland Soundscape Community (a regional affiliate of the World Forum for Acoustic Ecology). Operating at the intersection of acoustics, urban design, sound art, soundscape studies and experimental music, Drever's practice represents an ongoing inquiry into the affect, perception, design and practice of everyday environmental sound and human utterance. He has a special interest in soundscape methods, in particular field recording and soundwalking. Deriving from his study on the noise impact of high-speed hand dryers, he has recently been focusing on the experience of everyday hearing from a nonnormative perspective that he calls aural diversity.

György Fazekas is a Lecturer in Digital Media at the Centre for Digital Music, Queen Mary University of London, where he obtained a PhD in Semantic Audio. He is leading the QMUL team for the EU-funded project Audio Commons as Principal Investigator. He was coinvestigator of five collaborative projects, program cochair of the International Workshop on Semantic Applications for Audio and Music (SAAM2018), general chair of the International Audio Mostly conference (ACM-SIGCHI, 2017) and paper cochair of the AES 53rd International Conference on Semantic Audio (2014). He is guest editor of two *JAES* issues and published over 100 peer-reviewed papers related to semantic audio.

Roberto Filoseta (PhD) is a composer, sound artist, performer. He is currently a Principal Lecturer at the University of Hertfordshire, UK, with the role of Programme Leader for the MSc Music and Sound for Film and Games, and the MSc Music and Sound Technology.

Frederic Font is a Postdoctoral Researcher at the Music Technology Group of the Department of Information and Communication Technologies of Universitat Pompeu Fabra, Barcelona. His current research is focused on facilitating the reuse of audio content in music creation and audio production contexts. Alongside his research, Frederic is leading the development of the Freesound website and coordinating the EU-funded Audio Commons Initiative.

Paul Geluso's work focuses on the theoretical, practical and artistic aspects of sound recording and reproduction. He has been credited as recording and mix engineer on hundreds of music titles and film/video soundtracks that include Grammy-nominated CDs and an Oscar-winning film. He also coedited *Immersive Sound: The Art and Science of Binaural and Multi-channel Audio* published by Focal Press–Routledge and currently serves as an Assistant Professor of Music and the Associate Program Director of Music Technology at New York University.

Andrew Knight-Hill is a composer of electroacoustic music, specializing in studio-composed works both acousmatic (purely sound based) and audiovisual. His works have been performed extensively across the UK, in Europe and in the United States, including performances at Fyklingen, Stockholm; GRM, Paris; ZKM, Karlsruhe; London Contemporary Music Festival, London; New York Public Library, New York; San Francisco Tape Music Festival, San Francisco; Cinesonika, Vancouver; Festival Punto de Encuentro, Valencia; and many more. His works are composed with materials captured from the human and natural world, seeking to explore the beauty in everyday objects. He is particularly interested in how these materials are interpreted by audiences and how these interpretations relate to our experience of the real and the virtual. He is Senior Lecturer in Sound Design and Music Technology at the University of Greenwich, program leader of BA (hons) Sound Design and tutor on MA Digital Arts. For more information, see www.ahillav.co.uk.

Neil Hillman joined UK broadcaster Central Independent Television in 1982 after graduating in electronics from Aston University, leaving in 1990 to become a freelance sound supervisor/mixer/recordist. He founded UK audio postproduction facility The Audio Suite in 2002 to better serve his increasing workload as a sound designer and re-recording mixer. He continues that work alongside acting as a supervising sound editor; outside of the mixing theater, he operates as a sound supervisor/mixer/engineer on outside broadcasts and as a drama and documentary sound recordist on location. He was awarded a prestigious Royal Television Society award for "Best Production Craft Skills" in 2010, is a past multiple winner for sound design from the New York and Los Angeles Festivals and he has amassed over 700 individual IMDb film and television credits across more than 100 separate titles. He gained his PhD from the Department of Theatre, Film and Television at the University of York with a thesis entitled 'A new sound mixing framework for enhanced emotive sound design within contemporary moving-picture audio production and post-production' and his principle research interests continue to concern emotion in sound design.

George Kalliris is Professor of audio and audiovisual media technologies at the School of Journalism and Mass Communication of the Aristotle University of Thessaloniki (AUTh) Greece. He holds a five-year MSc equivalent degree from the School of Electrical Engineering—specialty in Telecommunications, Faculty of Engineering, AUTh, with a scholarship from the Cyprus Government for all years of study 1984–1989, a doctorate degree from the same school, Lab of Electroacoustics & TV Systems 1990–1995. During and after completing his doctoral studies, he worked in several research, development and innovation projects, as well as a part-time higher education teacher. In 1998 he was elected Lecturer of Electronic Media Technology at AUTh. In his current position, he is the Deputy Head of the School, master program director and the head of Electronic Media Lab. He has also taught and/or is teaching in four master's degree programs to the Film Studies School and as a visiting professor of Frederick University Cyprus. His current research interests and publications include audiovisual technologies for the new media, radio and television studio design; digital audiovideo processing-production-broadcasting-webcasting; multimedia content, restoration, management and retrieval. He is a member of the Audio Engineering Society (AES), member of the Association for Computing Machinery (ACM), board member of the Steering Committee of AUDIOMOSTLY yearly international conference and board member of the Hellenic Institute of Acoustics. For more information, see http://kalliris.blogspot.com/.

Maria Matsiola received her five-year MSc equivalent degree from the School of Electrical Engineering—specialty in Telecommunications, Faculty of Engineering, Aristotle University of Thessaloniki (AUTh) in 1996 and her PhD on New Technologies on Journalism, from the School of Journalism and Mass Communications of AUTh in 2008. She serves as Tenured Senior Teaching Fellow and Instructor in the Electronic Media Laboratory of the School of Journalism and Mass Communications of AUTh. Her scientific interests include New Media technologies, web radio and TV; audiovisual content management; e-learning; audiovisual application systems in distance education.

Davide Andrea Mauro is currently Assistant Professor in the Computer & Information Technology Department at Marshall University (USA). His research interests are in areas of Sound and Music Computing, in particular related to 3D audio technologies, sound design and synthesis and encoding of music information.

Adam Melvin is a composer and lecturer in Popular and Contemporary Music at Ulster University, Derry–Londonderry, Northern Ireland. A great deal of both his compositional and research practice is concerned with interrogating the relationship between music, sound, site and the visual arts, particularly the moving image. He has received numerous international performances and broadcasts of his music; his research has been published in *The Soundtrack, Short Film Studies* (Intellect) and the *Palgrave Handbook of Sound Design and Music in Screen Media*. He is a member of Dublin's Spatial Music Collective.

Sandra Pauletto is Associate Professor in Media Production at the Division of Media Technology and Interaction Design at KTH Royal Institute of Technology in Stockholm, Sweden. Previously she was Senior Lecturer in Sound Design at the University of York, UK. She has a background in physics, music and music technology. Her research focuses on sound design in various application areas including cinema and interactive systems. She has worked on a number of research projects supported by the British Academy, EPSRC, EU and Wellcome Trust. She has published widely in both science and arts and humanities journals on topics ranging from the design of cinematic voices to the use of sound in health communication applications. For more information, see https://www.kth.se/profile/pauletto

Sarah Pickett is an Associate Professor of Sound Design & Music Composition at Carnegie Mellon University. In addition to teaching, Ms. Pickett works as a freelance composer and sound designer in theaters across the United States. She is a founding member of the sound design collective District 5 Sound, which specializes in sound design for immersive and interactive environments. She has worked as a composer, sound designer, sound engineer, music director, musician and performer. Ms. Pickett began her professional career in sound as a location sound recordist on an independent film shoot and later went on to compose music for the film. The interactive nonlinear film by Diego Bonilla has won awards at film and media festivals in Russia, Brazil and the United States. Performance has been a lifelong passion, and Ms. Pickett continues to incorporate live elements in her design and composition work. She graduated with an MFA from Yale University and was awarded the Frieda Shaw, Dr. Diana Mason OBE and Denise Suttor Prize for Sound Design.

Andrea Valle is currently Aggregate Professor at the Humanities Departments, Università di Torino (Italy), where he teaches audio programming and semiotics of the Media in the Performing Arts program (DAMS). His main research interests are Sound and Music Computing, Semiotics, and Theory of Audiovision, particularly in relation to cinema.

Anna Xambó is Associate Professor in Music Technology at the Norwegian University of Science and Technology, Visiting Lecturer at the Centre for Digital Music at Queen Mary University of London working on the EU-funded project Audio Commons, and a co-founder and chair of Women Nordic Music Technology (WoNoMute). Her PhD dissertation is entitled *Tabletop Tangible Interfaces for Music Performance: Design and Evaluation* (2015, The Open University, UK). Her research focuses on interactive real-time systems for music investigating live coding and generative computer music, collaborative and participatory interfaces and real-time MIR.

Series Preface
Foundations in Sound Design:
A Multidisciplinary Approach

Edited by Michael Filimowicz

This series organizes topics in sound design that combine multidisciplinary perspectives across linear, interactive and embedded technologies. Such an approach is needed as today the practices of sound design are diversifying beyond what could adequately be captured by any single author or discipline. Today's sound designers need to be prepared just as much for games as for films, for programming in coding languages as for mastery of proprietary industry software, and for prototyping web applications or new industrial designs as for traditional occupations in film, television, music and audio.

The volumes are designed to be more future proof than most volumes on media technologies, by focusing on high-level concepts that can be easily put into practice. The first three volumes in the series are sequenced as follows:

Volume 1: Linear Media covers traditional topics such as audiovisual preproduction, production and postproduction but adds other important aspects of linear media as well, such as electronica music production, basic music theory for sound designers, as well as artistic compositional practices such as soundscape design and electroacoustic music.

Volume 2: Interactive Media expands the cinematic soundtrack developed in Volume 1 by developing interaction approaches through consideration of gaming technologies, music programming, installations, spatial audio, real-time synthesis, performances and web-based interfaces and databases including mobile and locative media.

Volume 3: Embedded Media brings much needed coverage to emerging areas such as auditory display, data sonification, the role of sound in the Internet of things, wearables and multimodal interaction by integrating physical computing technologies and product development in contexts ranging from toys to automobiles.

This approach to the foundations of sound design does justice to the ever growing uses, content variations, audiences and professional roles that sound design is brought to bear on in the contemporary context. Each volume in itself constitutes an introductory text to its respective area of sound design: linear, interactive and embedded media. Taken together, they comprise a comprehensive introduction to the many forms, technologies and practices of sound design.

These first three volumes set up the possibility for other books that can expand upon the foundational topics for deeper explorations of specialized topics. Prospective authors are encouraged to send ideas for other volumes that can be added to the initial three-volume set, either single- and coauthored monographs or edited volumes, by submitting a proposal to the series editor, Michael Filimowicz (michael@filimowi.cz). Finally, feedback from readers on the content is always welcome.

Michael Filimowicz

Volume Introduction

Michael Filimowicz

Foundations in Sound Design for Linear Media introduces key concepts for the production of linear media forms, such as film, video and sound composition. A linear medium is commonly understood as one in which the progression of events is fixed according to the duration established by a timeline. Such media are sometimes called "traditional" but also form the background of general audiovisual cultures upon which the more computationally interactive "new media" often draw upon for their expression. The chapters in this volume cover listening and recording; audio postproduction; key musical concepts and forms such as harmony, conceptual sound design, electronica, soundscape and electroacoustic composition; the audio commons; and sound's ontology and phenomenology.

Chapter 1—"The Nature of Sound and Recording" by Andrew Knight-Hill—begins the volume by providing a high-level introduction to acoustic and psychoacoustic phenomena. The physics of sound and psychology of listening form an empirical core for the practices of recording, which are put to use for a variety of aesthetic and narrative considerations.

Chapter 2—"Invisible Seams: The Role of Foley and Voice PostProduction Recordings in the Design of Cinematic Performances" by Sandra Pauletto—focuses on the contribution of Foley sound effects and voice postproduction recordings to the creation of the cinematic performances we enjoy on screen. It features extracts from interviews with sound professionals, and it reflects on the creative aspects of these processes.

Chapter 3—"Media Management, Sound Editing and Mixing" by George Kalliris, Charalampos A. Dimoulas and Maria Matsiola—details the basics of media management and setting up audio postproduction projects for editing and mixing. The differences between analog and digital signal chains is covered, as well as the main hardware and software components used today.

Chapter 4—"Audio Effects in Sound Design" by Brecht De Man—dives into the dynamic, spectral and sonic properties of sound by examining key audio effects processes. The most common creative goals associated with each effect type are detailed, e.g. from simulating first-person perspective to the creation of imaginary spaces.

Chapter 5—"The Mix Stems: Voice, Effects, Music" by Neil Hillman—analyzes the use of mix stems for structuring rich audio mixes comprised of a multitude of elements into a coherent outcome. Systems management and workflow are discussed for the production of mix outputs used in theatrical, broadcast and online distribution.

Chapter 6—"Mixing and Mastering" by Paul Geluso—develops a philosophy for mixing, covering topics such as critical listening, session preparation, monitoring, metering, target levels, signal flow, panning, equalization, dynamics, balancing and mastering. Stereo and immersive effects are discussed in relation to sound design, film sound, installation, theater, multimedia and music.

Chapter 7—"Designing Sound for 3D Films" by Damian Candusso—discusses the creative and technical challenges specific to soundtracks for 3D films. Three-dimensional films use cinema space differently, and the chapter provides an overview of the industry's adoption of 3D and its ramifications for sound design and sonic immersion.

Chapter 8—"Compositional Techniques for Sound Design" by Sarah Pickett—mines music compositional theory, since sound design today can include "note-for-note" notation for live musicians, sequencing for virtual or electronic instruments, plundering and reconstructive sampling, *musique concrète*, aleatory and algorithmic music or any combination of these.

Chapter 9—"Music Theory for Sound Designers" by Adam Melvin and Brian Bridges—addresses key principles from music theory. It includes discussion of metaphors for musical motion, analogies drawn from other art forms, and principles of musical construction informed by psychoacoustical studies and music cognition.

Chapter 10—"Leveraging Online Audio Commons Content for Media Production" by Anna Xambó, Frederic Font, György Fazekas and Mathieu Barthet—introduces the role of the Internet as an audio commons, where hundreds of thousands of audio files are available to sound designers who have forged a culture of sharing under the Creative Commons legal framework.

Chapter 11—"Sound Ontologies. Methods and Approaches for the Description of Sound" by Davide Andrea Mauro and Andrea Valle—highlights the need for a rigorous definition of categories so that sounds can be organized effectively into databases and classification schemes. Semantic, phenomenological, perceptual, psychoacoustic and ecological features are used as abstraction layers to create hierarchies in the organization of sound collections.

Chapter 12—"Electroacoustic Music: An Art of Sound" by Andrew Knight-Hill—explores sound design practices in the histories and genres of electroacoustic composition. Electroacoustic music has much to offer sound designers especially since it has inspired and led to many of the technologies and techniques used in audio postproduction today.

Chapter 13—"Electronic Dance Music in Narrative Film" by Roberto Filoseta—gives an in-depth discussion of how EDM (electronic dance music) approaches have changed film scoring practices, by examining closely two films that were critically important for the current trends: Tom Tykwer's *Run Lola Run* (1998) and Darren Aronofsky's *Pi* (1998).

Chapter 14—"Soundscape Composition—Listening to Context and Contingency" by John L. Drever—connects sound design practices to the genre of soundscape composition. In doing so, sound designers are encouraged to be less reliant on storehouses of ambiance recordings and to take a more compositional approach to rendering sounds of the world.

Finally, Chapter 15—"From Feeling Vibrations to Building Audiovisual Scenes: The Perceptual Practice of Storytelling With Sound" by Isabelle Delmotte—outlines exercises for the creative practitioner that are grounded in phenomenological insights, bringing out the richness of sonic environments, human ways of living in sound and audiovisual storytelling.

Acknowledgment

The chapter summaries here have in places drawn from the authors' chapter abstracts, the full versions of which can be found in Routledge's online reference for the volume.

The Nature of Sound and Recording

Andrew Knight-Hill

1.1. Introduction

This chapter explores sound as material and phenomenon. It is vital to understand the properties of the medium with which we work so as to be able to achieve the greatest level of control and to be best able to articulate your ideas and intentions through sound design.

We will cover an introduction to the physical and acoustic properties of sound, the processes of hearing and listening, and we will explore the tools: microphones and recording technology. These topics are expansive areas, so you will find that this chapter provides a wide breadth of knowledge, with links to additional resources that can be taken advantage of in order to further extend your knowledge.

The question of what sound is and how it operates brings together ideas and approaches from physics and philosophy, biology to engineering and design. All of these approaches culminate in our understanding of sound and how we can capture and manipulate it creatively.

By the end of this chapter, you will have begun to understand how sounds are created, how the physical properties of sound relate to the sounds that we hear and how best we can begin to approach the process of capturing and recording sounds.

1.2. Physical and Acoustic Properties of Sound

1.2.1. The Physical Sound Source

Sound is caused by objects that vibrate. Vibrating objects move back and forth, and as they do so, they impact the medium[1] around them. The vibrating object is often described as the *sound source*.

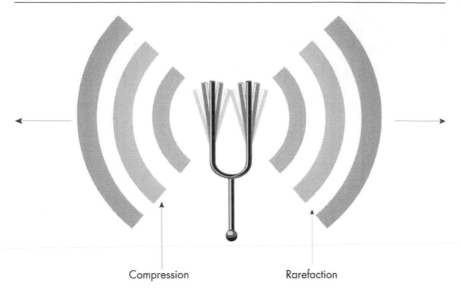

Compression   Rarefaction

Figure 1.1   Compression and Rarefaction: A vibrating sound source creates pressure fluctuations in the air around it.

Imagine a plucked string. At the moment it is plucked, the string begins to vibrate back and forth. As it does so, the air surrounding the string is impacted, with the molecules being pushed away (as the string moves toward them) and then pulled back (as the string moves away again). This push and pull is repeated over and over again many times each second, creating a propagating wavefront that radiates out from the string. When the air molecules are pushed together, they create areas of *compression* (higher pressure), and when the air molecules are pulled away, they create areas of what is called *rarefaction* (lower pressure). The back-and-forth motion of vibration, therefore, sets up subtle pressure fluctuations in the air that spread outward from the sound source.

The physical properties of each sound source dictate how it will vibrate, and these variations result in different wave patterns. Our string vibrates in a very different fashion to a bell or a motor, and so each of these objects creates very different sound waves when excited. The ability to infer how different objects might vibrate when struck is an incredibly useful tool for the sound designer. It can help you to select appropriate sound sources and to identify the most suitable locations to position a microphone for capturing those sounds. The latest physical modeling synthesis software works in exactly this way: physical parameters for an object are input, and an algorithm calculates the likely patterns of vibration and subsequent sound waves.[2] This process can synthesize the sound of familiar objects but also of fantastical objects that could never (or would be unlikely to) exist in the real world.

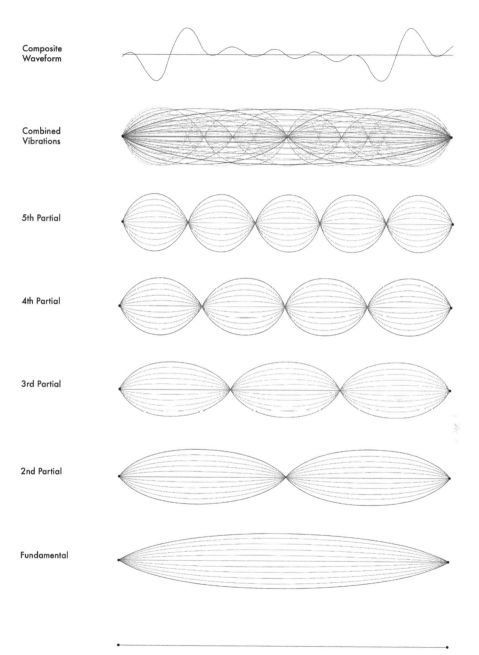

Composite
Waveform

Combined
Vibrations

5th Partial

4th Partial

3rd Partial

2nd Partial

Fundamental

Figure 1.2 Vibration Patterns of a String: Individual patterns combine to create the composite vibration that we can see with the naked eye. The resulting sound wave is more complex, with the individual partials both adding and subtracting from one another to create the resulting complex wave.

Image credit: Deborah Iaria and Luca Portik.

The physical material properties of an object are not the only factor that will affect its sound. How the object is made to vibrate (excited) will affect the sound, as will any modification to the object such as *dampening*. A string, cymbal or glass may be plucked, bowed or struck, and its vibration may be dampened by adding a cloth, by touching with light pressure or by filling any voids in the object. Each of these interventions cause the object's patterns of vibration to be altered thus to create different sounds. Familiar objects can be taken and excited in different ways to create a wide array of new sounds.

Figure 1.3 Same Glass, Different Sounds: Exciting the glass in different ways will create different sound textures, while filling up the glass will alter the pitch.

Image credit: Deborah Iaria and Luca Portik.

The shape and structure of the sound source dictate how the sound waves *propagate* (spread) out from the object. When you speak, sound waves travel from your vocal chords, up your throat, out through your mouth and nasal cavity and are projected forward across your tongue and past your lips. Movements of your tongue and lips alter the shape of the vocal cavity, thus modulating the sound that results. You can still hear someone talking if they are facing away from you, but the sound is much clearer if they are facing you. This is because the majority of the sound energy is projected forward out of the mouth. Every sound source will be different in this regard. A bell or glass will resonate completely and propagate sound in every direction, but the voice, a guitar or a violin will project sound out of their face.

The position of microphones around an object can radically alter the relative balance of sound captured, and this is largely the result of the way in which sounds propagate out from the sound source.

1.2.1.1.   A Note on the Visualization of Sound

There are many different ways to visualize sound. Each form of visualization makes certain decisions about the kind of information that is being

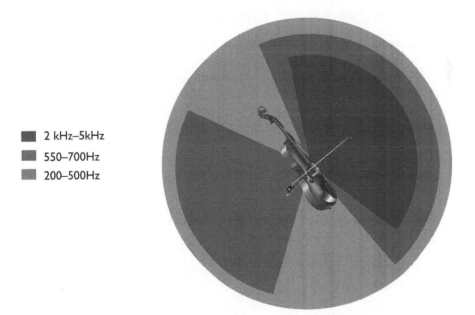

2 kHz–5kHz
550–700Hz
200–500Hz

Figure 1.4 Directional sonic propagation: Low-frequency sounds travel out in all directions while higher-frequency sounds are projected from the face of the instrument.

Image credit: Deborah Iaria and Luca Portik.

presented. In Figure 1.1 the sound is visualized as a longitudinal compression wave; this is how sounds operate and can provide details about the dispersion of sound, but this format of visualization is difficult to read and gain information about the parameters of sound. Most commonly, for example in digital audio workstations (DAWs), sounds are rendered as *transverse waves*. This format is ideal for detailed time-based work, it is very easy to see where sounds begin and end, as well as how strong the signals are. Transverse waves display the pattern of compression and rarefaction of a longitudinal sound wave, with the points of compression above a center line and points of rarefaction below (you can see this in the Complex Waveform at the top of Figure 1.2).

However, transverse waves are poor at rendering frequency information, and so another form of visualization is required when we want to visualize pitch information. Spectrograms display sound in terms of its component frequencies. It is possible to see clearly the frequencies present, as well as roughly where they begin and end, but it is more difficult to accurately interpret loudness and exact timings. Visualizations within this chapter will shift depending on the concepts or information under discussion. But it is important to understand how each of these visualizations are related and how each can be a useful tool for imagining sound.

1.2.2.   Propagated Sound: Sound in Spaces

Once a source has been excited, sound waves propagate out and away from the object. As they do so, the energy of the wave becomes more and more diffuse. At the source, the vibration energy is concentrated, but as the wave moves away, it spreads out in three dimensions. The same amount of energy becomes distributed over a much wider area. In fact, the relationship is exponential. At 2 m from the source, the sound energy is spread over four times the area and is therefore a quarter of the volume that it was when only 1 m from the source. At 3 m from the source, the sound energy is spread over nine times the area and thus is nine times quieter than at 1 m from the source. This exponential dissipation of energy is known as the *Inverse Square Rule*, and it is the reason why closer sound sources appear louder and why a microphone needs to be positioned relatively close to the sound source in order for it to capture a strong signal.

As sound waves propagate, they do not act in isolation. At some point in their journey, they will come into contact with either an object or another sound wave.

When sound waves impact with an object, they can be either reflected or absorbed. Hard materials—such as tiles, glass or concrete—reflect sounds. Soft materials—such as carpet, foam or flesh—absorb sounds. The relative propensity of an object to absorb sound is described by its *absorption coefficient*. This is a mathematical value calculated to describe how

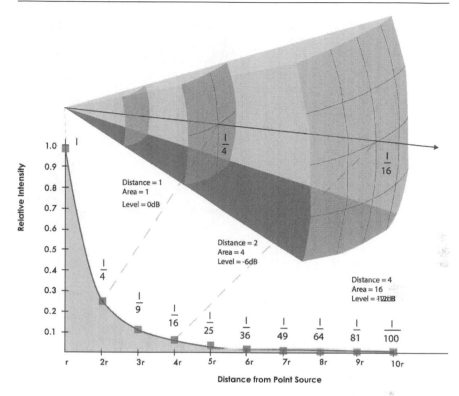

Figure 1.5 Inverse Square Rule = 1/Distance²: As sound waves travel away from their source, their energy becomes spread over an increasingly wide area. As it spreads, the intensity drops.

Image credit: Deborah Iaria and Luca Portik.

likely an object is to absorb sounds at specific frequencies. Absorption coefficients are used by acousticians to calculate the acoustic properties of a space. As sound designers, we may need to engage with the detailed math when designing acoustic treatment for a studio or listening environment, but most often we can leave the detailed calculations and instead take advantage of the general principles of sound absorption. These can inform our actions when recording sounds or in mixing and editing sounds to simulate the impression of a particular acoustic space.

The most common acoustic feature that results from the reflective properties of surfaces is reverberation (or reverb). Reverberation accompanies almost every sound because it is a physical result of the acoustic properties of the space within which the sound is made—artificial reverberation is a common effect plugin in digital audio workstations, often titled *Reverb*. When a sound source is excited, the wave travels away from the source and

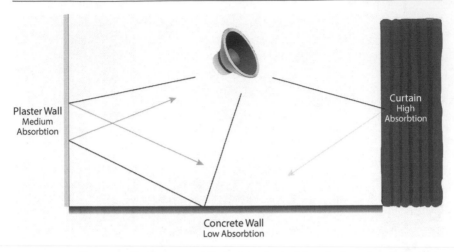

Figure 1.6   Plaster Wall, Concrete Wall, Curtain: Materials reflect or absorb sound to a different extent depending on their properties.

Image credit: Deborah Iaria and Luca Portik.

impacts on the surrounding surfaces. Depending on the material properties of these surfaces (how reflective/absorbent), a proportion of the sound will be reflected back. These reflected sounds combine with the original "direct" sound to embellish it with a tail of reverberation. Because these reflected sounds have traveled a longer distance to reach the listener, they are delayed and arrive slightly later than the direct sound. Because the delay time is short, the reflected sound is not heard independently from the direct sound. In fact, the two merge to create an overall impression of the sound in the space. As the room size changes, so too does the delay time on the reflected sound (as sounds must travel either a greater or shorter distance before being reflected). And if the material properties of the surfaces in the space are altered, so too are the types and qualities of the reflected sound (more or less absorption at specific frequencies, diffuse or coherent reflections). The ability to adapt these properties as parameters is a common feature of artificial reverb plugins.

Direct sound travels straight from the source to the listener; it is not reflected in any way. *Early reflections* are the first layer of reflected sounds; they impact with surrounding surfaces and are bounced back to combine with the original "direct" sound. Early reflections give the listener an impression of the spatial location and proximity of the surfaces around them. In contrast, *late reflections* have been reflected multiple times and provide more information about the overall acoustic properties of the space but much less specific directional information. These different

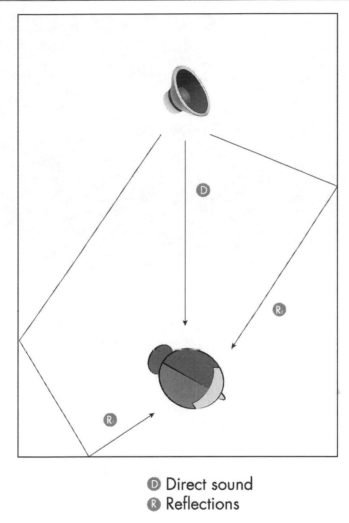

○D Direct sound
○R Reflections

Figure 1.7 Direct and Reflected Sound: Note the different distances that reflected sounds travel.

layers of reflected sound are sometimes available for individual editing and control within reverb plugins. Control of these independent layers allows you to mold the reverberation in detail, for example to balance an impression of nearby surfaces while retaining clarity within the original sound (achieved by mixing direct sound with early reflections and attenuating the late reflections).

Reverberation is often confused or conflated with echo. However, the two are quite distinct phenomena. An echo is the result of a reflected sound

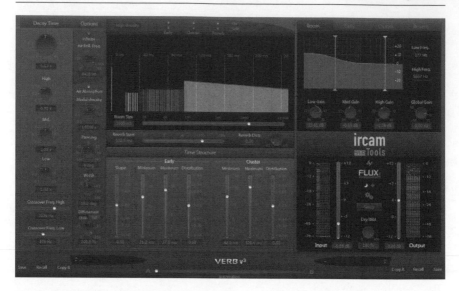

Figure 1.8 Screenshot of a Reverb Plugin: Note the independent control over room size, early reflections (split here into early and cluster), late reflections (called reverb) and the independent equalization control over each. The combination of these parameters builds up a highly accurate and adaptable simulation of acoustic reverberation.

Source: FLUX–IRCAM Tools Verb v3

that reaches the listener as a distinct wavefront; there is no blending with the direct sound, as occurs in reverberation. Instead you hear a distinct facsimile of the original sound, audible as a separate entity.

The surfaces within a space also impact on the type and character of reverb. Sounds reflecting off of flat and regular surfaces retain more cohesion than sounds impacting upon uneven or irregular surfaces. Without recalling too many of the horrors of physics/science classes, it is useful to reflect on the science of waves. For all reflecting waves, the *angle of incidence* (the angle at which the wave impacts with the surface (a)) is equal to the *angle of reflection* (the angle at which the wave leaves the surface (a)). On flat and even surfaces, this means that sound waves will reflect consistently and thus retain a more coherent wavefront (this is known as specular reflection). Irregular surfaces make for a much more complex and erratic pattern of reflection, as adjacent portions of the wavefront may be reflected in completely different directions.

An incoherent reflection pattern creates diffuse reverberation. Angled and irregular foam panels are designed to create this diffusion effect within studios. They diffuse any reflected sounds in order to break up coherent reflections and manage the studio's acoustic.

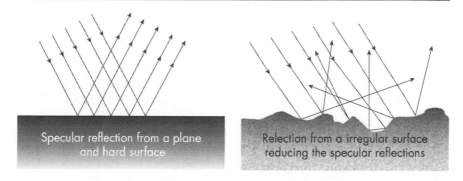

Figure 1.9 Flat and Uneven Surfaces: Flat surfaces result in coherent (specular) reflections, while uneven surfaces diffuse and scatter the reflections.

Figure 1.10 Diffuser Acoustic Panels: These are designed to both absorb and scatter reflections, reducing room tone and reverberation.

Source: Auralex Acoustics—Metro Diffusion Panel Promotional Image

1.2.3. The Sound Wave

We have already discussed the physical processes involved in the creation of sounds. but, to gain an even deeper understanding, let us zoom in to explore the physical properties of sound in detail.

The main properties of a sound wave are its amplitude and frequency. The *amplitude* of a sound wave reflects the amount of displacement that occurs within the medium and therefore the pressure gradient that is established (the difference between the highest point of compression and the highest point of rarefaction). Amplitude is often correlated with loudness.

Frequency is the rate of vibration, measured as the number of oscillations per second. Low-frequency oscillations are generally perceived as low-pitched sounds, while higher-frequency oscillations are often perceived as high-pitched sounds.

The parameters of amplitude and frequency can be measured against standardized scales and so can be described mathematically in a standardized way, even though the physical phenomena that we experience do not always correlate directly (section 3.2 explores this in more detail).

The purest type of sound is a *sine wave* oscillation. This wave produces a perfect pattern of compression and rarefaction with the result that all of the sound wave energy is concentrated at a single frequency. However, sine tones almost never exist in nature. Recorded sounds are almost always a complex of multiple vibrational waves that add together to create a composite waveform.

1.2.3.1. Partials and Composite Waveforms

If we return to our string example and observe its vibration patterns (Figure 1.2), we can see that, when struck, it vibrates simultaneously at a number of different frequencies. The string does not simply vibrate back and

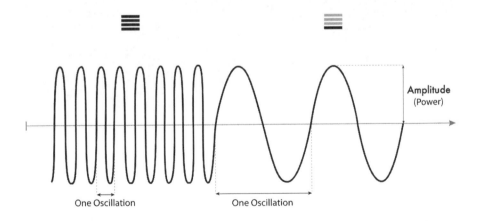

One Oscillation   One Oscillation

Amplitude (Power)

Frequency is the number of oscillations per second.

Figure 1.11 Sine Tone, the Most Basic Waveform: Frequency is the rate of oscillation and can be correlated with pitch (left = high/right = low), while amplitude represents the extent of pressure fluctuation and can be correlated with loudness.

forth along its total length; being flexible, it bends and also vibrates at other wavelengths. Being regular, these wave patterns create a series of concentrated sound energy bands that are referred to as partials. In the case of the string (a harmonic oscillator), the additional vibrations are whole number fractions of the total string length, i.e. half, third, quarter, fifth, sixth, seventh. The partial that results from the string oscillating along its whole length is called the fundamental. The overall pitch that we hear when listening to a composite sound wave of many different partials is often that of the fundamental.[3]

In the string, the partials are all evenly spaced, in consistent ratio with the fundamental frequency. The sound that this kind of vibration creates is

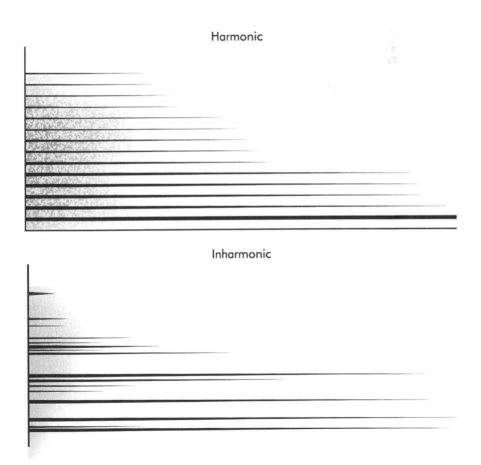

Figure 1.12 Harmonic and Inharmonic Partials: Note the even spacing of the harmonic partials compared with the inharmonic sound. The guitar is harmonic.

described as *harmonic*. Many instruments have been specifically designed to create this kind of harmonic sound—stringed instruments, wind instruments, brass instruments etc.[4]

But not all sounds contain partials that are regularly spaced. A bell, for example, contains distinct partials, but these are not evenly spaced. The resulting sound is pitched, but the relationships of the pitches within the sound are described as *inharmonic*. Metallic objects are characterized by their inharmonic sound spectrum.

However, not all sounds are distinctly pitched. *Noise* is a random fluctuation of sound wave energy and is therefore completely opposite to the focused concentration of the regular oscillations that result in pitched partials. Nearly all "natural" sounds will contain a noise component, and the proportion of noise will affect the character of the sound. An instrumental example can be found with the snare drum. When the snares are released, the drum possesses an open and ringing pitched sound; the dimensions of the drum amplify specific vibrational oscillations, giving it a certain pitched quality. However, when the snares are set, they interrupt the vibration of the drumskin, creating randomized sympathetic vibrations that are noisy in character.

Transients are high-amplitude, often noise-based, sounds that occur at the beginning of a waveform. They are short-lived—hence the name—oscillations that result from the excitation of the sounding object. They will often contain a higher proportion of high-frequency energy in the form of additional partials or noise.

Individual wave patterns add together to create the complex vibrations that make each sound unique. Nearly all sounds have composite waveforms, being made up of partials and/or noises. The relationships and balance between these components create the variety of unique timbres that allow us to distinguish between different sound objects.

The term *timbre* is used to describe the texture and character of a sound. It allows us to tell a piano apart from a flute (even when they play the same note), a moped from a motorbike, and one human voice from another. Our impression of timbre is influenced by both the spectral information present in the sound, as well as its shape over time (see section 1.2.3.2, Envelope). An object's timbre is influenced by a wide range of factors, and timbres of the same object can even change at different loudness levels; for example, think of the textural difference between whispering and shouting.

Manipulating timbre can be a vitally important tool. A harmonic musical underscore might completely obscure or clash with harmonic sound design material, while an inharmonic ambient soundscape might stand out distinctly from it. Layering up different timbres of harmonic and inharmonic sound can create rich multilayered soundscapes and sounds with complementary components. Alternatively, within sound design, it may be necessary to blend multiple sounds together to create a new sound object. By tweaking EQ (equalizer) and bringing the timbres of the objects more closely together, a fusion between the distinct sounds can take place, hiding their original independent origins and creating a new coherent sound object.

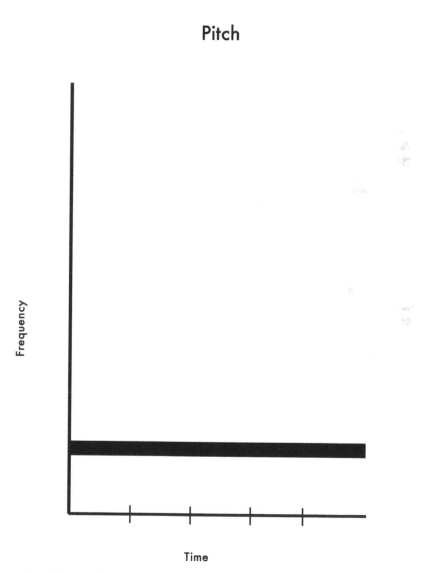

Figure 1.13a Pitch and Noise: This pitch example shows a sine tone, where all of the sound energy is concentrated at a single frequency.

Figure 1.13b   Pitch and Noise: Noise, by contrast, is a random fluctuation spread across the spectrum.

1.2.3.2. Envelope

Beyond the properties of sustained sound wave vibrations, the other factor that has a major impact on timbre is the *envelope*: the shape of the sound over time. As alluded to in our discussion of transients, the shape of a sound over time can have a major impact upon the perceived character. One of the first realizations made by Pierre Schaeffer in his initial

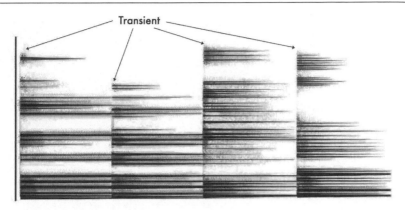

Figure 1.14   Transients Identified in a Spectrogram.

experiments with recorded sound[5] was that a recorded sound can be completely transformed when played back in reverse. Schaeffer demonstrated that the shape and disposition of sound energy over time play a major role in the perceived sound event. Of course, musicians had long realized that different sonic results could be achieved by playing their instruments in different ways, but they had never been able to play their instruments in reverse. Sound recording and editing allowed sounds to be heard backward, at different speeds and with parts of their envelope separated out.

The envelope of a sound can be described as consisting of an attack/ sustain/decay.[6]

The *attack* is the initial portion of a sound's envelope. String players may use their bow to initiate a slow and smooth beginning to the note or start with a fast and decisive stroke. Each gesture creates a very different type of attack. The second portion of the envelope, *sustain*, is the period in which the sound remains relatively stable in state. There is little change to the sound within this portion, and extended sustain allows the listener to focus upon the internal characteristics of the sound. The final portion of the sound, its *decay*, is the time it takes for the sound energy to dissipate and fade away. Manipulating the relative durations of each of these parameters can radically alter the character of a sound.

A popular technique in sound design and composition is to edit and splice recorded sounds so as to separate and reorder, or eliminate, various portions of the envelope. For example, to remove the attack portion of a sound and isolate only its sustain and decay, as in the piano drones created within Jóhann Jóhannsson's score to the film *Arrival* (2016), or to combine

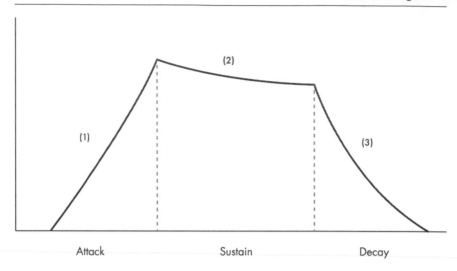

Figure 1.15 Envelope Diagram: Note the distinct attack, sustain and decay sections.

individual attack, sustain and decay components to form entirely new and novel sounds, as in the work of Gilles Gobeil (Gobeil, 1994).

1.2.3.3. Phase

Sound waves not only add together but can also cancel one another out. Because sound waves are patterns of compression and rarefaction around a fixed resting state, we can describe them as moving in cycles. As the sound moves into the compression state, the pressure increases, and as it moves past the resting state to the rarefaction phase, the pressure decreases. *Phase* is a value corresponding to a particular point within a cycle.

Where two identical waves are perfectly aligned, with their peaks and troughs in sync with one another, we would describe these sound waves as in phase. When *in phase,* the peaks of the waves and their corresponding troughs will add together to create a single wave with double the original amplitude of each input wave.

However, if the alignment shifts, we describe the sound waves as being *out of phase*. Should the peaks of one wave align precisely with the troughs of another identical wave, then the two waves will cancel each other out. This perfect negative alignment is sometimes called antiphase. We can hear this effect if we stand two loudspeakers 1 m apart, facing directly at one another. When both speakers play a 343-Hz sine tone, an antiphase relationship is established, with a single cycle of the sine tone projected into the distance between them.[7] However, when you place your head between the speakers, instead of hearing a loud sound in between

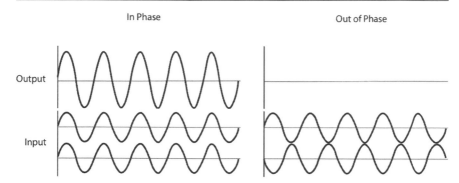

Figure 1.16   In Phase, Out of Phase: On the left, the input tones are in phase; their peaks add together to create an output sound with twice the amplitude. On the right, the input tones are out of phase; the peak of one coincides with the trough of the other, and so they cancel each other out, leaving no output.

them, you'll actually hear a marked drop in volume. The two waves are out of phase and so cancel each other out.

Phase cancellation can be used constructively to control or transform sounds; for example, it is designed into shotgun microphones to give them their focused directionality and is the key component in phasing and flanger effects. But phase cancellation can also cause unwanted problems, for example in stereo recordings. Imagine that you have two microphones recording a single source. When the source is positioned equally between the microphones, the sound waves radiate out and reach the left and right microphones after having traveled the same distance. Because the distances are equal, the sound waves will have passed through exactly the same number of cycles and thus will be in phase with each other when they excite the microphone diaphragms and are subsequently recombined within the recording medium. However, if the source moves so that one microphone is closer and the other farther away, then the distances over which the sound waves travel has changed, and so have the number of cycles that each sound wave will have undergone as it makes its journey. In this situation, the sound waves will be out of phase when they reach the microphone. This phase cancellation in recordings can result in the unwanted elimination of certain frequencies.

1.2.4.   Key Points

We have explained a wide range of physical principles to provide an overview of how sound works. By understanding the properties of sound waves, we begin to understand more about the subtleties of how sound

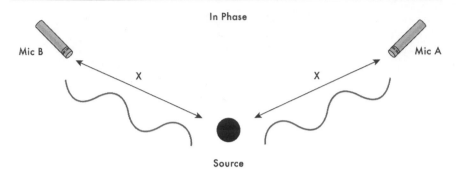

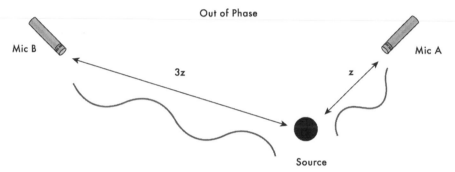

Figure 1.17   Phase Cancellation: This can occur when microphones are positioned at different distances from the sound source.

operates. This, in turn, helps provide a foundation for understanding how listeners hear sounds and how we can develop tools and techniques to best create and capture sounds.

- Sounds are caused by objects vibrating in a medium.
- Physical space has a significant impact on the way that we hear sounds.
- Sound waves can be combined together.
- Sounds are often, complex composites, made up of multiple vibrations.
- Sound waves can add or subtract from one another, depending on phase.

1.3.   Hearing and Listening

This topic extends all the way from the physiological and biological through to the conceptual and the philosophical. As workers in sound, it is vital for us understand the processes of listening so that we can effectively achieve our sonic goals. In the following sections, we will explore the physical organs of listening and begin to unpack the psychological and psychoacoustic dimensions that influence or affect what it is that we interpret and hear. We will reveal how some parameters of sound are not as absolute as generally assumed and highlight how this flexibility can actually be to the advantage of the informed sound designer.

1.3.1. The Ear

The ear is a specialized and sensitive organ that has evolved in order to facilitate the transfer of sound wave energy into neural signals. Its outward appearance on the side of the head is only part of the whole complex mechanism of the ear, much of which is safely situated within the skull.

The ear is often divided into three distinct sections: outer, middle, inner.

1.3.1.1. Outer Ear

The *outer ear* is called the pinna, this shaped disk of cartilage funnels sound waves into the almost cylindrical ear canal. The ear canal passes through the skull into the head and is sealed at the inner end by the eardrum. The outer ear is responsible for focusing and transferring sound waves into the more sensitive portions of the middle and inner ear.

Having two ears provides us with our perception of directionality in sound. Depending on the position of the sound source and the orientation

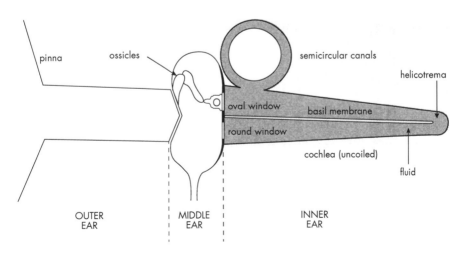

Figure 1.18 The Ear: A schematic diagram showing the three distinct sections. (Campbell and Greated, 1987, p. 41)

of the head, one ear will be slightly closer to the sound source than the other. This will result in the sound reaching one ear sooner than the other—defined as interaural time difference—and also results in a phase difference between the two signals (see section 1.2.3.3). The shape of the pinna itself and the presence of the shoulders and head act to filter and attenuate certain frequencies reaching the ear, thus adding more cues for the brain to interpret the relative spatial position of sound sources.

So if we want to create the impression of spatial location or spatial richness within the sound design edit process, we can reverse-engineer these effects. Take two copies of a monophonic sound and pan them apart. Nudge the second copy to delay it by a few milliseconds. (N.B.: The delay time must be short; too long, and you'll begin to introduce phase cancellation). Then pull back on the gain −6 dB, and use EQ to attenuate some of the higher frequencies. You should now have a pseudo stereo impression generated from a mono signal. By tweaking the parameters, you can manipulate the extent of the spatial depth, all the time taking advantage of the brain's understanding of space through differences in time, filtering and phase.

1.3.1.2.   Middle Ear

The *middle ear* is a small, air-filled, cavity that focuses and amplifies the vibrational movement reaching the outer ear, conducting these vibrations into the inner ear. Bounded by the eardrum toward the outer ear and two apertures toward the inner ear—the oval window and the round window—the middle ear is bridged by the ossicles. These three small bones (very small, ~ 3 mm in height and respectively called the hammer, anvil and stirrup) act to focus the sound wave vibrations reaching the eardrum and transmit these fluctuations into the oval window. The hammer is attached to the inside of the eardrum and thus moves in sympathy with it when activated by an incoming sound. This, in turn, causes the anvil to rotate around its pivot, transferring the movement onto the stirrup, which moves, piston-like, in and out of the oval window.

The ossicles also provide a mechanism for protecting the sensitive inner ear. Just as we blink when flashed with a bright light, the so-called *acoustic reflex* acts when a sudden loud sound is encountered. It causes a small muscle in the inner ear to contract, pulling the stirrup back from the oval window and reducing the amount of sound energy transmitted into the inner ear. The response time for this reflex is around 10 m/s, and it is only significant for frequencies below 1000 Hz. Therefore, it affords no protection against loud sounds with an attack time shorter than 10m/s, nor those at frequencies higher than 1000 Hz.

1.3.1.3. Inner Ear

The *inner ear* is another cavity but filled with fluid. It contains the *semicircular canals*, which provide our sense of balance, and the *cochlea*, which is responsible for our sense of hearing.

The cochlea is a small spiral tube, coiled like the shell of a snail. Running down the center of the cochlea is the basilar membrane, a flexible structure, the surface of which is covered by a carpet of tiny hair cells known as *stereocilia*; each hair cell is connected to a nerve fiber, and, when displaced, these hairs send electrical impulses to the brain.

As the stirrup pulses into the oval window, it displaces the cochlea fluid. This fluid displacement causes the basilar membrane to flex. Different frequencies of pulsation cause different displacement peaks along the membrane, thus causing different portions of stereocilia to fire. This process results in the complex incoming sound waves being separated into their component frequency bands, not unlike a spectrogram, and the corresponding neural signals transmitted to the brain.

1.3.1.4. Hearing Range

The physical makeup of the ear restricts the range of frequencies that humans are able to hear. The general rule of thumb is that humans can hear sounds between 20 Hz (hertz) and 20 kHz (kilohertz), but, of course, there are variations among individuals (see Table 1.1). As humans age, their sensitivity to higher frequencies is reduced; exposure to loud sounds can also damage the sensitive stereocilia of the ear, leading to partial hearing loss or sensitivity in specific frequency bands.

As a sound designer, your ears are your most important tool; therefore it is vital that you protect them from unnecessary exposure to loud sounds that otherwise might permanently damage them.

1.3.1.5. The Ear: Summary

Our description so far can sound like a purely mechanical process, which might give the illusion that the sounds that we hear are directly received. However, the reality of the situation is that these signals mean nothing until they are interpreted. This can result in some interesting phenomena.

1.3.2. Psychoacoustics: Perceiving Sound

Psychoacoustics is the study of sound perception. While acoustics deals with the physical action of sound in the external world, psychoacoustics is all about how sounds are interpreted and heard. Listening experiments

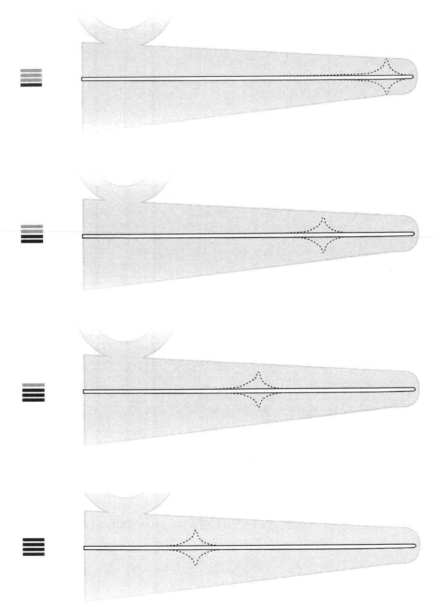

Figure 1.19   Inner Ear: The flexes of the basilar membrane, with different frequencies, create displacement peaks at different points along its length. Higher frequencies induce peaks closer to the oval window, while lower-frequency sounds induce peaks closer to the end of the basilar membrane (helicotrema).

Table 1.1   Hearing Frequency Ranges of Different Animals.

Animal	Hearing Range in Hertz
Humans	20–20,000
Bats	2000–110,000
Bottlenose dolphin	90–105,000
Cat	45–64,000
Dog	67–45,000
Chicken	125–2,000
Parakeet	200–8,500

Source: Adapted from Denny and McFadzean (2011, p. 180)

have demonstrated again and again that some of the sonic properties we might think of as absolute, such as loudness and pitch, are actually variable.

1.3.2.1.   Sonic Tricks

A prime example of this is the Shepard tone, a psychoacoustic phenomenon that demonstrates how our ears can be tricked. The aural equivalent of MC Escher's ever ascending staircase, Shepard tones appear to increase in pitch continuously, but they never actually depart from a limited frequency range. Multiple overlapping glissandi, spaced an octave apart, slide either upward (or downward) at the same rate. The glissandi fade in and fade out, masking their start and end points, and direct our ears toward the constantly repeating rising pattern in the middle of the scale. This effect is frequently used to create the impression of constant rising tension, for example in the crash sequence of the film *Flight* (2012)[8] and through the score to the film *Dunkirk* (2017).[9] It functions by taking advantage of so-called *pitch chroma*, the phenomena in which certain pitches appear to resemble each other, even when they possess a different frequency. The musical octave is an archetypal example of this phenomena. When you travel up the piano keys or a guitar string, the pitch height of each note continues to rise. But at a certain point, you will reach a note that has the same chroma of the note you started with. These notes are at different frequencies but have a similar character. The phenomena can be represented by a spiral, with musical octaves aligned vertically. Pitch chroma isn't just used in the Shepard tone, it is the basis for much musical construction and is frequently used to develop musical pattern chord progressions and phrases.

The musical notation of pitched notes as letters is a reflection of chroma. For example, all notes ascribed "G" have a similar chroma but different pitch frequencies. This can be observed in Figure 1.20. But remember, this

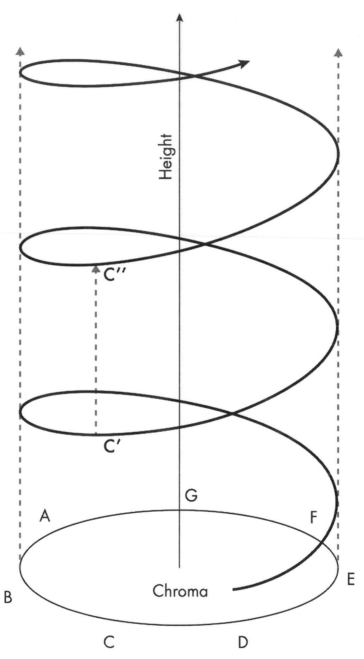

Figure 1.20 Spiral of Pitch Chroma.

phenomenon doesn't apply just to musical notes; any sound with a pitch component can possess a certain pitch chroma.

1.3.2.2. Pitch

The circular and looping natures of both the Shepard tone and pitch chroma reveal that we can't simply think about pitch as being a simple linear trajectory from low to high. Yes, pitches range from low to high, but pitch perception is actually a much more complex phenomenon.

We can describe pitch in scientific terms, ascribing specific frequencies according to the rate of oscillation (vibration), but, as we will explore shortly, these scientific definitions of pitch do not always match our perception. The standard unit for frequency is hertz, which represents the number of cycles per second in a periodic wave. Hertz can be a very useful reference scale and is frequently used within plugins and software (for example, equalizers).

However, while pitches can be scientifically described with a specific frequency value, our brain is unable to make such precise analytical judgments. For example, despite the objective value difference in number of cycles, a pitch at 980 Hz will be indistinguishable from a pitch at 990 Hz. This is due to the physiological limitations of the ear. As just explored, the basilar membrane becomes distorted at different points along its length in response to frequencies received. Two similar frequencies will activate a similar portion of the basilar membrane, thus sending common impulses to the brain and signaling two indistinguishable pitch phenomena. This limitation in pitch resolution is called *critical bandwidth*. Because of the physiology of the basilar membrane, critical bandwidth is much larger at higher frequencies than it is at lower frequencies.

This phenomenon also tells us that an increase in frequency is not directly proportional to an increase in perceived pitch. The frequency difference of an octave in a lower register is much smaller than at a higher one. For example,

In Table 1.2, both the jump from A3 to A4 and from A6 to A7 will be heard as a single octave, but the frequency changes are radically different

Table 1.2 In Both Cases, the Same Perceptual Difference (One Octave) but Radically Divergent Frequency Differences.

A3 = 220 Hz
One Octave Difference = **220 Hz**
A4 = 440 Hz

A6 = 1,760 Hz
Octave Difference = **1,760 Hz**
A7 = 3,520 Hz

in each. The same is true for smaller intervals such as between notes, while the perceived jump from one tone to the next is heard as the same pitch interval, the actual frequency distance between each will be different.

Because of differences in the critical bandwidth, a far greater frequency change is required in higher registers before distinct pitches become identifiable.

For example: C4/D4 = 262 Hz/294 Hz = **32 Hz** difference = one tone
G4/A4 = 392 Hz/440 Hz = **48 Hz** difference = one tone

This phenomenon might seem strange and nonsensical, but it actually tells us something important about how we hear pitch. Pitch is not absolute; it is a perceived phenomenon.

In most people, the perception of pitch and of pitch changes is established by a comparison, i.e. pitch A is higher than pitch B. There is no fixed reference point against which pitches are judged.[10] This means that we can recognize a common melody even if we begin at different pitch heights (see Table 1.3).

So why is this the case? Why do we hear the same semitone intervals in each version when the frequency differences are not the same? The answer is that, while the absolute frequency values are different, the ratio difference between the notes is the same in both cases. In both examples in Table 1.4, the ratios of change between the second and third notes, as set out in Table 1.3, are shown to be identical.

Table 1.3 Frequency differences for the opening phrase of "Twinkle, Twinkle, Little Star" and how these differ depending on the starting note. In both cases, the perceived interval steps are the same, and the melody can be clearly identified in each case.

"Twinkle, Twinkle, Little Star"—Starting at C4						
Note:	C4	C4	G4	G4	A4	A4	G4
Semitone intervals:		7 semitones		2 semitones		2 semitones	
Frequency:	262 Hz	262 Hz	392 Hz	392 Hz	440 Hz	440 Hz	392 Hz
Frequency difference:		**130 Hz**		**48 Hz**		**48 Hz**	

Table 1.4 Interval Ratios Calculated by Dividing the Frequencies.

C-G = 3:2 = **1.5**	392/262 = **1.496** = 494/330	**1.5** = 3:2 = E-B

In musical terms, differences between each note are described as an *interval*. These are often expressed in relation to whole tone or semitone steps. It can be useful to understand the musical terminologies used to describe pitches and intervals because many plugins and audio software programs will refer to these within their interfaces or modes of operation (Table 1.5).

For a long time in history, there was no standardization of pitch; each ensemble would tune to a different common pitch. However, despite this variation in tuning, the same melodies could still be performed by different ensembles because the ratio differences between the notes remained constant. It has only become possible relatively recently for ensembles to tune to an absolute common pitch (yet there is even variation in this between the United States (440-Hz) and Europe (442 Hz)).

What is most important to remember is that pitch is not absolute; when working with pitches, it is often more useful to think in terms of pitch change and the variations between pitches rather than some absolute reference point or scale.

1.3.2.3. Loudness

Loudness is also a relative phenomenon. You will find many different scales of loudness in the audio tools that you use, each of which defines volume in relation to a fixed reference point. However, the fixed reference point changes in each loudness scale. This is due to the fact that loudness is perceived as a variation in intensity, not against a fixed point, and that perceived loudness is also affected by the frequencies present within the sound.

Therefore, while all of the terminology around the scales and measurements for loudness might at first make it seem as though it is a scientifically quantifiable effect, we can quickly see that these scales have been applied by humans in order to try and approximate the flexible ways in which we hear and interpret sounds.

If we identify a sound as being 90 dB "loud," what we really mean is that this sound is 90 dB louder than another sound. This "other" sound is

Table 1.5 Range of Different Terminology for Describing Pitch.

Cent = 100th of a tone
Semitone = ½ of a tone = 50 cents
Whole tone = 100 cents
Unison = identical tones = 0 cents
Octave = same chroma different pitch = +/−1,200 cents

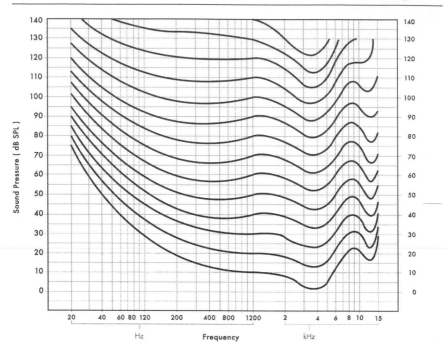

Figure 1.21 Fletcher-Munson Curve: These display how our sensitivity to different frequencies affects how we hear. Our ears are most sensitive to sounds in the 3- to 5-kHz range, and thus we are able to hear them at a very low level. However, 40-Hz frequencies need to be ten times louder before we can hear them at the same intensity.

generally taken to be the quietest sound that a human being could normally be reasonably expected to hear (otherwise called the *threshold of hearing*). However, as is visible from the Fletcher-Munson Curve (Figure 1.21), human sensitivity to sound changes across the frequency range; therefore sounds at 30 Hz need to be a lot louder to become audible than sounds at 3,000 Hz do. This is the reason that white noise sounds so harsh. Scientifically, white noise is made up of random fluctuations of sound all at the same intensity. However, because our sensitivity to sounds is higher in the midrange, we perceive white noise to be louder in this portion of the frequency spectrum. Pink noise, on the other hand, has been logarithmically attenuated in order to follow the contour of human loudness sensitivity; therefore, scientifically, the loudness varies across the frequency range, but perceptually pink noise sounds equally loud at all frequencies.

Some audio equipment can be adjusted to take human loudness perception into consideration. With sound pressure level (SPL) meters, you can

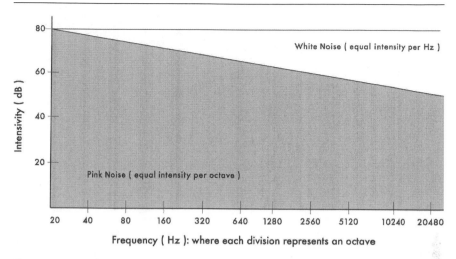

Figure 1.22   White Noise versus Pink Noise.

often switch between A and C weighting scales. *A weighting* represents the scientific measurement of loudness and is used to test the response of loudness within audio systems and equipment. While *C weighting* follows the human loudness contour and therefore relates to the listeners' perception of sounds.

Decibels are the common unit of loudness measurement, but there are a range of variants. These variants are usually identified with appended acronyms. For listening, loudness is usually defined in relation to the threshold of hearing, identified as 0 dB SPL (sound pressure level). Loud sounds are then measured against this, for example the loudness of an average conversation is measured at 60 dB SPL, while an underground train in motion creates a sound that reaches 110 dB SPL.

Within audio systems, loudness is measured from negative infinity (silence) upward. The maximum amplitude level that digital systems can accommodate before clipping is 0 dB full scale (0 dBFS). *Clipping* occurs when the amplitude of the incoming signal exceeds the dynamic range of the system; when this happens, the peaks and troughs of the sound wave stretch beyond the range that can be encoded, and so these extremities are truncated and lost. With the tips of the waves removed, the waveform is altered, with distortion and unwanted artifacts introduced. For example, a sine wave that is clipped becomes a square wave, thus acquiring additional harmonics not present within the original sound (Figure 1.24).

Clipping can occur at any gain stage within the signal chain and in either analogue or digital systems; however, digital clipping is far more problematic. Each track in your DAW possesses a level meter and will

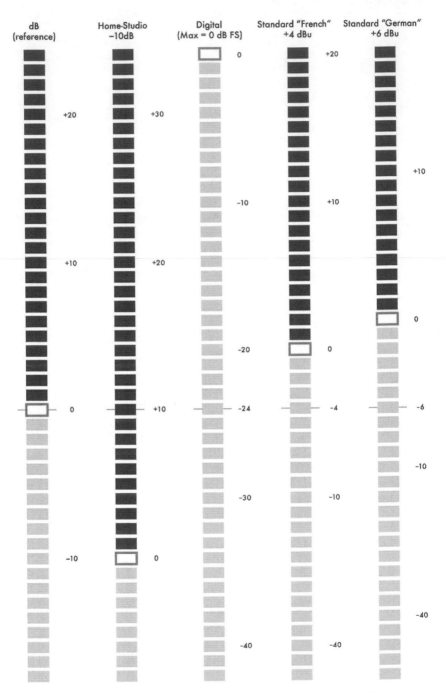

Figure 1.23 Comparison of Different Decibel Scales of Loudness Measurement: Note the differences in available headroom.

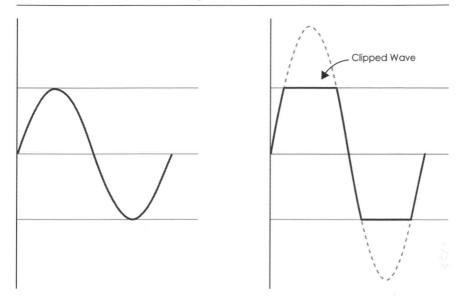

Figure 1.24 Clipping: This occurs when the amplitude of a wave exceeds the range of the system. This truncates the waveform, introducing distortion.

identify where clipping occurs. Any clipping should be avoided on individual channels as well as on the master fader, which represents the final combination of waveforms.

Within the decibel scale, an increase of +6 dB equates to a doubling in intensity.[11] The logarithmic nature of the loudness scale means that this remains consistent wherever you are in terms of overall loudness. Therefore, the increase from −6 dB to 0 dB is a doubling in loudness, as is the increase between −24 dB and −18 dB. The most important thing to remember is that whenever you move a fader up +6 dB, you are doubling the sound output and that when you reduce it by −6 dB, you are halving the loudness of that sound.

If you are seeking to make something appear loud but your faders are already maxed out, you should consider playing with the psychoacoustic options available to you. Because the ear is more sensitive to frequencies within the 3-kHz range (see Figure 1.21), boosting amplitude in this frequency range will make sounds appear louder to humans without causing clipping within the analogue/digital signal chain. Another solution might be to introduce greater contrasts in loudness, by dropping the level of other elements in the mix or adding moments of silence. Quieter contexts will foreground your louder sounds. Think, for example, of how it always gets very quiet in horror films before the monster crashes abruptly into the

scene. One last solution can be to introduce very light levels of deliberate distortion into the mix. Our ears recognize distortion as a cue that we
are listening to very loud sounds, and therefore introducing subtle distortion can trick listeners into thinking that the sound they are listening to is
louder than it actually is.[12]

Exposure to loud sounds can cause permanent hearing damage. The
longer you are exposed to loud sounds, the more likely you are to suffer
permanent hearing damage. The louder the sound, the less time you can
be exposed to it before hearing loss occurs. Note in Figure 1.25 how the
exposure time halves for every increase in 3 dB.

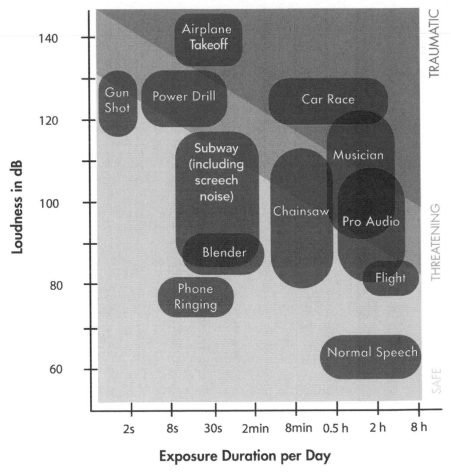

Figure 1.25   Chart of Loudness Exposure and Risk.

1.3.2.4.   Duration

All sounds possess *duration*, and the spectrum and character of the sound may shift and change over this duration. Just as with pitch, our hearing has limits in terms of perceiving duration. Some sounds may last so long that they begin before we arrive and end long after we have departed. Conversely, our ears are unable to resolve distinct sounds when they occur less than 10 ms after one another. Two beeps 12 ms apart are likely to be heard as separate sounds, but the same two beeps played 8 ms apart will be heard as one continuous sound.

This fact has led a number of contemporary musicians and composers to suggest that pitch and rhythm, often situated as distinct characteristics, are actually on the same continuous time scale. If you take a recording of a pitched sound and slow it down, eventually you will reveal a rhythm, the oscillations of the vibration. And if you take a rhythmic sound and speed it up, you will eventually reach a continuous pitch. This is a key feature of the hardcore techno style Speedcore, where the tempo ranges between 300 bpm and 1,000 bpm, thus converting the drumbeats from rhythms into tones. Karlheinz Stockhausen created the term Unified Time Structuring to explain this confluence of pitch, duration and timbre (see Chapter 12, "Electroacoustic Music: An Art of Sound").

1.3.2.5.   Timbre

Timbre is a composite phenomenon, affected by a wide range of parameters. This makes timbre one of the trickiest characteristics of sound to define, but it is also one of the most significant. Timbre is the color, or character, of a sound, the aspect that allows us to distinguish between a glass and a bell or between the crunch of a boot on gravel and the clop of heels on concrete. Timbre is one of the main parameters that we sculpt in sound design.

The spectrum of frequencies present within the sound, as well as its envelope, will affect the overall timbre. As the loudness of a sound changes, so often does its timbre (often with an increase of higher-frequency content). The relative balance of pitched to noisy material within the spectrum will define the timbre of the sound. Any resonant characteristics of the source will also contribute to modulate the overall characteristic of the timbre through formants. *Formants* are created when the harmonics of the sound reinforce themselves within a congruent physical space. Where the dimensions of a physical space match the frequencies present in a sound, these frequencies will be reflected back in phase with the original sound; where they meet, they will multiply together and thus increase their loudness. In special situations, the scale of the resonant space can be altered to morph the timbre of the sound. This is exactly what happens with the mouth, where changing the shape and the size of the cavity modulates the timbre of the sound. If you imitate the sound of a passing car, then you are performing this modulation, changing the size of your vocal cavity to shift the timbre and pitch of the resulting sound.

Playback speed also modulates timbre, shifting the character of a recorded sound and affecting the perceived scale of the sounding object. Objects dropped in pitch tend to sound as if they emanated from larger physical objects, and this feature is often used to enhance and make recorded sounds appear more impressive and impactful. Parameters of a synthesizer are all designed to shift and sculpt the resulting timbre of the sound, from the character and balance of the various oscillators, to the cutoff frequency and resonance of the filter, to the ADSR (attack/decay/ sustain/release) envelope; all of these parameters are designed to mold and change the timbre of the resulting sound.

Timbre's composite nature means that it is influenced by a wide array of factors, some of which we have explored here. New tools are constantly being developed in order to allow greater control and manipulation of timbre, and these allow us to access timbre as a primary sonic and musical characteristic—morphing between timbres, transforming and shifting the character of the sound. *Timbral fusion* can be used to merge sounds into one mass, while *timbral difference* can be used to ensure that different sound materials remain distinct and do not clash.

1.3.3. Key Points

In this section, we have explored the physical and acoustic properties of sounds. These characteristics influence the way in which we hear sounds, but, as we have discussed, there is more going on than simply receiving these sounds. Our brain interprets and seeks to make sense of complex information reaching us. In a creative context, we can take advantage of these processes to create compelling and engaging soundscapes to help drive narrative and atmosphere.

These characteristics of listening, in combination with an understanding of the physical properties of sound, can be vital guides for understanding how others hear the sounds that we create. They also define and direct the development the audio tools and technologies with which you will work.

- The ear is a complex organ designed to interpret sounds within a specific frequency range.
- The brain can be fooled by sonic tricks to hear things that do not exist physically.
- Pitch is relative (not absolute).
- Loudness is relative (not absolute).

- +6 dB equates to a doubling in loudness.
- Timbre is one of the key parameters for sound design.
- Your hearing is your most important tool as a sound designer—take care of it.

1.4. Microphones and Recording Technology

This section provides an overview to the tools and processes of recording sound. As it does so, it will highlight practical considerations and concerns that can help to direct and guide your decision-making process, as you select the best ways to capture different sounds.

1.4.1. Microphones

Microphones are instruments designed to capture sounds. The characteristics and properties of sound (explored in the first section) inform their design and development, and while there is continual ongoing research to deliver improved sensitivity and frequency response, the basic mechanics of the microphone have not changed.

Microphones have been specifically adapted to capture sound traveling through the air. There are a number of different types, but, fundamentally, each type works to transform the kinetic vibrational energy of the sound wave into an electrical impulse. The microphone capsule is the component that makes this transformation, and its delicate electronics are often protected by metal grids and sometimes also by foam shielding.

1.4.1.1. Dynamic Microphones

Dynamic microphones are rugged and able to withstand high sound pressure levels. They are often used in live music contexts where robustness is essential. But they can be highly useful for recording loud sounds and in situations where other microphones might be too delicate.

A diaphragm moves in response to sound wave vibrations, driving an attached copper coil back and forth over a magnet. An electrical voltage is induced through the process of electromagnetic induction, as the magnetic fields of the copper coil and magnet cross. The induced voltage directly relates to the movement of the diaphragm and coil and therefore to the incoming sound wave.

Because the dynamic microphone utilizes the process of electromagnetic induction, it does not require its own power to operate. However, because the mass of the diaphragm and copper coil is relatively large (in comparison to other capsule designs), the dynamic microphone struggles

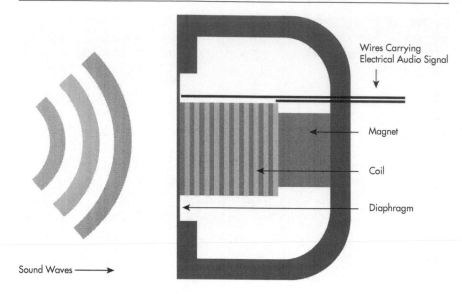

Wires Carrying
Electrical Audio Signal

↓

Magnet

Coil

Diaphragm

Sound Waves ⟶

Figure 1.26   Cross Section of a Dynamic Microphone.

to capture subtle or low-energy sound waves. Therefore, where more detail is needed, a different instrument is often called for.

1.4.1.2. Condenser Microphones

Condenser microphones are able to capture higher levels of detail than dynamic microphones. They are the ideal choice for studio recording contexts but are often used outside of the studio as well. They operate through the principle of capacitance (and thus are sometimes called capacitor microphones). A thin electrically conductive diaphragm membrane is held in close proximity to a fixed metal plate, while an external power source provides a voltage to set up an electrical charge between them. As the thin conductive diaphragm moves back and forth against the fixed metal plate, the capacitance is changed. This variation in capacitance reflects the incoming sound wave as an electrical signal.

Because the mass of the conductive diaphragm is extremely low, it is able to move more freely and thus can reflect considerable subtleties in the incoming sound wave vibrations. As a result, condenser microphones offer a much higher level of sonic detail and are particularly excellent at capturing detailed transient responses. They also have the widest frequency responses of any microphone type.

The downsides to the condenser microphone are that their high sensitivity often makes them vulnerable to wind noise or large changes in sound pressure level[13] and that they also require an external power source.

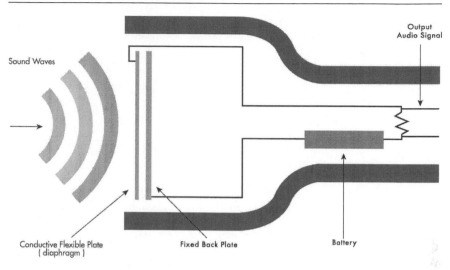

Figure 1.27 Cross Section of a Condenser Microphone.

However, power for microphones has been conveniently standardized since the 1960s through a technology called Phantom Power. Most mixing desks and location sound recorders will provide Phantom Power to condenser microphones, and you will sometimes find it abbreviated +48 V or Ph48. representing the 48 V that pass back down the standard 3-pin XLR cable in the opposite direction to the incoming electrical sound signal. *Note*: Without phantom power (or batteries), condenser microphones will not operate.

1.4.1.2.1. Small or Large Diaphragm?

Condenser microphones can be further divided into those that have larger or smaller diaphragms. Small diaphragm condenser microphones are often pencil shaped, long and thin, with the diaphragm at one end. Large diaphragm condenser microphones are larger and side addressed.

Large diaphragm condenser microphones have a much better signal-to-noise ratio, as the larger diaphragm is able to capture more sound energy, but they present an uneven frequency and polar pattern response, meaning that the sound they capture is often less "realistic." However, this coloration of the sound can actually be desirable. The widening of the polar pattern at low frequencies helps to avoid the proximity effect and results in a rich low-frequency boost to recorded sounds. Large diaphragm microphones can make sound objects appear vibrant, rich and "larger than life."

Small diaphragm condenser microphones are more accurate in almost all other aspects. They have a much better transient response—because a smaller diaphragm can more accurately and consistently respond to incoming sound waves—they possess an extended frequency response, and they have a highly consistent pickup pattern across different frequency ranges. Therefore, small diaphragm condenser microphones are the best choice for capturing transparent and highly detailed recordings.

1.4.1.3.   Contact Microphones

A *contact microphone* is designed to capture sound moving through solid surfaces. Therefore, unlike the dynamic and condenser microphones, it

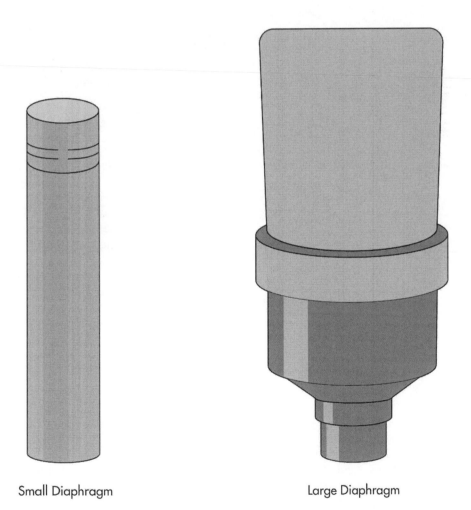

Small Diaphragm   Large Diaphragm

Figure 1.28   Examples of Small and Large Diaphragm Condenser Microphones.

does not have a diaphragm but instead uses piezoelectric materials that generate electrical current in response to surface-borne vibration or tactile stress. They are passive but often require a preamplifier or high-impedance transformer to capture the fullest range of frequencies.

Contact microphones can be attached to instruments but can also be excellent for capturing rich material or otherworldly sounds. They were used extensively in the soundtrack to the 2013 film *Gravity* to capture the structure-borne vibrations of the spaceships. They can be made inexpensively and are great tools for discovering new sound worlds.[14]

1.4.1.4. Hydrophones

A relative of the contact microphone, a *hydrophone* is designed to capture underwater sound. Just like the contact microphone, it uses piezoelectric materials to capture vibrations traveling through liquid. Their response is also greatly improved by the use of a preamplifier or high-impedance transformer to balance out the frequency range.

The ocean is excellent at conducting low-frequency and infrasonic sounds, and so many sounds recorded with hydrophones can be sped up to create materials audible to humans.

1.4.2. Polar Patterns

As we explored in section 1.2, sounds travel in three dimensions, and the construction and function of different microphones mean that they each capture sounds traveling in from different directions. This focused directionality allows us to position microphones to capture certain sounds while avoiding others. Sound sources can either be described as *on axis* (meaning that the polar pattern will pick up their sounds) or *off axis* (meaning that the polar pattern will not pick up these sounds). The four basic polar patterns are:

- Omnidirectional.
- Cardioid.
- Bidirectional (figure-of-eight).
- Hypercardioid.

1.4.2.1. Omnidirectional

Equally sensitive to sound coming from all directions, omnidirectional microphones are perfect for capturing ambiences and are also less susceptible than other, more focused patterns to both wind noise and the proximity effect (see section 1.4.3). Lavalier microphones, attached to individuals, are often omnidirectional for precisely these reasons.

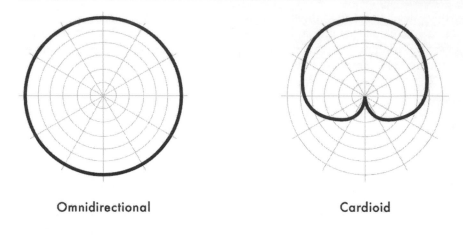

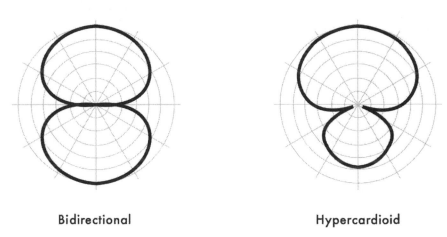

Figure 1.29   Common Microphone Polar Pickup Patterns.

1.4.2.2.   Cardioid

The *cardioid* pattern is heart shaped and captures sound primarily from the front, while sounds from the sides are somewhat quieter and sounds from the rear are rejected completely (these are sometimes called *unidirectional* microphones).

Cardioid microphones are best used to capture specific sound sources and as a result are often the most frequently used microphones. They are excellent for recording anything that is desired dry and close.

1.4.2.3. Bidirectional (Figure-of-Eight)

A *bidirectional* (or figure-of-eight) microphone is sensitive to sounds coming from the front and the rear but rejects anything coming from the sides. This microphone is most often applied in combination with other microphones, where its unique abilities combine to provide a wide range of advanced multichannel possibilities.

One such example is the mid/side (M/S) pairing, in which a figure-of-eight microphone is combined with a cardioid, thus providing a wide "side" image on one channel with a focused frontal "mid" sound image on the other. In postproduction, this allows the editor to mix in the stereo width of the recorded sound, either focusing in on the frontal "mid" section or the wide "side" section.

1.4.2.4. Hypercardioid ("Shotgun")

In some situations, standard cardioid microphones are simply not directional enough. The *hypercardioid* microphone takes advantage of phase cancellation principles to create a very focused and directional pattern. Often referred to as Shotgun microphones because of their long barrel, these microphones are used on film sets to capture actors' dialogue and are ideal for recording in noisy or busy environments.

The barrel of the microphone contains side slots into which the off-axis sound passes. These off-axis sounds are out of phase with the on-axis sounds that pass directly down the barrel, and therefore, through a process of phase cancellation and interference, the off-axis sounds eliminate one another. This provides increased directionality but can result in some uneven phase characteristics. These microphones can struggle in particularly reverberant spaces, where the delay time of the early and late reflections (out of phase with the original signal) are a key part of the sonic characteristic.

1.4.2.5. Multichannel Microphones

As previously mentioned e, each of the main microphone types can be combined to create hybrid multichannel patterns, but some microphones are specifically designed to capture multichannel sound on their own. Stereo microphones, such as those on many handheld recorders, are common, but you may come across more complex multichannel microphones such as ambisonic microphones. These often record multiple axes of sound and take advantage of complex mathematics to decode and interpolate the spatial image of the sound.

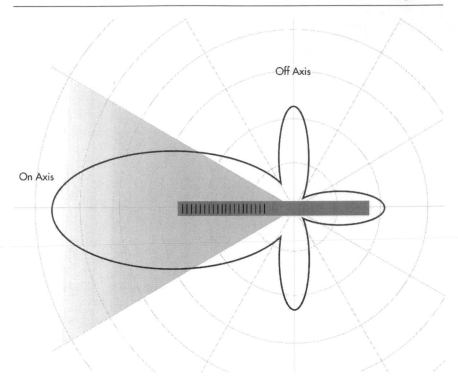

Figure 1.30 Polar Pattern of a Shotgun Microphone, Demonstrating Its High Directionality: Sounds traveling off axis are eliminated through phase cancellation as a result of the holes along the barrel of the microphone.

1.4.3. Microphone Positioning

The positioning of microphones must take into consideration the sonic properties of the primary sound source—how it resonates, any movement—and any additional sound sources and reverberation or ambience in the recording location. There are no hard-and-fast rules because individual recording situations and desired goals vary widely. Rather than presenting specific microphone setups that you might copy, the following section highlights some key factors that you should consider in selecting where to position your own microphones. The intention is that this will help you to focus on the primary consideration, which should always be the sound that you want to capture.

1.4.3.1. What Are You Recording? (Primary Sound Source)

First, we need to consider the object of our focus, its properties and how it vibrates. As explored above (section 1.2.1), each physical sound source

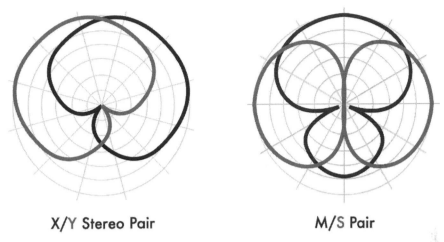

X/Y **Stereo Pair**   M/S **Pair**

Figure 1.31   Two Common Stereo Microphone Setups: The X/Y stereo pair utilizes two cardioid microphones with capsules aligned but set pointing away from each other at a 90° angle. One channel captures sound from the right, while the other captures sound from the left, thus providing an impression of stereo space. The M/S (mid/side) pair combines a hypercardioid microphone, which captures sound directly ahead (mid), with a bidirectional (figure-of-eight) microphone, set at right angles, to captures sound to the left and right (side).

possesses unique properties. The microphones that we choose and how we position them must relate to the properties of the vibrating object. Some objects may radiate sound in all directions, while some may have a more focused frontal propagation pattern. Therefore, moving the microphone around an object may radically change the qualities of the sound captured. It is often worthwhile experimenting by shifting the microphone around your desired sound source, adjusting its relative angle, height and distance in order to find the position that highlights the qualities you desire the most. Try to imagine how the object vibrates in order to work out where you might best place your microphone. Through experience, you should build up an impression of which microphone positions work best for specific sound object types.

Once we have considered the sound source itself, we need to consider its context and location. Are other sound sources nearby that are undesirable? What sonic perspective are we seeking to capture? Is the reverberation a key part of the sonic characteristic that makes this sound appealing?

The closer we position a microphone to a sound source, the greater the signal (desirable sound) to noise (unwanted sound) ratio will be. If we try to record dialogue across a room with the microphone a long way away, we will actually capture a mixture of the desired dialogue along with any other sound sources in that room and the reverberant sounds of the space. If, on the other hand, we position a microphone in close proximity to the speaker, then the dialogue sound will be recorded at a much higher level than any of the other sounds within the space; thus we will have a strong signal-to-noise ratio.

As we learned from the Inverse Square Rule (section 1.2.2), sound dissipates exponentially as it radiates from the sound source. Therefore, it is often desirable to position the microphone as close to the source as possible for a strongly recorded signal. However, the ideal distance between the source and the microphone will be dictated by the sonic perspective that we are seeking to capture.

1.4.3.1.1. Proximity Effect

The proximity effect is a coloration of sound caused by positioning the microphone very close to the sounding object. It causes a boost in the lower frequencies, creating a richer and fuller low-frequency sound. Sometimes this is undesirable, making instruments or objects sound boomy, while in other cases this can actually lend weight to vocals and make key objects sound more impressive.

Omnidirectional microphones are not affected by proximity effect, while cardioids are somewhat susceptible and bidirectional (figure-of-eight) microphones are highly susceptible. You can therefore use decisions over polar pattern type in order to either promote or dissuade proximity effect in your recordings.

1.4.3.2. Where Are You Recording?

The context within which you record sound often plays a big part in its character and coloration. If the space is a key part of the sound that you are seeking to record, then you should aim to capture it with a slightly more distant microphone position or the application of multiple microphones. The M/S rig is highly useful within this context as it provides a central channel that can be focused upon the sound source and a separate channel of wide ambience that can capture spatial context. These independent channels can then be mixed within postproduction to deliver the final desirable balance of central and ambient sound.

Whatever you are seeking to record, you should aim to capture individual sound sources on separate audio channels. This will give you a high level of control within the postproduction process. Pay particular attention to other sound sources within the space and how they might interfere with the recording of your desired sound. Often you may need to position microphones in a specific way that can keep unwanted sound off axis.

1.4.3.3. Does the Sound Source Move?

If the source is moving, then its trajectory needs to be taken into consideration, this is not just applicable to fast moving objects such as a car or train but can also be important for performed sounds, such as Foley or ADR (automated or automatic dialogue replacement), in which the performer may need to make physical movements in order to achieve the desired performance. Microphones should be positioned in such a way that the sounds of the performance will still fall within the polar pattern, even with the gestures and movement of the performer.

For larger movement impressions, it may be suitable to use an array of multiple microphones. The coincident XY pairs found on many handheld recording devices (e.g. Zoom H5, Tascam DR100) offer a balance between a reasonable stereo spread for capturing some lateral movement, while retaining a clarity and focus for individual sound sources. The coincident capsules are positioned as close to one another as possible and thus receive the same wavefront signal at almost identical times. In contrast, spaced arrays offer a much wider stereo pattern that can capture a greater impression of movement, but they are less suited to capturing individual sound sources. This is because the capsules are distant from one another and so therefore will receive very different wavefront patterns, thus they are susceptible to phase cancellation effects when the signals are combined within postproduction.

It is important to take distance from the sound source into consideration when using multiple microphones in a spaced array because uneven distances between the sound source and the microphone can result in phase cancellations caused by the different lengths of time that the sound wave will take to reach each microphone. This was explored above in section 1.2.3.3.

1.4.4. Signal Chain/Gain Staging

A well positioned microphone is only the start of the signal chain. One of the most important and misunderstood aspects of recording is the bit that comes after the microphone, the signal chain, which transfers and amplifies the electrical signal induced in the microphone.

Microphones deliver relatively modest output signals (though some are stronger than others) and these signals therefore need to be boosted and amplified so as to make the signal usable. Preamplifiers boost the level of an input signal up to "line" level, a standard used in a wide range of studio equipment. The quality of the preamplifier can thus greatly affect the quality of the signal from the microphone. Many recording devices come with their own inbuilt preamplifiers, but these often pale into comparison with more professional or bespoke units.

However, it doesn't matter what preamp you are using if the levels are set incorrectly. There are three main variables to consider:

1. Loudness of the sound source
2. Distance between the microphone and the sound source
3. Sensitivity of the microphone

These parameters will help to guide you in setting the correct level on the preamplifier. A loud source will obviously induce a higher signal from the microphone than a quiet one, as will a microphone that is placed closer to the sound source. Condenser microphones are generally more sensitive, offering a much higher output level than a dynamic microphone, and thus they require less gain.

The most common error in the signal chain is to boost the incoming signal too much and to introduce distortion or digital clipping (see, sections 1.3.2.3 and 1.4.5). Distortion and clipping irretrievably damage the input sound, and you should seek to avoid them at all costs.

There are often a number of different points along the signal chain where the level can be boosted. So you should take great care to ensure that the settings are consistent throughout the signal chain. There is no point in boosting the signal at point A if it is cut at point B before being boosted again at point C. Take advantage of your highest-quality preamplifier to do the majority of the work and allow the other gain stages to pass the signal without affecting it. Some preamps possess both an input gain and an output level. These dual-level controls can allow you experiment creatively by forcing the preamp to create analogue dynamic range compression that will color the sound.[15]

For maximum clarity, it is usually most appropriate to keep the gain setting as low as possible while using the output level to boost the signal to the desired amount.

The low noise levels of digital recording mean that signals can be cleanly boosted in postproduction, and there is therefore no need for incoming signals to be boosted to such maximized volume levels within the input.

1.4.4.1. Headroom

Headroom is buffer space in the dynamic range that can accommodate for any unexpectedly loud sounds in a recording. You should always aim to leave about 10 dB of headroom when you set the gain on a preamplifier. This provides a relatively strong signal, while allowing for unexpectedly loud transients or flourishes to still pass through without clipping. Some hard disk recorders include limiters or compressors in the signal chain, but these affect the dynamic range of your sound and therefore should only be used as a last resort.

Remember to check the audio level at each point in the signal chain. Clipping can occur at each point (gain stage) in both analogue and digital domains.

1.4.4.2. Dynamic Range Compression: Compressors, Limiters, Expanders and Gates

These dynamic range processes can be found both within recording and postproduction studios and digital audio workstations, as well as in many location sound recorders.

Dynamic range compression is a process by which the dynamic range of an audio signal is reduced. It decreases the difference between the highest and lowest amplitudes of a sound wave, making louder things quieter and quieter things louder and in the process reducing the range in volume between the loudest and quietest sounds. It can be highly useful in smoothing out amplitude changes and creating a more consistent audio signal, with fewer variations in loudness over time. For example, compression is often applied to recordings of speech within postproduction.

However, excessive or incorrect usage can radically alter sounds and introduce unwanted noise into your mix. Therefore, when recording, it is often best to not rely on the tools that create dynamic range compression by default but to instead allow for appropriate headroom. This doesn't preclude you from undertaking any dynamic processing later in the postproduction phase, and it ensures that you capture a recording of the full dynamic range of a sound source.

Compressors have four main parameters:

1. *Threshold*—the amplitude at which the compression process is triggered. Usually measured in decibels, any sound at a level below the threshold will not be affected.
2. *Ratio*—the extent to which sounds above the threshold will be attenuated. Usually expressed in a ratio e.g. at a ratio of 2:1, the amplitude above the threshold is halved; at a ratio of 4:1, the amplitude above the threshold is quartered.
3. *Attack and Release*—the amount of time it takes for the compression process to take effect (usually measured in milliseconds (ms)). These parameters can radically affect the resulting sound. A *short attack* will introduce the compression suddenly, as soon as the threshold is passed. While a *slow attack* will require an input level that consistently exceeds the threshold for the compression process to kick in.

Likewise, a *short release* will immediately cease the compression process once the input level falls below the threshold, while a *long release* will continue to compress the audio signal even when it no longer exceeds the threshold.

4. *Knee*—allows the user to smooth out the onset of the compression by increasing the softness setting. A *hard knee* will engage the compression as soon as the threshold is reached, while a *soft knee* will increase the compression ratio as the input level increases, thus smoothing out the onset of the compression and, in some cases, masking its operation.

Limiters are special types of compressors that have a fast attack time and a high ratio (often 60:1 or above, sometimes expressed as ∞:1). These can be useful in emergencies for controlling unexpectedly loud sounds above a certain threshold, which would otherwise introduce clipping and distortion.

Expanders function in the inverse way to compressors. They increase the dynamic range of a sound by further attenuating the amplitude of any input signal that is below a defined input threshold. Therefore, only sounds that pass the threshold become audible. Expanders can be used as noise gates to reduce or eliminate portions of silence between distinct loud impulses.

Compression and expansion can be used creatively to enhance sound but can also create unwanted effects within the recording process.

1.4.5. Digital Audio

The majority of the recording that you'll do these days is with digital technology. Therefore, it is vitally important to understand how sounds are encoded digitally so that you can most effectively work with materials and deliver the correct formats to your clients and collaborators. Analogue recording formats can still be found within creative contexts, but digital is the current industry standard. Digital allows for a much greater signal-to-noise ratio and greater flexibility in duplicating, storing and transferring data.

Because computers work in a binary fashion (something is either a 1 or 0), sound wave information needs to be converted into this format before it can be understood and processed. Sound waves are continuous streams that must be converted into a series of finite units within the digitization process. This is accomplished by taking discrete samples of the continuous waveform, with each sample point representing the amplitude of the wave at a specific point in time. These samples provide a rough approximation of the smooth waveform that can be understood by the computer.

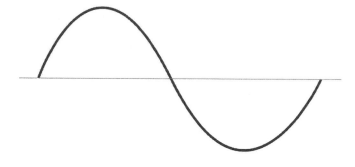

Analog Sound Wave

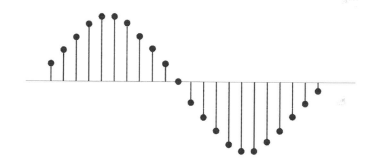

Digital Sound Wave

Figure 1.32   Continuous Analogue and Discrete Digital Waves.

1.4.5.1.   Sample Rate

Sample rate is a measure of the number of samples taken each second. Higher sample rates provide more accurate approximations of the incoming wave but require greater amounts of data storage to hold this information. The standard sample rate for work with moving image is 48 kHz (48,000 samples per second), but there are a number of other standards in use for different media.[16]

The decision to settle on the 48k-Hz sample rate is twofold. First, the maximum frequency that can be encoded into a digital signal is defined

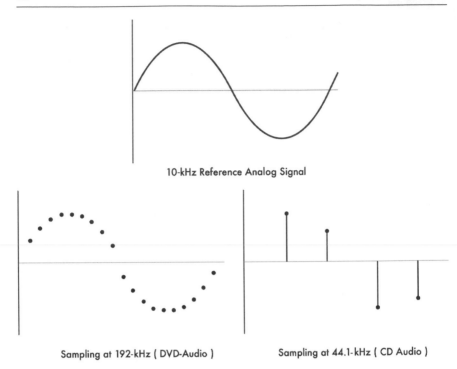

10-kHz Reference Analog Signal

Sampling at 192-kHz (DVD-Audio)   Sampling at 44.1-kHz (CD Audio)

Figure 1.33 The Impact of Sample Rate Difference in Encoding an Analogue Signal: Higher sample rates provide a more accurate approximation of the analogue wave.

by the rate of sampling. In order to capture a wave in discrete terms, you need to be able to record both a maximum and minimum points. At lower sample rates, the sound wave can be misencoded, resulting in aliasing and the rendering of wavelengths not actually present in the original signal (Figure 1.34).

Therefore, in order to accurately encode frequencies up to the limit of human hearing (20 kHz) you need a sample rate of at least 40 kHz. Thus, the sample rate for CDs was established at 44.1 kHz giving a maximum recorded frequency of 22.05 kHz.

Second, in order to retain consistency with video and avoid potential sync issues (picture and sound out of alignment), a standardized rate was selected that was concordant with the 24 fps of standard film. Thus, the 48 kHz sample rate allows sound and image to be locked together within a single project and not go out of sync.

Each session that you load in a DAW will have a setting for the sample rate. Most contemporary audio software will recognize when you try to import audio at different sample rates and adjust accordingly. However, in

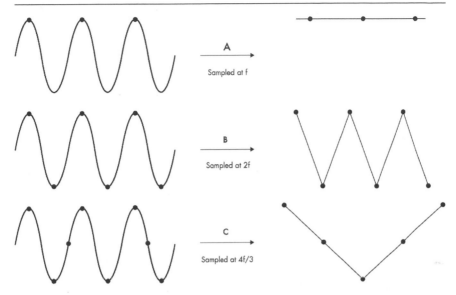

Figure 1.34 Sample Rate and Frequency: Sample rate needs to be twice that of the incoming frequency (f) in order to encode a sound wave (example B). Where the sample rate is the same as that of the frequency, no wave will be recorded (example A).

some cases, the software will simply read the audio that it is given at the rate of the project session. The result is that 44.1-kHz sounds loaded into a 48-kHz session would be read at the session rate and therefore play back faster than originally intended, resulting in an increase in pitch. The inverse is true for 48-kHz sounds being imported into a 44.1-kHz music session: the 48-kHz audio will be read more slowly by the 44.1-kHz session and therefore will be stretched and decreased in pitch. This change not only alters the pitch of the sound but will also affect the timings and therefore can result in your dialogue being both transposed and shifted out of sync.

1.4.5.2. Bit Depth

Bit depth defines the detail in amplitude and thus the ratio between the loudest and quietest sounds that a digital file can encode. The higher the bit rate, the better the signal-to-noise ratio of the encoded sound.

The most common bit depths that you will encounter are 16 and 24 bits. Twenty-four bits most closely approximate the sensory response of the human auditory system and therefore provide the best solution. Again the trade-off is between the quality and the size of the encoded file.

The real benefit to increasing bit depth is in the quieter portions of the sound wave, increased resolution enables many more subtleties to be

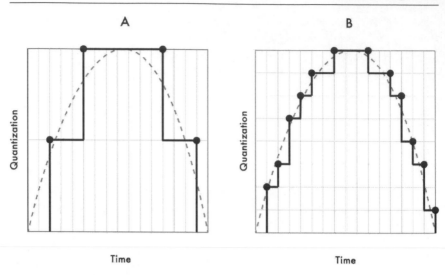

Figure 1.35 Bit Depth: In encoding information about the amplitude of the sound wave, higher bit depths encode a greater dynamic range (B) and thus can represent more subtle changes in loudness than lower bit depths (A).

encoded in the digital waveform and for the full dynamic range of sound sources to be accommodated.

1.4.5.3. Digital Compression: File Sizes

Common consumer formats compress file information to improve convenience for the end user in terms of downloads and storage.[17] But loss of information means loss of quality. When working with compressed formats, you will often encounter the term "bit rate," which should not be confused with bit depth.

Bit rate is a computer term that defines the amount of information being processed each second. For a 48-kHz 24-bit stereo sound file, the bit rate would be 2,304,000 bits (2.3 MB per second) (Table 1.6).

1.4.5.4. Digital-to-Analogue Conversion

In order to hear sounds resulting from digital processing or synthesis, we need to convert the digital information into an analogue waveform. This process is done through digital–to–analogue conversion (DAC), which converts discontinuous digital sample rate information into a continuous electrical current used to drive a loudspeaker.

Table 1.6 Calculating Bit Rate from Sample Rate and Bit Depth for a Stereo File.

Sample rate × (Bit depth × 2 (for L and R channels)) 48,000 × (24 × 2)	= Stereo bit rate at 48 kHz = 2,304,000 bits

As with conversion into the digital domain, *aliasing* can occur where the frequencies of sounds in the digital file exceed the Nyquist frequency. This can result in the introduction of frequencies in the output signal that are not present in the original file. Filtering processes are automatically applied in order to eliminate these unwanted subharmonics, but in some situations these filters can introduce their own distortions as they remove essential harmonics above the cutoff frequency.

Because amplitude information is encoded at a relatively lower resolution than that of frequency (through sampling), there is a greater capacity for error. When a signal is digitized, there can be a potential for banding at certain points where subtle variations are averaged out within the production of discrete values. This banding can cause audible disruption to signals, and therefore the technique of dithering is applied to eliminate it. *Dithering* introduces a factor of randomization into the signal in order to break up any undesirable grouping at specific intensities and thus acts to smooth out any aliasing within the output signal.

1.4.5.5. Key Working Practices in Digital Audio

It is vitally important to understand and keep track of the sample rate and bit depth with which you are working and to remain consistent throughout your project. Once you downsample to a lower rate, you discard details and information about the sound wave that you can never retrieve. Moving from a lower sample rate to a higher one does not input any new information; it simply remaps the lower resolution across a higher rate format. Compressed audio formats, such as AAC or mp3, eliminate large proportions of the audio data, sacrificing information about the sound wave and instead relying on psychoacoustic tricks to give the impression of a full spectrum. You should take care when employing such compressed files within projects as they will be much less accommodating for transformation and processing, due to being much more susceptible to digital errors, artifacts and aliasing.

For creative purposes, working at a higher sample rate provides greater flexibility in terms of transformation and processing. The trade-off is always that working at higher sample rates requires both a machine that can process information at a higher data rate and larger amounts of storage space. Audio transformations rendered at 96 kHz have much more data to process and therefore are able to deliver much more detailed and nuanced results. Stretching a 96-kHz audio file to be twice as long still leaves the equivalent original information of 48-kHz sample rate across the duration, so there will be no perceptible loss in quality for the listener.

Obviously, when the digital file is time stretched, its sample rate is not reduced; instead the computer redistributes the samples that it already has to cover the extended duration and then approximates additional sample points to complete the waveform. The higher the resolution of sample points, then the more accurate the resynthesized or interpolated data. However, at extreme stretch ratios or with low initial sample rates digital artifacts and aliasing are introduced where the computer is unable to make an accurate approximation of the sound and thus must make something up. This is often audible as a metallic or glassy timbre and is an indicator that sounds have reached their limit for transformation.

1.4.5.6.   Rendering and Reality

As we have explored the recording process (and that of listening), we have demonstrated how sound wave information is converted from vibrations and pressure fluctuations in the air into voltage differences in analogue circuits (or electrical impulses in the brain). These voltage differences can also be converted into digital information through processes of sampling in analogue-to-digital conversion to deliver the binary data that can then be visually represented as waveforms in our DAW. We then edit and manipulate these waveforms before reconverting this data back into analogue electrical signals through digital-to-analogue conversion and once again into sound pressure vibrations in the air via loudspeaker.

These processes of transformation reveal how our sound design practice is built upon the traces of events and objects. There is a disconnection from absolute reality: we do not capture a specific sound object when we record it; rather we record a trace of it. And this fact is vitally important when we come to constructing and developing worlds of sound. We do not engage with reality itself but with something constructed, naturalistic perhaps, but fundamentally unreal.

The sounds that we layer together need only to approximate the visual images that they accompany because the human brain will project itself upon the materials that it is provided in order to fuse and combine the two. Therefore, our attention should be not on what is the correct sound for that visual action we are seeking to articulate but instead on how our sounds might further the story and the world of the film. Our craft is not built upon simply inserting the sound of a generic footstep in sync with an actor's motion but for the footstep to be light/heavy/dark/smooth/soft/rough/stilted. These characteristics will project themselves into the minds of the audience to infer associations and motivations behind the characters and the objects in the film.

The choices that we make when recording can help to highlight specific characteristics and change the relationships that we have with sounds. Michel Chion refers to this characteristic of sounds projecting into the scene through his concept of materializing sound indices:

> A sound of voices, noise, or music has a particular number of materializing sound indices, from zero to infinity, whose relative abundance or scarcity always influences the perception of the scene and its meaning. Materializing indices can pull the scene toward the material and the concrete, or their sparsity can lead to a perception of the characters and story as ethereal and fluid.
>
> (Chion, 1994, p. 114)

Chion argues that we are often encouraged to eliminate material characteristics striving toward the impossible perfection of capturing pure sound, freed from the grubby physicality of production. But it is exactly these noisy elements, these flaws of tactile and material cues, which give sounds their character and provide them with their potential to elaborate and elucidate a rich soundscape that enriches the associated world of film.

Every decision within the recording and signal chain process should take into consideration the creative intentions and the goals for the final output. Microphones can be chosen and positioned to capture exactly the right type and extent of material character. A cardioid microphone might provide excellent transient response but lack the ability to capture the appropriate resonances and reverberation. So this cardioid might be supplemented with an omnidirectional microphone, thus capturing both the well-defined transient as well as a balanced and open sustain and decay. Microphone positioning can be even more powerful, highlighting tactile qualities and enhancing timbres though proximity and perspective, tuned to the desired balance of creative intentions and application of sound.

These creative applications have less to do with reality and much more to do with narrative; highlighting the impact of the physical, psychological and metaphysical within the story of the film. It is here that the creative potentials of sound design lie, extending characters, objects and locations beyond their realistic and concrete visual forms by projecting onto them symbolic meanings.

Key Points

- There are many different microphone types, each of can fulfill a specific creative purpose.

- Experiment with different microphone placements and positioning.
- Avoid distortion and clipping within the signal chain; remember to check your entire signal pathway sequentially to eliminate the possibility of clipping at each gain stage.
- Know your sample rate and bit depth. Work at the highest rate you can afford.
- Microphones and recording technology are just tools. Don't let them take over. Make sure that your creativity is leading the way.

1.5. Summary

This chapter has been designed to provide you with a practical guide to the nature of sound and recording with which you can swiftly and affirmatively build your creative practice and continue your sonic explorations. While occasionally technical and scientific in subject matter, it has been the goal of this chapter to equip you with an understanding of how sound operates, through which you are able to develop your own strategies for creative sound design and recording.

The links to following additional texts can provide you with in-depth reading on each of the subject areas explored here and can provide more specifics details to expand upon the topics discussed. You should seek to use this chapter and the following resources as a jumping-off point to experiment and develop your practical and creative skills.

Equipped with an understanding of how sound operates as a physical phenomenon—how it is interpreted by listeners through the psychoacoustic through to techniques of recording and production that allow you to capture and highlight specific characteristics of sound—you are now well placed to continue your journey of sonic discovery, to imagine, to innovate and to communicate your intentions to audiences through whole new worlds of sound.

Acknowledgments

Special thanks to Luca Portik and Deborah Iaria for their graphic renderings within this chapter. Thank you also to Michael Filimowicz for his constructive comments on an earlier draft.

Notes

1. The medium is the material surrounding the source through which the sound waves travel. The medium that we most frequently refer to when discussing sound is air, but water and solid surfaces can be mediums for sound too.

2. One such example is the Next Generation Sound Synthesis (NESS) project run by Edinburgh University. www.ness.music.ed.ac.uk/

3. In the example of a guitar string, the fundamental has the greatest amount of sound wave energy as bending the string with one point of extension along its length takes much less energy than to bend it multiple times (as is the case with the higher partials). However, sometimes the listener will perceive the pitch of a sound even when the fundamental frequency is not present; see Murray Campbell and Clive Greated's discussion of the Bassoon's "missing" fundamental (1987, p. 88).

4. Campbell and Greated, describe the acoustic properties of traditional instruments in great detail within their *Musician's Guide to Acoustics* (1987).

5. See Chapter 12, "Electroacoustic Music: An Art of Sound" in this volume.

6. In order to confuse things, synthesizers utilize a slightly different nomenclature. In 1960s, Vladimir Ussachevsky was assisting Robert Moog to improve his synthesizers and identified four components to describe the shape of a synthesized waveform: attack, decay, sustain, release. The attack and sustain portions remain largely the same, but in order to factor in attack transients into synthesis, they adopted decay as the second portion of an envelope, defined as the rate of reduction from an initial attack peak level to the sustained steady state. Thus, the end of the synthesized envelope needed a new name and was defined as the release, the rate at which a sound decays after the control input has been released. It is important to keep track of the context in which decay is being used but also note that even some synthesizers use different terms.

7. A sine tone at 343 Hz has a wavelength of exactly 1 m.

8. See Soundworks, 2012.

9. See Hooton, 2017.

10. Some people experience what is called absolute pitch, or "perfect pitch." These individuals are able to identify or recreate a specific note from their memory. The occurrence of absolute pitch is thought to be one in 10,000 people.

11. This applies to all of the dB scales—dB SPL, dB U, dBFS.

12. On the flip side, high-quality loudspeakers have very low levels of distortion and therefore must be driven harder and louder before any distortion occurs, this can result in the volume level being turned up to dangerous levels because the undistorted sound simply doesn't seem to be that loud.

13. Which can cause interruptions to the signal through distortion, clipping and overload.

14. www.zachpoff.com/resources/alex-rice-piezo-preamplifier/

15. See section 1.4.4.2, "Dynamic Range Compression."

16. Audio standards: CD = 44.1 kHz/16 bit; DVD = 48 kHz/24 bit; Audio-DVD = 96 kHz.

17. *Digital compression* refers to the shrinking of file sizes. This can result in the loss of resolution in terms of sample rate and dynamic range, but it should not be confused with processes of dynamic range compression discussed in section 4.4.2. File formats like AAC eliminate bit depth, instead seeking to encode the audio information in as few bits as possible.

References

Arrival. (2016). [DVD]. Directed by D. Villeneuve. USA: Paramount.

Chion, M. (1994). *Audio-vision: Sound on screen.* New York: Columbia University Press.

Dunkirk. (2017). [DVD]. Directed by C. Nolan. UK: Syncopy.
Flight. (2012). [DVD]. Directed by R. Zemeckis. USA: Paramount.
Gobeil, G. (1994). *La Mécanique Des Ruptures*. Empreintes DIGITALes—IMED 9421.

Additional Materials

Physical and Acoustic Properties of Sound

Campbell, M. and Greated, C. (1987). *Musician's guide to acoustics*. Oxford: Oxford University Press.
Howard, D. and Angus, J. (2017). *Acoustics and psychoacoustics*. 5th ed. London: Routledge.
Manning, P. (2004). *Electronic and computer music*. New York: Oxford University Press.

Hearing and Listening

Bregman, A. (1990). *Auditory scene analysis*. Cambridge, MA: MIT Press.
Denny, M. and McFadzean, A. (2011). *Engineering animals: How life works*. Cambridge, MA: Harvard University Press.
Gelfand, S. A. (2004). *Hearing: An introduction to psychological and physiological acoustics*. 4th ed. New York: Marcel Dekker.
Gravity. (2013). [DVD]. Directed by Alfonso Cuarón. USA: Warner Brothers.
Hooton, C. (2017). [Online video]. *The Shepard tone: The auditory illusion that makes Hans Zimmer's Dunkirk score so powerful and even shaped the screenplay*. Available at: www.independent.co.uk/arts-entertainment/films/news/the-shepard-tone-dunkirk-hans-zimmer-video-scale-score-soundtrack-christopher-nolan-inception-a7862211.html [Accessed May 2018].
Moore, B. C. (2004). *An introduction to the psychology of hearing*. 5th ed. London: Elsevier Academic Press.
Murray, L. (2019). *Sound theory from sound practice*. London: Routledge.
Smalley, D. (1997). Spectromorphology: Explaining sound shapes. *Organised Sound*, 2(2), pp. 107–126. Cambridge: Cambridge University Press.
Soundworks. (2012). [online video]. *The sound of flight*. Available at: http://soundworks collection.com/videos/flight [Accessed May 2018].

Microphones and Recording Technology

Chion, M. (2009). *Film, a sound art*. New York: Columbia University Press.
Huber, D. (2017). *Modern recording techniques*. 9th ed. London: Focal Press.
Katz, M. (2014). *Mastering audio: The art and the science*.3rd ed. London: Focal Press.
Rumsey, F. and McCormik, T. (2014). *Sound and recording: Applications and theory*. 7th ed. London: Routledge.
Viers, R. (2012). *Location sound bible: How to record professional dialogue for film and TV*. Studio City, CA: Michael Wiese Productions.

2

Invisible Seams

The Role of Foley and Voice Postproduction Recordings in the Design of Cinematic Performances

Sandra Pauletto

2.1. Introduction

A film actor's performance is a jigsaw. Not only because different shots are edited together to create an idea of cause and effect and an idea of time that, in reality, never existed but because their voice and the sounds of their body could be produced at completely different moments in time, in different parts of the world, by different objects, and even by different people. Thirty-five minutes into a film we could be seeing and hearing actors as they were performing in Hollywood during the shoot, but at 36 minutes perhaps hear their voices from a performance in London three months after that shoot, and at 37 minutes the sound of their footsteps could be produced by someone called John walking on the spot on a Foley stage in California. We would never know.

The processes used to re-record voice and body movements' sounds in postproduction are called ADR (automated or automatic dialogue replacement) and Foley (from the name of sound effects pioneer Jack Foley). They are often used to substitute noisy or unintelligible production sound, but it would be misleading to think that their role is limited to this corrective function.

ADR and Foley are integral parts of what makes cinema different from the simple recording of an acting performance. They contribute to the construction, down to the fine details and subtle modulations of the final character that we see and love on screen. Extreme care, knowledge and skill are involved in producing these sounds and in sewing them seamlessly into the film.

This chapter explores the contributions of these two sound postproduction processes to the creation of cinematic performances. It focuses on the work of the sound experts at the center of this creative process rather than the actors or the directors, as this aspect is often the least explored and its implications on the final performance the least understood.

In the spirit of finding "a way to account for the role of technology in performance" (Wojcik, 2006), this chapter features excerpts from original interviews with ADR mixer Nick Kray and ADR recordist James Hyde (Kray, 2018; Hyde 2018), Foley artist Gareth Rhys Jones (Rhys Jones, 2018) and Foley engineer Julien Pirrie (Pirrie, 2018), as well as reference to existing online interviews with renowned ADR mixer Doc Kane (Kane, 2018) and Foley artist John Roesch (Roesch, 2018).

2.2.   Cinematic Voices

"In movies actors do not just speak, they are recorded" (Sergi, 1999, p. 126), and this reproduction, which is articulated by sound experts, is the basis of the final performance we experience. Cinematic voices are the result of a combination of processes that start with the actor's performance in production to the final mix in postproduction (Pauletto, 2012). They are the result of a design: "In fact, the voice is often the most aggressively manipulated element of the film sound track" (Whittington, 2007, p. 174).

There are many types of voices in film and TV—dialogues, voice-overs, internal monologues—that function within storytelling differently and do not have the same need for postproduction voice replacement. Human dialogue is recorded in production and draws us, the audience, into the middle of the scene, into the middle of the stage. The illusion is that that conversation is happening here and now: the characters are thinking those words, and they are reacting as they speak and move their lips.

As a spectator, we can almost "participate" in the performance, take sides, and reply in our head (or perhaps out loud!) to them. All seems to be happening now, immediately, and there is no premeditation to it. Anything breaking that illusion of immediacy and presence would take us out of the story.

Dialogue in animated pictures is slightly different. It does share the immediacy of the dialogue between human characters, but, due to the nature of the visuals, the audience is never fooled into thinking that this could have been somehow produced in a "real" location. As a consequence, the performance in animations is constructed differently. The voices are recorded first, and even though they are recorded in a studio rather than on location, they represent the "production shoot": a shoot without the camera and a set and therefore with very few sound problems and little need for ADR. In animation, actors often record dialogue together rather than separately, so that they can respond to one another's performances, which they would not be able to experience otherwise. The visuals come after the vocal performances, and the lip synchronization is created adapting the visuals to the recorded voice rather than vice versa. Therefore, although dialogue in animation is immediate and synchronized to the image, it shares many aspects of "displaced" voices such as voice-overs.

Internal monologues, hearing what someone is thinking, is one step removed from the immediacy of dialogue. Internal monologues expose the "recorded" nature of film. In real life, we do not hear what other people are thinking, although we might hear our own voice in our own mind. When we listen to internal monologues in films, an artifact is revealed: that voice has been made "visible" to us, the audience, via technology, and it has been added on top of what resembles reality. The lack of synchronization between the voice and the character's lips is not perceived as an error. We are happy to accept the lack of synchronization as a convention that signals that this voice needs to be interpreted as thought rather than as dialogue. Finally, a voice-over, a narrator's voice, tells us that cinema is not reality and that design, premeditation, can be part of it. An additional space, a metaspace outside the story, or diegesis, can exist that is far removed from immediacy. In this new space, a voice without a visible body can exist and provide us with additional information (Chion, 1999).

These differences between cinematic voices and their different degrees of immediacy and presence are reflected in the way these voices are created in production and postproduction and in how immersed these voices are in the space of the story. While we expect internal monologues and voice-overs not to be recorded in the same space as the story because they represent different, imaginary spaces, human dialogue needs to be immersed in the story world. Dialogue is therefore perhaps the most fragile of the cinematic voices and indeed the voice that needs the most careful attention by the sound crew. In this chapter, we will focus on how dialogue is re-recorded, partially or completely, in ADR, rather than on other cinematic voices that have been discussed in literature more frequently (Kozloff, 1989; Doane, 1980; Chion, 1999; Marcello, 2006; Wojcik, 2006).

Acting is to be able to respond to a situation as if it were real, as if it were happening for the first time, and as if the actor really felt those emotions. A large part of this performance is expressed physically, visually, but is also, crucially, expressed sonically through the sound of voice and body movements. Great efforts, therefore, are made to record these performances both visually and sonically during production on location. However, there are times when the captured soundtrack is not suitable for use, be that for qualitative or artistic reasons. Fortunately, the malleability of our perception of sound grants us flexibility in regard to how we treat sonic performances, and in postproduction we can re-record fragments of vocal and body performances and embed them into the film by employing the expertise of ADR and Foley professionals.

2.2.1. The Need for ADR Recording

There are many reasons for needing ADR. These vary between improving intelligibility, removing noise and the need to alter performance. While some noise issues can be eliminated using digital noise-removal tools, broadband sounds that overlap with the voice cannot be reduced without affecting the overall signal. The only option is to re-record the voice so that the signal-to-noise ratio can be improved. Sometimes a performance is not quite as strong as the director originally thought or the words need changing. Finally, a film can be completely revoiced to create different foreign language versions. Essentially, ADR is always an opportunity to improve the overall effect of a scene.

Lines of dialogue are re-recorded in loops, in synchronization to the picture, in a recording studio. The actor watches the scene, sometimes listening to the original dialogue through headphones, and reperforms those lines a number of times while the recording equipment is rolling. Dialogue lines are usually recorded separately as it is rare that two actors can be in the same place at the same time.

Before digital sound, loops of film used to be created and played back in the studio for the actor. The process was called "looping," and it was both time-consuming and costly. Nowadays digital technology has made some aspects of ADR very efficient as loops are created simply defining markers in Pro Tools (Avid, 2018), the most common editing and mixing software in film sound, and they can be cued up very quickly. Crucially, it is possible to immediately preview how well the recorded lines can be embedded in the film. Preliminary synchronization, tone and space matching can be executed in the ADR studio thanks to the expert skills of the ADR team.

ADR mixer Nick Kray says:

> ADR has got the technology behind [it]. Recording and editing dialogue has got so good, and it's so much easier for us to shoot a lot more in a shorter space of time with the actor that there is a tendency now, and it has been going on for a while, for the director to say: "Don't worry about it, we'll sort it out in ADR". So they get to have all the elaborate sets, and stuff that is going on, or even a lot of characters that don't even exist on the set—they make them in CGI and they haven't got a voice yet—so you have to do them in ADR.
>
> (Kray, 2012)

The ADR process involves both performance and technical aspects and, therefore, in addition to the actor, the following personnel will often be present: film director, dialogue editor, ADR mixer and ADR recordist.

2.2.2. Performing in ADR

Producing a good performance in ADR is not simple. The actors are asked to be in the studio for ADR sessions often weeks and months after the production shoot has ended. They are usually working on new projects and are immersed in other characters and stories. Not only do the actors have to reimmerse themselves in the original story, they must do so under completely different conditions to those in which the original performance was captured: they are not in the right environment, they are not reacting to other actors, and they are not wearing the right costumes or moving the right props. Therefore, the most important aspect of the session is to allow the actors to reenter the appropriate headspace to be able to perform at their best.

On this Kray says:

> What an actor finds probably most difficult to do is not actually re-record the lines, it's trying to find the emotion again in a different environment [. . .] We [director, dialogue editor, ADR mixer and recordist] have to make them feel comfortable enough to forget that they are in this room [the recording studio].
>
> (Kray, 2012)

The actor will receive performance directions from the film director and, when necessary, technical directions from the dialogue editor or the ADR mixer.

> [The director] will give them an emotion, they'll give them a reason to feel the way they need to feel again to match what is going on on screen.
>
> (Kray, 2012)

The dialogue editor and the ADR mixer will give technical directions such as asking the actor to run on the spot to reproduce the physical effect of running out of breath if that is what is happening in the scene, or to turn their head slightly left or right in order to be very slightly off the microphone's axis when the scene is dynamic, or to swing their shoulders a bit if, in the scene, they are walking, or to produce some very subtle lip smacks if their character is supposed to be nervous and their mouth is supposed to be quite dry.

Kray mentions another technique to help enhance the naturalness of the recorded performance:

> Say you have got a three minute scene, and there are ten lines of dialogue in that scene that they need to re-record. Depending

on what's happening in that scene, if it's a quite quiet, emotional [scene] between two people, you are going to have all these different subtleties in there, in the way they perform that dialogue, but also around that you've got all that sort of lovely, well, what would be considered silence, but it isn't, there is stuff there, you just can't hear it yet. So what we tend to do is something called "breath pass", which is: we'll cue up the whole scene, and we'll say [to the actor] [. . .] "We want you to go through the whole scene, and what we would like you to do is, just kind of follow the action of what you are doing on screen and whisper. Don't worry if your dialogue is late, whisper yourself through the dialogue and then out of it, and then watch yourself, and watch if you are breathing, watch what's going on, and watch your lips stuff, and don't worry if it is late, just get something in there", and eventually what we'll do, we'll do a couple of passes of that, and then we'll take the pieces and [. . .] we'll lay that underneath [the previously recorded dialogue] [. . .] and the difference it can make is incredible. Suddenly you bring this [dialogue] alive.

(Kray, 2012)

Physical changes due to body movements or emotional variations affect the sound of the human voice. The human brain is highly attuned to picking up on these tiny variations. If they were missing, the audience would feel that something is wrong: they would probably not know what was missing, but they would perceive a lack of "naturalness." Voice and body movement, and therefore ADR and Foley, are two sides of the same coin: the performance. Even when the body is not visible, in voice-overs or animation, the movement participates in the performance. David Kaye, the voice of Megatron, one of the Transformers, used the Laban method (a method developed by dancer and choreographer Rudolf Laban to express realistic human movement and behavior) in his voice performances when auditioning for cartoons (Bevilacqua, 1997).

Artistic and technical directions, together with the possibility of watching the original performance on screen, support the actor's performance during ADR.

When they [actors] are [re-recording the lines] to picture, to sync, [they are] thinking about themselves in that scene again, and so you can get the most wonderful performances, in perfect sync coming out, and the director goes: "Wow, that's interesting. Play that back, that's really interesting", and I'm thinking "Tonally, is this matching?", and the [dialogue] editor and ADR recordist will look at the waveform: "Yes, it is in sync, tone is matching acoustically, I can get it in there, it sounds good", and when all those come together they go: "Wow, that's perfect, that's brilliant". You play it back to the actor and they go: "Yeah, yeah, cool. Can we just try one more, can we just try one more?", and so you get this lovely positive rhythm going on, and you try and keep that going on.

(Kray, 2012)

Similarly, American ADR mixer Doc Kane describes the actors' positive approach to ADR when they know that they can improve their overall performance. He recounts Meryl Streep, for example, using ADR to "smooth that cut out with my vocal" or "tie those two lines together a lot better" (SoundWorks Collection, 2018).

One of the most efficient techniques to record ADR is "listen and repeat," but it has its disadvantages. In this approach, the actor watches the scene and listens to the original production lines, via headphones, and then is asked to repeat those same lines. While this technique is very effective when replacing a line for clarity, it is unhelpful when attempting to create a new or improved performance.

[Listen and repeat] can change a performance quite drastically. The problem is when you hear something, and then you repeat it, what you are doing is, you are forgetting what you have been told to do in the first place.

(Kray, 2012)

It becomes difficult to follow new directions and create new performance alchemy: there is a risk that the process becomes mechanical rather than creative.

2.2.3. ADR Roles and Tasks

Thanks to the portability and efficiency of digital technology, it is possible to preview to the director and dialogue editor how the new lines will fit together with original production lines in the film and other sounds in the ADR session. Three main sonic aspects need to be matched very precisely in order for the new lines to fit: synchronization, voice tone and space.

The ADR mixer matches tone by choosing the appropriate microphones, positioning the actor, giving technical directions, and, at times, using equalization (EQ). Then they match the space of the production lines adding room tone and appropriate reverb. The ADR recordist very quickly live-edits and synchronizes the recorded lines to picture. While most actors will be able to perform lines almost in synchronization to the picture, there will always be some adjustments to be made.

ADR recordist James Hyde says:

> I have to try and grab the audio within a couple of seconds and try
> and fit it in quickly. And I do that by looking at the original guide
> track we got off set [. . .] The first thing I learnt was from [dia-
> logue editor] Tony Currie actually. He basically said: "It always
> takes the human mind a certain amount of time to pick up whether
> something is in sync, so" he said, "you do a very quick edit, back
> sync it, so back sync the end of the sentence, it's the very first
> thing you can do, and then you can start moving back through it".
> It's learning where you can cut into sentences, if there is a nice
> gap you want to cut it, but if there is a nice breath you don't want
> to cut the breath, you want to keep that. And then it's whether you
> want to keep it at the end or the start of the word. And then it's also
> extended vowels, you can cut into them and extend them.
>
> (Hyde, 2012)

However, it is worth remembering that if too much editing is needed, it is
likely that the editor and director would want to re-record the line again
with more natural synchronization.

The dialogue editor, who is keeping notes throughout the process, will
receive the raw recorded lines and will be able to fine-tune the synchronization. When necessary, they will take screenshots of any Pro Tools setting
that can be useful in the later stages of postproduction. An important task
of the ADR recordist is also to organize, label and run the ADR sessions as
efficiently as possible. It is essential that the dialogue editor is able to pull
out any segment of line as quickly as possible when finalizing the dialogue
edit and during the mix, so the labeling and organization of the Pro Tools
sessions need to be precise.

The ADR mixer and recordist roles are complementary: their work is
very focused and efficient. A key skill is to be able to anticipate what the
director and editor would most likely prefer to listen to, and at the same
time being able to make technical adjustments without interfering with
the performance and the discussions. A lot of nonverbal communication,
made up of looks and signals, goes on within the ADR team. This supports
the natural flow and pace of the session, which is dictated by the actor.

The tonal match of the voice starts with the choice of microphones. The
ADR mixer will consult the specification sheet given by the production
team and select the same microphones as those used in production. These
could be one or two shotgun microphones (positioned at different distances to record different perspectives) and a personal radio microphone.
If exactly the same microphones cannot be used, then the ADR mixer will
choose the microphones that best match the production microphones' polar

patterns and frequency responses. For narration or animation (when there is no production sound to be matched), a large diaphragm microphone, like the Neumann U87, will be used. Choosing the correct microphones goes a long way to matching the tone of the voice. Physical and performance techniques also help make it similar to the original sound. There are times when, for particular voices that seem to resonate differently in different spaces, tonal adjustments using EQ may be required.

On this Kray's says:

> Some people have a very strange structure to their vocal tone, and it doesn't react well to actually being recorded [in the studio]. So when you try and marry it in again, nothing but the environment they were in seems to actually work.
>
> (Kray, 2012)

When this happens, EQ might be used.

> We start carving off the bottom end. And then suddenly you find that [the voice] sort of drifts into a similar tonal quality [to the production lines], and then the editor will go: "Ah, ok, cool, I'm going to make a note of that."
>
> (Kray, 2012)

It is also crucial to maintain the dynamics of the performance when recording. Doc Kane recalls how Robin Williams, for example, could go from a whisper to a scream very quickly, and, in order to maintain the performance and avoid distortion, in cases like this a mixer needs to have exceptionally good anticipatory skills and ride the faders: become a "human limiter and compressor" (SoundWorks Collection, 2018).

When the tone is right, the ADR mixer will add room tone and reverb. They will select between a series of reverb settings that would match a number of small, medium-sized and large spaces. More recently, convolution reverbs have started to be used. A convolution reverb combines a dry recording with an impulse response of a space. An *impulse* is a very short, click-like sound, and an *impulse response* is the recording of an impulse played in a room with objects and walls reflecting the sound and creating reverb. An impulse response is basically a temporal representation of the reverberation of a space: a sort of signature of a space. It contains all the information about its geometry and the objects and materials present in that space at the time of the recording. The result of the combination of a dry sound with an impulse response is the sound immersed in that specific space where the impulse response was first recorded, i.e. the sound with the reverb.

Kray favors convolution reverbs because of their naturalness and detail:

> The physics aren't missing in a convolution reverb: the physics
> exist within that impulse response. [. . .] Convolution reverbs are
> exceptionally good at very low [audio] level.
>
> (Kray, 2012)

Another sign of how the process of ADR is evolving thanks to these new
digital tools is that ADR teams are starting to record their own impulse
response libraries of interesting spaces.

> There are some nice stairwells in this building, and we always do
> ADR on stairwells. And even though the convolution reverbs for
> medium spaces are good, they just don't sound like that scatter
> effect that you get in a stairwell, so when we get a chance we are
> going to do one in the stairwell, here, and just load it in and try
> it out.
>
> (Kray, 2012)

2.2.4.   Creative and Technical Challenges in ADR

ADR recording requires exceptional sound knowledge, technical skill and
attention to detail, as well as the ability to quickly adapt to new situations
and personalities and to resolve difficult problems with creativity.

As an example, Nick Kray recounts one of the most satisfying and chal-
lenging works he was involved in with dialogue editor Tony Currie and
ADR recordist James Hyde. They were working on the four-part TV series
Titanic (2012).

> There were 90 individual speaking parts, and the crowd on top of
> that. And we did all of it here in this room. It took me a month,
> a whole month. [. . .] We then had to create the sound of 1500
> people on the boat with the ship going down, out of the crowd.
> And the maximum I can get in here, and still make it work, is 23,
> so it was a hell of challenge. [. . .] So we came up with a way of
> doing this in here, and it was just different ways of miking things,
> and using the room in a way that I could make things differently,
> acoustically, in the room. I have done things before where I had
> made very few people sound like a lot, but I had to have a lot of
> people sound like 1500 people, and it was tricky . . . For Brave
> (2012) I had 17 people sound like 300, I didn't know if I could
> get 23 sound like 1500, but we did, and nobody knew that those
> 1500 people that make up the soundscape, especially the most

horrendous ending when the boat is going up and everyone is slid-
ing down, they are in the water, and there is all this people . . . It
was made of 23 men and women in this room, and multiple passes
of things, and a couple of different techniques to keep it dynamic.
And we got them to change their voices as much as we could as
well, because crowd actors are very very good voice actors, so
they have great range, so there were little techniques used. And
we did it, 1500, and we "made" 1500 people, and we ADRed 87
main cast members, and we did it all in four weeks. I had never
ever done that before or since.

(Kray, 2012)

2.3. The Art of Foley

A performance is the combination of an actor's visual presence, facial
expression, vocal performance and body movements. A lot of it is portrayed visually, but we tend to underestimate how much, as human beings,
we use the sound of body movements and interactions with objects to
judge someone's intentions and emotional state or to simply understand
the material nature of the world around us. Fast footsteps on the pave
ment behind us would make us turn around with suspicion; a soft knocking
at the door might be the prelude to receiving an apology; the slamming
of cupboard doors might mean that someone is angry; the sound of children playing with a ball followed by the sound of broken glass probably
means that someone's window is not intact anymore. We understand a lot
by hearing the sound of people moving and interacting, and filmmakers
who understand this can utilize Foley sounds to create incredibly charged
scenes: "Foley sounds provide pivotal sonic anchors to the body and unify
the space in which the body moves" (Whittington, 2007, p. 159).

Sound effects have accompanied film throughout its history, both in
the silent era and later in the time of synchronous sound. All manner
of mechanical machines were used to create sound effects for films in
real time, in the same way as music was played live to accompany the
story. Mechanical sound effects machines, such as rotating wooden wind
machines, thunder metal sheets and rain machines, were used in theater
and in cinema to accompany the story and give a "body" to its world.

In the 1920s, when the synchronization of sound to picture became possible, additional sounds, such as footsteps and body movements, could be
added for the first time to the repertoire of sounds accompanying a film.

Jack Foley was one of the pioneers who developed techniques to record
sound effects in synchronization to the picture, and that is where the term
"Foley effects" originates (Wright, 2014; Ament, 2014; Viers, 2014;

Whittington, 2007). Since that time, audio technology has advanced at tremendous speed. Nowadays we can record ice melting with hydrophones, a tropical forest atmosphere with an array of microphones; we can create the most interesting combinations of recorded and synthesized sounds or produce completely new sounds with a digital synthesizer.

However, footsteps, body movements and interactions between characters and objects are still recorded, fundamentally, using exactly the same technique as Jack Foley did almost one hundred years ago. Why are these sounds different?

Foley sounds usually accompany characters' activity. In many ways they are the "voice" of the characters' "bodies." They seem mundane and less remarkable than, say, the sound of a car speeding or a gunshot or a storm, but it is their performative nature that makes them often difficult to record on location, and almost impossible to synthesize digitally (Pauletto, 2017).

Foley sound, more often than not, needs to sound "natural." While there are types of films, such as animations, comedies and horrors, where Foley sounds can be highly stylized and diverge from our everyday experience of sound, this chapter specifically focuses on the more common, naturalistic use of Foley, which is perhaps the most difficult to achieve because it allows for little error. While we would accept a wide variety of sounds for the footsteps of a monster, we are a lot less forgiving if the footsteps of a human character do not match our expectations.

Foley sounds are recorded in a studio in synchronization to the picture, similarly to how we record the voice in ADR. Interestingly, even though, as previously mentioned, we are very attuned to hearing the subtleties of these sounds and would immediately recognize when they sound "wrong," we are very happily fooled by hearing a very different person—the Foley artist rather than the original actor—performing these sounds using different objects from those used on screen. The synchronization and quality of the sounds need to be perfect, but the ways in which this is achieved can be very far removed from what we see on screen. Once again, invisible seams need to be sewn in the construction of the cinematic performance to allow this illusion to be successful.

2.3.1.   Foley Recording

Foley artist Gareth Rhys Jones and Foley mixer Julien Pirrie have worked together on many films, TV programs and animations—*Carol* (2015), *Blue Bloods* (2018), *Shaun the Sheep* (2007) to name a few—at the Foley Barn studio (The Foley Barn, 2018).

In their experience, the relationship between Foley artist and engineer is absolutely pivotal to the quality of the resulting work.

Jones says:

> The communication has to be right, the taste has to be similar, because otherwise you would be clashing all the time. You spend eight hours a day with each other, talking all of the time, under pressure, [it is] very intense sometimes, so it's vital: if the relationship is not there, it is not going to work. And in every sense, you would be unhappy at work and what you produce won't be good either.
>
> (Jones, 2017)

It is often the case that Foley artist and engineer form a partnership that continues across many projects. The engineer cues the sections of Foley scenes to be recorded, corrects the synchronization live, and assesses with the artist whether the quality of the sound and the performance are appropriate.

Pirrie says:

> Knowing what you want to hear is an important skill for a Foley Engineer, which is developed over experience. [. . .] The skill, in my opinion, is knowing what fits the scene and understanding what a client wants. It is also important to be creative and think outside of the box in order to achieve the sounds you want.
>
> (Pirrie, 2018)

Depending on the project, directions and notes can come from a number of sources:

> Sometimes we have specific notes for a scene of what a director or mixer wants or where there are problems, alternatively sometimes we are left to complete the Foley as we see fit. It is a team effort, with Foley engineer and artist, but I also have a strong idea of what I want to hear and will give the Foley artist notes. I believe this is important, as the Foley artist skill is finding and creating sounds but they don't necessarily hear how a sound is translating on the mic or how the sound is fitting in with the production sound. So it is important to work as a team to find the ideal sound for the project.
>
> (Pirrie, 2018)

Jones and Pirrie refer to a good recording as sounding "natural," as opposed to sounding "Foley," i.e. somewhat artificial, constructed. The Foley mixer will also be able to assess how "mixable" a recording is and whether a performance should be slightly altered to improve that aspect of the resulting sound. This depends very much on how well the artist and mixer know the piece they are working on: the story, the music, ambience and other sounds that will likely be used in that scene.

In a big action film, for example, Jones says:

> I might make a surface [for footsteps] grittier and harder than it actually is, because then I know that that will come through in certain places where the mixer will want it, and other times I might do completely the opposite, if I know that the ambience is, literally birdsong and a bit of wind, then I soften everything up so that it fits in with that.
>
> (Jones, 2017)

American Foley artist John Roesch describes in similar terms the working collaboration between himself and Foley mixer Mary Jo Lang:

> You know, Mary Jo is responsible for the correct recording of elements in the right places, having it set up correctly, and then having it delivered correctly. But, that's really doing her job an injustice, because her ear is just as important as mine or Alyson's [Alyson Moore, additional Foley artist]. Just today, there was a character that is very skittish. They came in off stage to talk to another character. I made them kind of come in fast, but then Mary Jo reminded me that this is a character that's coming in to apologise. Giving it some thought, Mary Jo's right, she would not come in fast, she'd really come in very slowly. So, we [all] kind of share that; in other words, if Alyson saw something like that, she might mention it, or I might mention it to her, there's a nice symbiosis amongst all of us. She also understands layering, that is, a good overall Foley job is not unlike mixing for a song. You know, you lay down the drums, then you lay down the keyboards, put all those things together, and when you play them back, hopefully you have a nice harmony so you don't really have to move a lot of the faders. It just all plays at a particular level. And of course, if we've done our job correctly, you don't know we've done it.
>
> (Ejnes, 2013)

The artist will usually use two condenser microphones to simulate different distances and slightly different tone qualities. They will adjust the microphone placements in relation to the sound source and perform while watching the scene on screen. Props will be available to use in the studio. Often Foley recording studios are fascinating junkyards full of completely normal objects (where one might find a feather duster that can become a bird or a baseball glove that can become an animated character's footsteps (Wired, 2016)), interesting sound instruments (for example, the water-phone (Newstalk, 2013)), water tanks and perhaps even a car door.

2.3.2. Performing Foley

A Foley artist needs to have two great skills in equal measure: timing and the ability to choose the right prop to perform a sound with appropriate sonic qualities. Both skills are essential, and they have to come together.

Foley engineer Pirrie summarizes the importance of these two aspects coming together:

> Quality of sound is always my focus when recording. These days, Pro Tools makes editing easier than ever, so if the quality is there, any problems with sync can usually be fixed. Whereas, a bad sound perfectly performed to sync will still be a bad sound. That is not to say sync isn't important though. Sync is what makes Foley still more economical than sound editing. A Foley Artist can perform the right sound to sync in the same time it takes to watch the action through, compared to an editor who could spend hours finding and editing a [sound] library to create the same effect.
>
> (Pirrie, 2018)

Foley artists often have a background in music or acting that will help them have a good sense of timing. Jones is a martial artist, used to play guitar, and has a great knowledge of materials and what they sound like from years of experience building sets. All of these experiences, in addition to having worked in the industry in many roles (writer, director, set designer, etc.), contribute to his ability to mimic someone's gestures with great accuracy and judge the appropriateness of the final recording.

Additionally:

> you have to be incredibly patient, because it is very meticulous [work]. [. . .] And you have to actually physically have a lot of stamina, because it's a long day, and you are picking things out, or maybe you are walking on the spot for three days solid, or running on the spot. And some [scenes need] big heavy props, like suits of armors and stuff.
>
> (Jones, 2017)

Jones points out how important it is, especially at the start of someone's career, to listen to your own work. It is essential to go to mixes and hear how the Foley recordings are embedded in the soundtrack. This helps develop the critical and analytical skills that allow judging how a sound should be performed in order to deliver a "mixable" recording.

Pirrie also says:

> Usually you know [when a Foley recording is easily mixable]
> through experience and by sitting in mixes and learning what does
> and doesn't work. I regularly playback and review our work as we
> go, playing it along with the production sound to make sure it sits
> in with the scene.
>
> (Pirrie, 2018)

A Foley artist will pay attention to mundane sounds happening in their
everyday life, which are usually completely ignored by others. Developing a "mental archive" of objects-interactions-sounds associations is key
to be able to quickly select an appropriate prop for a sound.

> For ages we couldn't do IV drips very well. We could do all the
> rattle and everything, but I had bad shoulders and I had been using
> a lot of ice packs, and I discovered that the ice packs, once they are
> melted, are the perfect IV bag. And so I have given them to other
> artists, and they were delighted, and said: "Oh, ok, that works
> really well" or "[Meat grinders] they just make the best door handles for rustic doors."
>
> (Jones, 2017)

Foley engineer Pirrie also recounts many unusual objects used during his
Foley sessions:

> I have used everything from pistachio shells as long fingernails, a
> mic stand as a police baton, a toilet roll holder for a creepy squeak,
> and have done strange things with dry ice and metal to create
> sounds for a car.
>
> (Pirrie, 2018)

An experienced Foley artist would usually be able to start performing as
they watch a scene for the first time. It often helps, however, to do the body
movements, i.e. principally the clothes sounds, first. This gives the artist
a chance to view the scene and gather information on how other sounds
should be performed.

Jones says:

> It is amazing if you do the footsteps without having done the
> moves you fall over yourself. [. . .] It's amazing how [watching
> the scene when doing the moves] sticks in your head.
>
> (Jones, 2017)

Foley artists, similarly to musicians and music composers, have the ability to "hear" in their heads the sound of a movement as they see it on screen. Foley artist John Roesch says:

> Detail is our business. We want to make sure that whatever we see on that screen, we are making the most honest representation thereof sonically. When I look at a scene I hear the sounds in my head, so all I am doing is figure out: "well, ok, that sound I hear in my head? How do I actually get that down on tape?" if you will.
>
> (Wired, 2016)

This ability to be able to mimic the sound of someone's movements evolves into almost a complete identification when performing the movements of a character over a longer period, for example, for many episodes of a TV program.

Jones mentions how after having done over 60 episodes of one particular program, "I can do [a particular character's footsteps] in my sleep, I just know when he is going to scuff. I can just tell by his head drops that he is going to do three feet and not two" (Jones, 2017).

Although all Foley artists need to have timing and an excellent judgment of the tonal quality of the sounds produced, there can be specific skills that make one artist better for certain works than others.

> I'm good for cartoons, for animation for some reason, I just think like a toy. I just find it really easy to be concise. I think if you are concise you are better at animation, so you are precise with your feet, and editors like me because of that, because it is easier for them to edit. I know other artists that are probably better than I am at costume dramas. There is one lady I worked with called Ruth Sullivan, her moves are mind blowing. I would do it [some of the Foley sounds] later. I would do hands, touches and pockets later. She would it all in one: she would just do now the whole thing in one.
>
> (Jones, 2017)

To produce the perfect performance, Foley artists are able to deconstruct in their head the physics of the action seen on screen.

Jones describes a few examples:

> Someone was captive inside a barrel, and we have barrels on the farm, but I have just emptied a filing cabinet, used the dustbin lid, and I can tell by looking at them that that would work, that the hollowness would work, if I mike it in the right way. [. . .] We had

to do a rock monster that was made out of rocks in a film called *The 9th Life of Louis Drax* (2016), and that was quite difficult to make it sound like it was a whole, like it was a body, and not just me dragging rocks across the floor, which of course that is exactly what it was, but we got there in the end. I filled sacks, pillow cases, full of stones and different size rocks and then just I gave myself a broken back just by shaking them. [. . .] A footstep sometimes sounds wrong but that's because, maybe, you just got the impact, but there is a lot going on underneath. So you have to, sometimes, lift things up, like that little bit of wood there [Jones points at a wooden prop in the studio], so that you have the impact but you also have the spreading out base underneath, and suddenly it becomes real, it becomes part of the character.

(Jones, 2017)

The tacit, embodied knowledge of Foley artists is still a very difficult aspect of this process to unravel (Pauletto, 2017), and the subtle nature of Foley sounds, the required invisibility of the technique, does not help the quest for more clarity. When asked how he knows when a sound is right, Jones answers without hesitance, "When you don't notice it [in the film]." When asked how he knows when a sound is wrong, he replies, "When you notice it" (Jones, 2017). Pirrie agrees: "You know when Foley is done well because it feels right and you don't notice it. [. . .] Bad Foley stands out and pulls an audience out of the scene" (Pirrie, 2018).

And that's absolutely correct. We would notice that something is wrong if we suddenly did not hear people's footsteps, despite the fact that we do not usually spend time consciously listening to people's footstep sounds.

However, for a Foley artist, who focuses on these apparently mundane sounds all the time, should it not become difficult to judge whether or not these sounds will be noticed by a general audience?

To this question, Jones answers:

I look at all the elements, I see the weight of the person, how they are walking, what shoes they wear, what they walk on, what I think the atmos is going to be, what I think the music is going to be, and that all goes through my head in seconds, and with the engineer, [. . .] and then I choose a pair of shoes, or I make a surface, I walk with a certain weight, or a certain speed, and then through experience I know that that's going to work.

(Jones, 2017)

So it is the experience and knowledge of the context in which the sounds are going to be played that allows an artist to judge how softly or loudly, roughly or smoothly they should perform the sounds in order to make them "disappear" in the film. Once again, a skill that could be considered typical of musical performance, rather than belonging to the technical domain, is at play.

2.4. Conclusions

Foley and ADR are pieces of the jigsaw that creates an actor's performance for film or television. They exploit the complex nature of our perception of sound and its ambiguity. Sound allows us to be blissfully fooled, while at the same time being completely unforgiving. We will probably be able to recognize our child among others from a single cough, but we are happily fooled into thinking that the snapping sound of a celery stick could be a bone breaking. It is this elasticity that allows us to collage together segments of different performances from different people and bring them together as a new whole that never existed before. This is filmmaking, just as different segments of video, edited together, can create a new world, a new story and new meaning.

The processes of ADR and Foley require a lot of technical skill, but they are not primarily technical processes. They are creative and essentially performative processes, inherently cinematic, in which newly created associations between sound and images produce, as if by magic, new meanings in the minds of the audience. It is for this reason that, so far at least, these processes cannot be fully automated through software or other means because at their core there is the complexity, not yet mastered by computers, of human performance.

References

The 9th Life of Louis Drax. (2016). Available at: www.imdb.com/title/tt3991412/?ref_=fn_al_tt_1 [Accessed July 2018].

Ament, V. T. (2014). *The Foley Grail: The art of performing sound for film, games, and animation*. New York: Focal Press.

Avid, Pro Tools. Available at: www.avid.com/pro-tools [Accessed July 2018].

Bevilacqua, J. (1997). Voice acting 101. *Animation World Magazine*, 2(1), Apr. Available at: www.awn.com/mag/issue2.1/articles/bevilacqua2.1.html [Accessed July 2018].

Blue Bloods. (2018). Available at: www.imdb.com/title/tt1595859/?ref_=nm_flmg_snd_12 [Accessed July 2018].

Brave. (2012). Available at: www.imdb.com/title/tt1217209/?ref_=nm_flmg_snd_97 [Accessed July 2018].

Carol. (2015). Available at: www.imdb.com/title/tt2402927/?ref_=nm_knf_i3 [Accessed July 2018].

Chion, M. (1999). *The voice in cinema*. New York: Columbia University Press.

Doane, M. A. (1980). The voice in the cinema: The articulation of body and space. *Yale French Studies*, 60, pp. 33–50.

Ejnes, S. (2013). *An Interview with John Roesch*. Available at: http://designingsound. org/2013/12/17/an-interview-with-john-roesch/ [Accessed July 2018].

Foley Barn Studio. Available at: www.thefoleybarn.com [Accessed July 2018].

Hyde, J. (2012). Personal interview with Sandra Pauletto.

Hyde, J. ADR Recordist, IMDB database. Available at: www.imdb.com/name/nm385 4267/?ref_=fn_al_nm_2 [Accessed July 2018].

Kane, D., ADR mixer, IMDB database. Available at: www.imdb.com/name/nm0437301/ [Accessed July 2018].

Kozloff, S. (1989). *Invisible storytellers: Voice-over narration in American fiction film*. Berkeley: University of California Press.

Kray, N. (2012). Personal interview with Sandra Pauletto.

Kray, N., ADR Mixer, IMDB database. Available at: www.imdb.com/name/nm2390390/ [Accessed July 2018].

Marcello, S. (2006). A performance design: An analysis of film acting and sound design. *Journal of Film and Video*, 58(1–2) (Spring–Summer), pp. 59–70. Champaign and Urbana: University of Illinois Press. Available at: www.jstor.org/stable/20688516 [Accessed July 2018].

Newstalk. (2013). *Emmy nominated Foley artist Caoimhe Doyle demonstrates movie sound effects*. Available at: www.youtube.com/watch?v=GrbgY6ajTgo [Accessed July 2018].

Pauletto, S. (2012). The sound design of cinematic voices. *The New Soundtrack*, 2(2), pp. 127–142. doi:10.3366/sound.2012.0034.

Pauletto, S. (2017). The voice delivers the threats, Foley delivers the punch: Embodied knowledge in FOLEY artistry. In: M. Mera, R. Sadoff and B. Winters, eds., *The Routledge companion to screen music and sound*. New York: Routledge, pp. 338–348.

Pirrie, J. (2018). Email interview with Sandra Pauletto.

Pirrie, J., Foley mixer, IMDB database. Available at: www.imdb.com/name/nm4350102/ [Accessed July 2018].

Rhys Jones, G. (2017). Personal interview with Sandra Pauletto.

Rhys Jones, G., Foley artist, IMDB database. Available at: www.imdb.com/name/nm04 28101/?ref_=fn_al_nm_1 [Accessed July 2018].

Roesch, J., Foley artist, IMDB database. Available at: www.imdb.com/name/nm0736430/ ?ref_=fn_al_nm_1 [Accessed July 2018].

Sergi, G. (1999). Actors and the sound gang. *Screen Acting*, pp. 126–137.

Shaun the Sheep. (2007). Available at: www.imdb.com/title/tt0983983/?ref_=nm_flmg_ snd_35 [Accessed July 2018].

SoundWorks Collection, Veteran ADR Mixer Doc Kane of Walt Disney Studios. Available at: http://soundworkscollection.com/videos/veteran-adr-mixer-doc-kane-of-walt-disney-studios [Accessed July 2018].

Titanic. (2012). Four-part TV series. Available at: www.imdb.com/title/tt1869152/ [Accessed July 2018].

Viers, R. (2014). *The sound effects bible: How to create and record Hollywood style sound effects*. Studio City, CA: Michael Wiese Productions.

Whittington, W. (2007). *Sound design and science fiction*. Austin: University of Texas Press.

Wired. (2016). *Where the sounds from the world's favorite movies are born*. Available at: www.youtube.com/watch?v=0GPGfDCZ1EE [Accessed July 2018].

Wojcik P. R. (2006). The sound of film acting. *Journal of Film and Video*, 58(1–2) (Spring/ Summer), pp. 71–83. Champaign and Urbana: University of Illinois Press. Available at: www.jstor.org/stable/20688517 [Accessed July 2018].

Wright, B. (2014). Footsteps with character: The art and craft of Foley. *Screen*, 55(2).

3

Media Management, Sound Editing and Mixing

George Kalliris, Charalampos A. Dimoulas, and Maria Matsiola

3.1. Introduction

As Hollman (1988, p. 14.1) states, "[T]he term *postproduction* loosely applies to all the processes needed to prepare finished picture and sound masters from the original-production materials ready for replication and distribution." In another perspective, Langford (2014) argues that editing methods can be corrective, creative or restorative in their techniques and goals.

Sound editing became a reality after World War II when magnetic tape became widely available (Eargle, 1992). Since then, large steps have been made in the field, most dramatically with the advent of digital technology, resulting in the contemporary digital audio workstations (DAWs), whose utilization provides to users almost unlimited capabilities. The digital transformation of audio signal into a *sound file* and its visual representation as *waveform* made postproduction accessible to everyone from novices to professionals. Numerous software applications are available on the market, providing the ability to visualize the audible signal to realize its levels, dynamics, weaknesses and progress in time and simplifying the perception of the needed processes. The parts of the audio signal can be quickly located, and time intervals are easily accessed since they can be seen with the accuracy of the individual digital samples, permitting even more precise elaboration. Sound faults are recognizable with a glimpse.

Nonlinear (reaching the desired point in a sound file at any time) and *nondestructive* (dealing only with pointers and not the original data itself) editing has released us from the anxiety of being right from the first attempt and has allowed audio editors to perform many trial versions without losing the original quality of the material. Another gain achieved with digital postproduction is the rectification of recordings that were not done properly and that are difficult, if not impossible, to do again, such as often occurs with news reporting.

3.2. Media Management for Audio Production

The process of audio editing and the associated management involves the selection and organization of all the necessary material, including scripts, recorded/archived audio and general audiovisual assets that are part of a project (Figure 3.1). Overall, media assets commonly include imaging and video content, i.e. in cases that the targeted sound design is related to audiovisual and multimedia productions; computer games (including audio-only games); and general digital storytelling products (Dimoulas, 2019; Dimoulas et al., 2014; Floros, Tatlas and Potirakis, 2011). In other cases, useful material might be associated with mobile captures and shared media that can be enhanced and properly labeled with the applicable geo-temporal and semantic tags, through applicable annotation propagation mechanisms (Dimoulas, Veglis and Kalliris, 2014, 2015, 2018; Dimoulas and Symeonidis, 2015). The availability of powerful processing tools

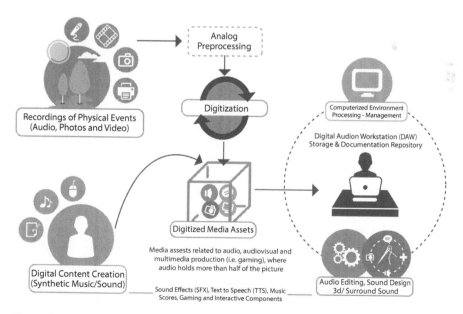

Figure 3.1 Typical Arrangement of Computerized Environment for Audio Production/ Editing, Sound Design and Overall Media Management.

Source: Dimoulas (2015)

allows for efficient restoration and enhancement of archived material, with quality improvements at expedient compression rates. In this context, proper documentation, with efficient content description and metadata management, propel knowledge acquisition and discovery, offering efficient content exploitation and reuse. Today, all the necessary processing and management are deployed through dedicated computerized environments, encompassing the digital audio workstation (DAW) systems along with the content-repository and management infrastructure (Figure 3.1). Again, there is a need for careful production organization by means of media documentation in all these aspects, so that content profiling and reuse can be delivered easily, efficiently and cost-effectively (Dimoulas et al., 2014).

Further, archived material like audiotapes, cassettes and vinyl records could be utilized. A crucial issue is the quality preservation status of archived media and the availability of the corresponding playback equipment that is required for capturing analog content in the corresponding analog–to–digital conversion (ADC) entry points (i.e. reel-to-reel audio tape recorders and vinyl records pickup playback devices, which have nearly disappeared in current use). Analog signal processing, adapted to the storage medium, might be applied for content restoration and enhancement, prior to the digitization process (Dimoulas et al., 2014). Hence, the task of processing needs detection is equally important, where related scientific know-how and technical experience can provide useful feedback in listing common degradation problems for various media and content types. For instance, additive broadband noise, hiss, wow and flutter, clipping and harmonic distortion or even reverberation are some of the common sound-related problems, which are strongly dependent on the conditions in which the initial recordings were conducted, as well as the storage media (vinyl records, tapes, etc.) and their preservation status (i.e. faults caused by frequency of use, physical-mechanical damage, protection against the already mentioned electrical, magnetic and chemical interferences, etc.) (Dimoulas et al., 2014).

The next questions that have to be answered from the very beginning of the project are related to compatibility issues and format selection. Thus, decisions over digitization parameters, such as audio resolution and quantization bit length, need to be made. These decisions will determine content quality (sound range and bandwidth richness), prescribing the applicable publishing or broadcasting formats that will eventually be selected. In general, attainment of maximum available and affordable quality during ADC is suggested, even if lower quality is used in the project or in the final product. The corresponding digitization parameters are determined by the source material status, the desired output format and the consideration that small-scale upsampling (both in resolution and quantization) is feasible without significant appearance of quality deterioration. Related poor quality and/or restoration/enhancement artifacts might be produced, which are usually more annoying than the initial degradation. In fact, generated artifacts are incongruous to the initial content qualities, which are very different in character from any physical degradation issues, which is very important. Analogously, lossless compression might be utilized during initial encoding (uncompressed audio is also attainable), considering that high-capacity storage media are currently available, featuring high data transfer speed and reliability at small costs. In any case, much care should be taken regarding the quality of the source material, in order to avoid using unreasonable digitization parameters and wasting storage capacity, with the risk of creating additional artifacts (Dimoulas et al., 2014).

In addition, new outdoor or studio recordings can be made, taking advantage of today's high-quality digital formats. Narrative audio and/or multichannel sound effects might also be deployed, taking into consideration aesthetic and informative reasons as well, i.e. depending on the targeted narration or interactive storytelling contexts. Typical amplitude, spectral and dynamics signal processing might be needed along with spatial enhancement and multichannel matrix encoding, especially in cases where high resolution and/or surround sound is involved in the sound design process. With audiovisual and multimedia productions, sound effects are inserted in order to match what is viewed on screen but also with ambient sound and sources that are not seen. This process is even more demanding in circumstances where the entire narration is delivered through the audio channels (i.e. radio drama), so it has to be even more descriptive and immersive to the virtual word of the story. For this purpose, besides the use of field recordings and clips from related sound-effect libraries, Foley techniques allow for specific sound effects to be artificially created and recorded in dedicated studios. In addition, sound effects are usually distributed throughout the channels of the corresponding surround sound, satisfying the spatial sound reproduction demands or even offering immersive audio experience. Finally, background music shapes the atmosphere, adds emotion and rhythm to the movie, but it also may have a diegetic nature, presenting actual sounds that were originated in the film word, whether their source is visible or not (Dimoulas et al., 2014; Dimoulas, Kalliris and Papanikolaou, 2008). Figure 3.2 presents the types of audio material that can be utilized throughout a demanding audio editing or sound design project.

As mentioned, an important issue in the AV production industry is content documentation. Nowadays, a plurality of content description and media asset management tools are available. Among them, MPEG-7 and MPEG-21 protocols provide standardized content description along with

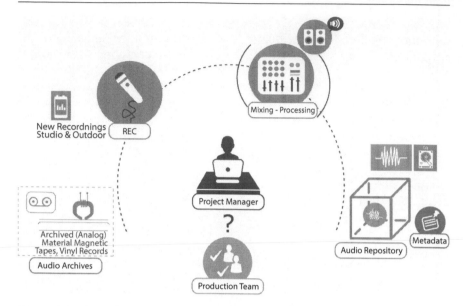

Figure 3.2 Typical Arrangement of Computerized Environment for Audio Production/Editing, Sound Design and Overall Media Management.

Source: Dimoulas (2015)

metadata management and media linking mechanisms, aimed at serving interoperability and transparent multimedia access (Figure 3.3). Intellectual property issues, including collaboration with digital rights management (DRM) systems might also be involved. Besides these, Dublin Core and MXF (Material Exchange Format) are other widely used multimedia metadata protocols that are intended to enable easier, cross-platform content distribution. MXF is a remarkable achievement of collaboration between manufacturers and major organizations, which is preferred by many experienced broadcasters. In addition, professional media assets management products are currently deployed, facilitating the media production, postproduction and broadcasting workflow. However, these solutions have not yet launched at an extended universal scale, an issue that is related to the accompanied difficulties, such as lack of uniformity, commercial competition, immaturity, know-how, user expertise requirements, purchase and maintenance costs and others. Nevertheless, a somewhat common ground already exists in the adoption of XML and other markup languages for data storage, transport and display (Figure 3.3). Apart from the core content, maintaining processing log files and edit decision lists (EDLs) on a central documentation server, along with basic

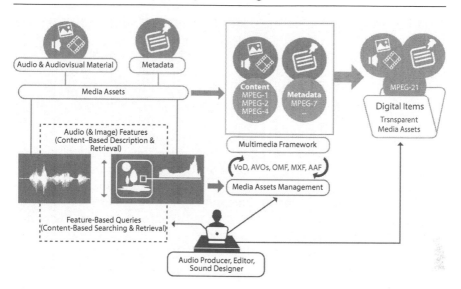

Figure 3.3 Content Documentation and Management as Part of an Audio Production.

Source: Dimoulas (2015)

user entries, could provide fully automated content description, with multiple format extraction capabilities, i.e. in case of a multiuser collaborative and networking environment. On the other hand, project coordinators can always supervise and deliver this kind of metadata translation and standardization, thus providing valuable semantic feedback and also contributing in the gradual implementation of suitable intelligent agents (Dimoulas et al., 2014).

3.3. Sound Editing

Editing is the aggregation of procedures elaborated so that the original audio material reaches its final optimized form. In the analogue era, editing and mixing were two distinct procedures. The main editing process was performed after the final mixing on the stereo mix tape in order that unwanted sounds such as breathing, counting before the beginning of the song etc. could be removed. Also, adding a reference tone and placing the songs in the right order before proceeding to the creation of the transfer medium (e.g. vinyl records) were among the editing tasks. Likewise, before completing a production with the addition of music or any other kind of sound elements in voice recordings made on tapes for radio broadcast use, processing was performed.

In the case of multitrack tape recorders, editing could not be executed with the use of the razor so the punch-in-and-out technique was applied for the necessary corrections. This was achieved by recording on a new track, for example vocals, while the other tracks were used for simultaneously playing back of the music. Another kind of editing procedure was the overdubbing of a small sound section that should be corrected on the already recorded track.

The advent of nonlinear editing (NLE) and subsequently digital audio workstations (DAWs) brought about the merging and consolidation of the editing and mixing procedures both at the device level and in the chronological order in which they are performed (hence, the nonlinearity). Perhaps one of the most significant capabilities of DAWs is that they provide editing for each separate channel of a multichannel recording. That is a feature that offered not only more advanced possibilities to the sound engineers but that led to new roads of musical synthesis for the musicians, since they are able to use techniques such as looping and small-scale interventions inside the audio tracks which, in some cases, result in major changes in the aesthetics of the musical outcome. Consequently, criticism has been made of these practices due to the conveniences that facilitate the musician's "life" provided by the NLEs and the DAWs. The main argument of this line of criticism is that at the end of the day "software makes music and not the musician" with all these available editing tools and tricks.

The utilization of contemporary digital NLE software and plugins with graphical user interfaces (GUIs), which present sound in its visualized waveform, allows the user to perform a series of tasks, such as rearrangement of the signal simply by cutting, copying and pasting parts, adjusting levels, adding signal processing etc., having at all times a visual representation of the audio form. These tools made editing a less time-consuming and potentially more experimental job. The simplest and initial procedure is to eliminate unwanted parts or to change other parts' positions in time. More advanced and complex operations regarding the signal processing that end up in the modification of the original material toward the creation of the desired outcome will be analyzed in the following paragraphs. In any case, listening correctly and carefully of the sound information that is about to be edited is the first and most important part of the work and should be maintained throughout all stages.

Since magnetic tape was the only possible medium for sound editing for a very long period, from the early days until the 1990s, it cannot be neglected and left without mentioning. The forthcoming paragraphs will pay a small tribute to sound editing with magnetic tape.

The three heads of the tape recorder (erase, record and playback) were engaged in the processes of recording and playing back the audio, and the sound editor had to listen and mark with a white pencil on the external side of the tape over the playback head the starting and the ending points of the segment to be omitted.

Shuttling (changing the speed to locate the right points) and jogging (controlling) were helpful features in this procedure, which was a mechanical one in contrast to the current digital manipulations. Subsequently, the capstan on the tape transport was released, and the tape was freed and afterward spliced with a special demagnetized razor at the appropriate angle to smooth the transition. According to the requirements of the editing process, the angle may be set to 90° for speech (cutting in the middle of a word for a quick transition) or in a long diagonal cut for a cross-fade or even in a zigzag (Benson, 1988). Finally, the two edges of the tape were reconnected with adhesive tape. In Figure 3.4, the whole procedure is presented with snapshots from educational material in Electroacoustics Laboratory of the Electrical and Computers Engineering Department at Aristotle University of Thessaloniki, Greece.

Today, signal processing may be achieved with either software or hardware found incorporated into consoles, as stand-alone units or as plugins. Digital audio workstations (DAWs), originally introduced in the late

Figure 3.4 Magnetic Tape Recorder Editing Procedure.

Source: Provided courtesy of Prof. George Papanikolaou

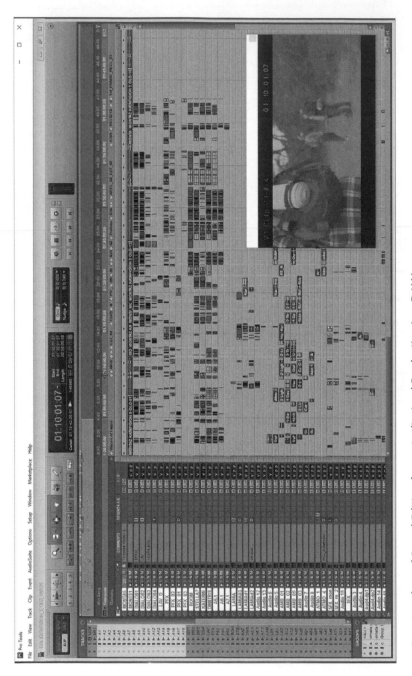

Figure 3.5 Interface of Sound Editing for an Audiovisual Production in a DAW.

Source: Provided courtesy of Avid Technology, Inc.

1980s by Digidesign, which launched the pioneering SoundTools software and shortly afterward Pro Tools, was the first digital multitrack system (Papanikolaou and Kalliris, 1991, 1992). Now more than 20 different DAWs can be found either as commercial products or as free and open source software.[1] They function as the conjunction of digital recording, storage devices and computers, and they carry out all of the procedures of recording, editing, mixing and playback. Most of them provide fully automated onboard mixing. Furthermore, integrated digital audio workstations also include a console that enables the sound engineer to become even more precise, vastly increasing the large number of channels that may be in simultaneous use. Figure 3.5 is a snapshot of the interface of a DAW where sound editing of an audiovisual work is performed and on which the editor may in parallel watch the rough or final cut of the video. The visual, easy, quick and zoomable access to any part of the recording along with the ability to audition all the commands before they are executed provides unlimited capabilities. Another very significant feature, common for all kinds of software, is the "undo" function that may lead to the preservation of multiple versions and their histories without any losses.

3.4.   Common Techniques

The concept of the atmosphere of the real-world conditions (interior or outdoor recordings) may be created either during recording or with post-production signal processing methods. Audio software offers numerous functionalities in the editing procedure through presets that also allow customization by the sound engineer. The processors fall under one or simultaneously under more than one of the following categories.

3.4.1.   Frequency Equalization (EQ) Processing

In the term of frequency processing, filters and equalizers that are applied to the audio signal are included as the most common ones. *Filters*, as their name suggests, reject, retain or emphasize a selection of the partials of different frequencies and amplitude (Dutilleux and Zölzer, 2002). Low-pass, high-pass, band-pass and band-stop are the broad categories of filters either permitting or rejecting the flow of low, high and defined bands of frequencies or boosting/attenuating them respectively by shaping the audio spectrum. They have functions that allow access to parameters like center/cutoff frequency, bandwidth and gain, via control of the associated coefficients. Regarding equalizers, the frequency band is divided in smaller bands based on octave intervals, and *equalizing* is the procedure by which a whole band or individual frequencies are increased or decreased. The bandwidth becomes narrower when the number of bands becomes larger. When the processing is applied in a whole band, there is a center frequency around which this takes place. The major categories for equalizers are graphic and parametric. The former one has vertical slider controls that can boost or cut specific frequencies, and the parameters involved in the latter category are the choice of frequency that will be affected and the degree of boost or cut (Eargle, 2003). In Figure 3.6, an incorporated parametric equalizer in a DAW is presented.

3.4.2. Time (Reverb and Delay) Processing

This is the kind of processing that affects the time interval between the signal and its repetitions such as reverberation and delay. In the real world, reflections on the surfaces of the enclosure surrounding the sound source cause reverberations that are audible as repetitions of the original sound. Many presets of physical spaces (churches, concert halls in various dimensions, variously sized rooms etc.) that simulate these repetitions are incorporated in DAWs allowing additional handling by the audio editor by adding depth and spatial dimension to the original sound. The main goal is to provide to the audience the acoustics of *what physically happens in the given space*, in which the sound was or should be recorded. For instance, recordings in the interior of a church should sound reverberant, and time processors can provide that capability through correct utilization of the parameters of the initial reflections and their decay. Furthermore, delay effects belong to time processing procedures as well. In Figure 3.7, the graphical interfaces of delay (a) and reverb (b) effect, along with customized parameters, are presented.

3.4.3. Dynamics Processing

The *dynamic range* of the signal amplitude is parametrically adjusted through the use of processors such as limiters, compressors, expanders and noise gates. *Compressors* amplify the lower levels of signal and reduce their louder parts, and *limiters* are compressors with a high (infinite) compression ratio and with fast attack and recovery (release) time. The parameters used in dynamic processing are the following: *compression ratio*, the parameter that defines the amount of gain reduction; *compression threshold*, the level at which compression is engaged; *attack time*, the amount of time needed for compression to reduce gain; and *release time*, the amount of time needed for the attenuated signal to reach its original level.

On the other hand, the dynamic range of a signal is increased through the use of an *expander*. The use of dynamic processing results in making less loud parts of a signal stand out above the "noise" of the other tracks without increasing the loud ones (Benson, 1988). Their applications lay among the

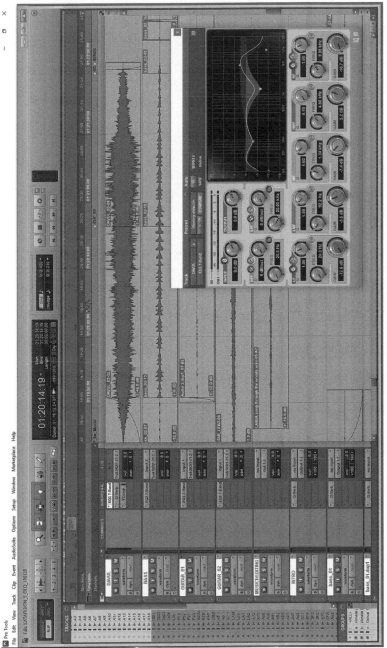

Figure 3.6 Parametric Equalizer Incorporated in a DAW.

Source: Provided courtesy of Avid Technology, Inc.

(a)

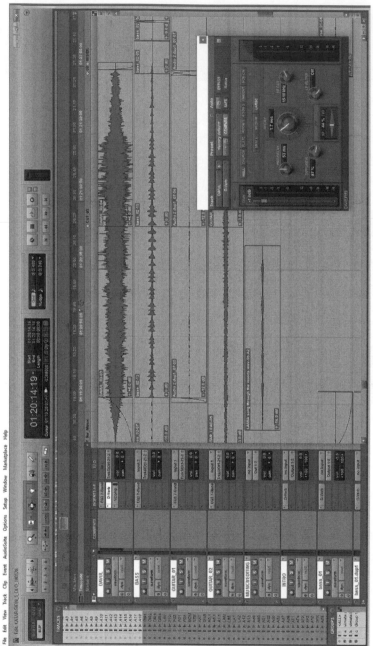

Figure 3.7 Interfaces of Time Processing: (a) delay effect, (b) reverb effect.

Source: Provided courtesy of Avid Technology, Inc.

variations caused due to movements of the performers or voiceovers to whole program compression for transmission uniformity purposes (Eargle, 2003).

3.4.4.   Restoration and Noise Reduction

Using these processes, unwanted sounds that may be recorded either from external sources such as traffic or any kind of ambient noise or from internal sources such as power supplies or fluorescent lights, as well as noise generated by the equipment and the media themselves such as hums, hisses, pops etc., are reduced or eliminated. Of course, as recording equipment and storage media evolve, the need for restoration is diminishing; on the other hand, sound recording takes place in increasingly difficult circumstances such as sports activities using action cameras, aboard drones, in crowded environments etc. The kind of noise that can be easily handled is the uniform noises such as the one produced by a ventilator or an air-conditioning system. On the other hand, traffic noise cannot be treated likewise since it contains highly differentiated frequency content.

The most common noise reduction procedure is to find a part of the audio signal that is purely noise and assume it as the data profile of the audio frequencies in the entire sound file that should be reduced or removed. This is due to the fact that discrimination between the signal and the noise has to be made manually and not decided by the algorithm. A number of presets, depending on the DAW, can be customized in the algorithmic process so that the final outcome fulfills expectations. In Figure 3.8, a noise reduction interface presenting the spectral de-noise procedure is shown. On the right vertical part of the figure, the restoration and de-noising modules are visible. The audio file is presented with a spectrogram view where elements such as noise, ambience and unwanted frequencies are visible (Alten, 2013). Spectrogram editing is used for a more refined view of an audio file. The sound editor handling the frequency display resolution becomes more analytical. Often, a compromise between the elimination of the noise and the alteration of the original signal is in order.

3.4.5.   Multiband Dynamics Processing

In this type of processing, the spectrum of the signal is divided and receives individual compression settings in more than one frequency band without the one affecting the other, and then they are combined into a single broadband signal (Eargle, 2003; Alten, 2013). The division usually comes up to a three- or four-band crossover, and the result is a more detailed, effective and optimized tailored processing. Multiband dynamics processing is mainly applied in the mastering procedure, which is thoroughly examined in Chapter 6 of this volume. Plugins and specialized software are used, and in Figure 3.9 the working environment of such software is shown.

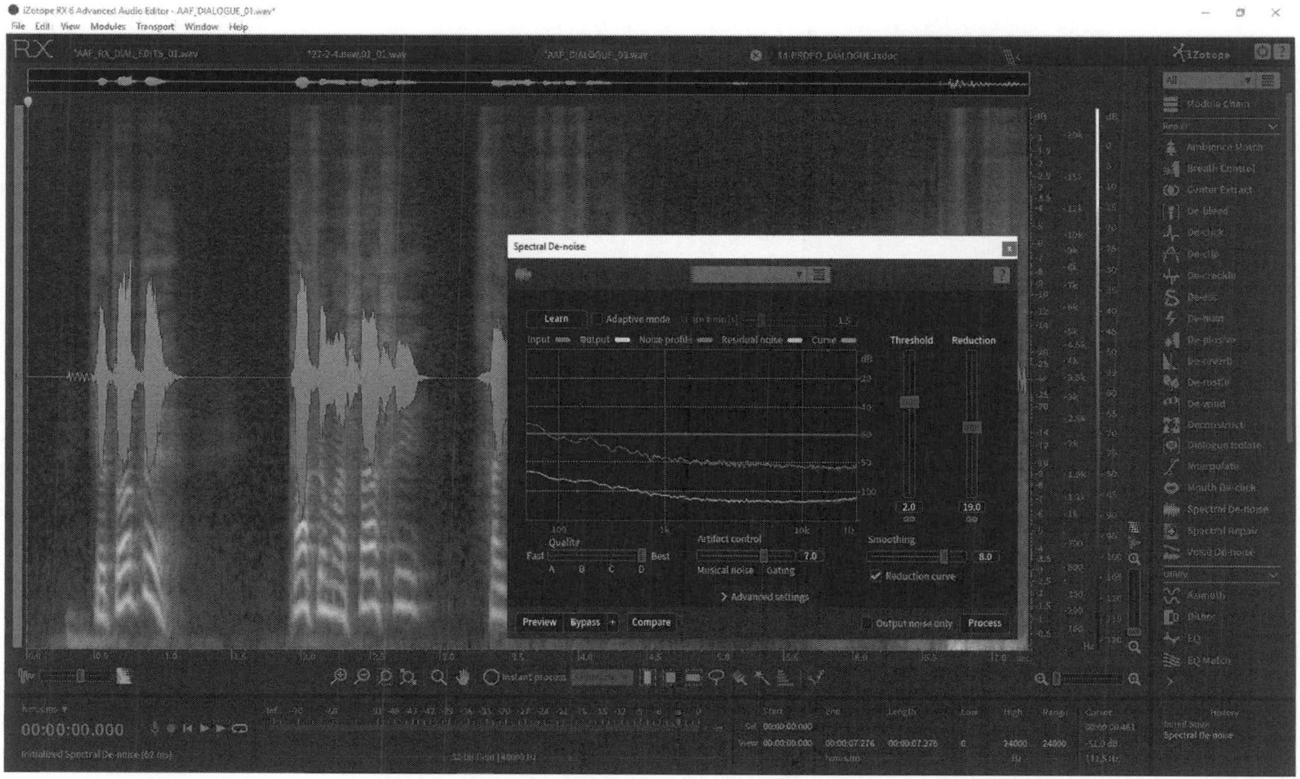

Figure 3.8 De-noising Interface with the Available Modules Shown on the right.

Source: Provided courtesy of iZotope, Inc.

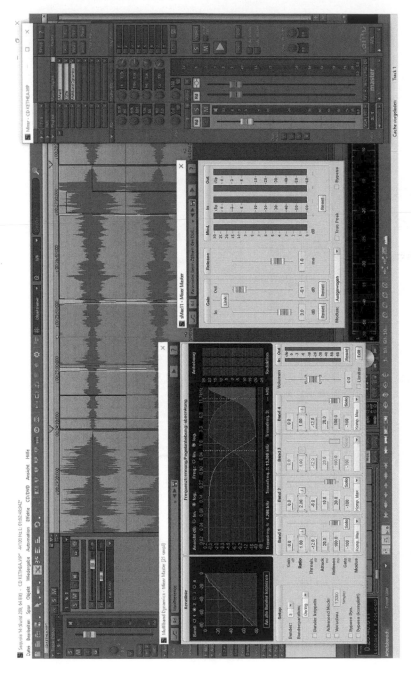

Figure 3.9 Working Environment of Software for Mastering Processing Applying Multiband Dynamics.

Source: Provided courtesy of Magix (Sequoia 14)

3.5. Postsound Recordings

3.5.1. Voiceovers and Overdubs

In studio recordings, as well as in large outdoor events, consoles are used to manage the number of sources. The console is commonly known as the "heart" of any recording system since it nests all the inputs and outputs, routes all signals to desired outputs (auxiliary and main), monitors each of the inputs, distributes outputs of its own and between the connected recording devices. Also it holds a number of equalizers and effects, especially in the latest digital consoles. The metering systems are essential, and they provide visual indications since sound passes through circuitry as voltage. Consoles or mixers in their smaller version can be found in analog, digital and virtual form on computers. Their size, features and operations vary according the needs of the work they perform in production, postproduction or broadcast.

The largest amount of console inputs are usually occupied by the microphones. Depending on the special needs and purposes of each recording, different types of them (dynamic, ribbon, condenser, electrocondenser), as well as of different pickup patterns (omnidirectional, cardioid, figure-of-eight) and of different sizes (shotguns, handheld, minimics), are used. The main features that are important when selecting a microphone are the operating principles, directional characteristics and sound response (Alten, 2013).

Especially in the case of voice recordings, narration or dialogue, where the main issue is intelligibility, the selection and correct placement of the microphone is crucial, guided according to the way that the written text will be pronounced. A number of factors determine the vocalization, such as emphasis and rhythm; however, the words should be clear and distinct whether they stand alone or are accompanied by music or other forms of sound, such as environmental or artificial. It is easier to perform editing on consonants since they give a hard sound compared to vowels, particularly diphthong vowels, which produce long sounds (Wyatt and Amyes, 2005). Clarity is a feature that should be achieved when recording or through postproduction since this provides intelligibility in the mix. One of the most common processors applied in voice editing is the de-esser, which is used to reduce the annoying sibilance caused by the consonant sounds of -s and -sh often encountered in speech recordings. Others are equalizing and compressing, especially in the cases of mixing with more sound elements since the human frequency range is not that wide in relation to that of instruments and ambient sounds. Another application that can be implemented is the "stretching" of sound in time in order to correct mistimed recordings (Langford, 2014). In Figure 3.10(a), voice editing for

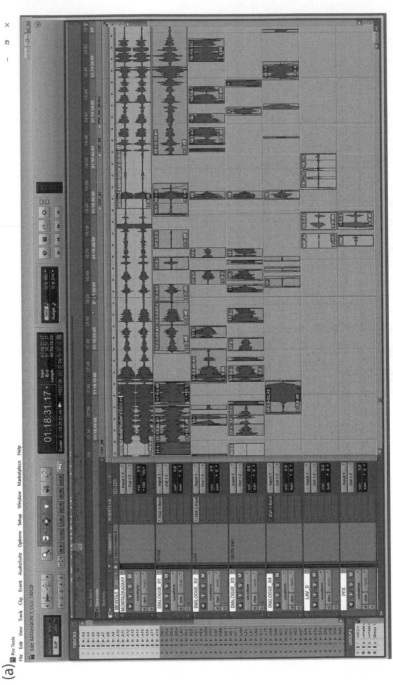

(a)

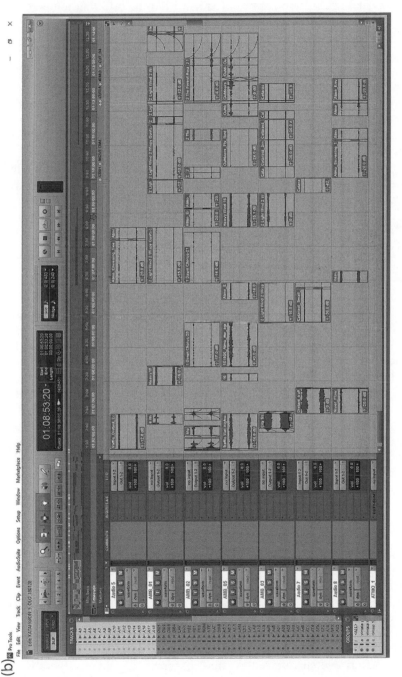

Figure 3.10 Examples of Mixed Multiple Background Sound Elements: (a) voice editing, (b) atmosphere sounds editing.

Source: Provided courtesy of Avid Technology, Inc.

an audiovisual production is presented in which the different tonal colors used can be helpful in organizing the process.

3.5.2. Sound Effects

A *sound* effect can be defined as any sound recorded or performed live for the purpose of simulating a sound in a story or event (Viers, 2008), and it can create the illusion of reality, continuity, spatial depth and space (Wyatt and Amyes, 2005). They give a powerful sensation and hold a central position to every production and they are not new. They have been used in filming since the 1920s, and in those days they were created through being recorded directly on the set, gathered outdoors in field recordings or created by a Foley artist in a studio (Flueckiger, 2009). They are used not only in the background since they have the power to create foreground spaces as well. Interacting with visual and musical design, they foster intense attention and lead expectation (Hanson, 2007). They even provoke humor and irony. Their utilization can be semiotic since they can be easily considered as signs, and as Jekosch (2005, p. 193) vividly states, "[E]ach acoustic event can be perceived as a sign carrier through which information about the world is communicated." In that sense, they may be processed in the human mind as something other than what they really are. An example of this semiotic dimension is the sound of bells that is received as a call for religious people to go to church, which in reality it is just the sound produced by the impact of two metal objects. Foley effects, previously mentioned, were named after Jack Foley who worked at Universal Pictures during the 1930s and are one of the categories of sound effects added to audiovisual productions. They are the kind of sounds performed by artists using common items and materials, and it is of great importance for them to be convincing since they add richness, gesture and realism (Viers, 2008; Wyatt and Amyes, 2005). Typically they are edited in absolute synchronization with the picture.

Sounds that give information about the location and the surrounding environment are called *background effects* or *atmos* (short for atmosphere), and they do not usually correlate with the action on screen. However, they create presence or ambience within a scene, filling a void in the soundtrack and identifying the location (Viers, 2008; Wyatt and Amyes, 2005). Traffic, birds and weather sounds are the kind of atmos effects. They are mainly mixed into the background, raising interest in the scene, and usually many tracks of sound are used in order to create the whole atmosphere of the supposed location in which the action takes place. In Figure 3.10(b), an example of mixed multiple background sound elements is presented.

Another term often referred in the broad area of sound effects is the soundscape, which is used to define an acoustic environment. According to

Raimbault and Dubois (2005, p. 340), "[S]oundscape encompasses sound variations experienced in space and time, grounded in the topography of the built-up area and accounts for the relationship between the individual experience and subjectivity with a physical and a socio-cultural context." Soundscape compositions are formed with a variety of electroacoustic techniques incorporating both narrative and abstracted elements into their language, and they may function as sound-based art (Truax, 2012).

Sound effects can be recorded in the field, creating a collection of sounds through a procedure that demands careful choice of the right place and time of the day. However, today large audio libraries compile a great variety of field sounds, organized in clever and useful categories, that are ready to use and easily accessible, thus offering the capability of formulating various aesthetics intended for audio or audiovisual productions.

3.6. Sound Mixing

Sound mixing involves a number of procedures of different categories, including musical decisions, technical and spatial renderings, that, when elaborated with precision, will eventually result in the desired outcome. As Langford (2014, p. 63) states, "[T]he end goal is usually a feeling or a mood rather than an absolute technical achievement." In the cases where moving pictures are involved, it is of absolute importance to maintain the synchronization of sound, music and image to engage the viewers in the media experience (Hanson, 2007). Mixing is ultimately an art.

Although multichannel audio signal processing used for sound mixing is typically performed manually and not automated by software, new systems are deployed serving to simplify some of the tasks aiming to improve the quality of editing and combining multichannel audio (Reiss, 2011). While mixing, signal processing is a procedure that works additively since more than one source is involved. New techniques propose that knowledge of previous practices, as well as of psychoacoustic studies and machine learning applications, can lead to automated adjustments of repetitive tasks (Reiss, 2011; De Man et al., 2014).

Editing involves procedures that deal with each sound file separately, while mixing weaves together all of the pieces; however, as already mentioned, contemporary hardware and software led to the consolidation of these two distinct processes within the same applications used in DAWs. Sound mixing, besides ensuring the continuity of all parts involved, as well as maintaining the intended volume, quality and sequencing, is also used to establish an aural perspective that is about the connection between the sound and the proximity and directivity of the place from which the sound is originating (Alten, 2013). Sequencing two sounds should include the kind of transitions that are appropriate for specific parts of the production. Sound editors may choose among cuts, fades in and out, cross-fades etc. The final outcome, after editing and mixing, of the distance and the direction of interrelated sounds should match their places in the physical environment and the desired aesthetics. This concerns instruments in an orchestra on stage as well as people and ambient audio in a film. The scene's visualization in the mind of the sound engineer can be very valuable.

Zettl (2011) argues that sound for video and film has four basic functions: information, outer orientation, inner orientation and structure. All forms of speech are included in the information function. Sound elements concerning the space, the time, the situation and the external event conditions lay in the outer orientation function. Inner orientation includes mood and internal conditions, and last but not least the role of structural function is to supplement the rhythm. These come together in an outcome where all the audio material sounds right, and no element distracts the audience at the expense of the others is a well mixed soundtrack.

Mixing procedures should commence with balancing the sources and deciding which of them stand in the foreground, which in the middle and which in the background. Of course, the equilibrium may change as the scenes evolve, e.g. the foreground may become background and vice versa. Especially in audiovisual productions, sound elements exist in many layers and are alternating in and out of focus at precise timings to stand out and create the appropriate impression. On the other hand, mixing music often demands an editor who is also a musician since music is a form of a "language" that has its own requirements. Knowing the instruments and their sonic elements is critical for deciding how they should be contained in a mix. In a live concert, vocals and each of the instruments hold their balanced role under the control of the conductor, as some are leading and some are accompanying. These roles, which are of equal importance, must be kept intact during the postproduction following a flow driven by the mixing process.

Mixing in multiple channels should result in clear recognition of the sound information contained in each channel and not losing segments along the way. The integrity of the whole composition should be preserved regardless the number of the sources involved. Today, the utilization of DAWs offer sound engineers the ability to perform mixing procedures with absolute precision with the use of the computer mouse and other interfaces. A waveform can be partitioned, and each segment may receive a different treatment (amplification, attenuation, etc.). At the point where they are merged, a cross-fade is applied either by the user or by the software itself to avoid unwanted clicks where waveforms might otherwise merge at points other than the zero crossing. Another practice is to perform a manual graphical "management" of the signal, creating an envelope that sets the boundaries of the sound amplitude. In Figure 3.11, interfaces of these processes are presented. In (a), the envelope technique is shown, and in (b), in the second from the top channel, the partitioning appears.

All the aforementioned led to the limited use of the console as hardware, even when embedded with significant software, since in the past those processes were partially achieved using the mixing consoles and especially the faders but are more easily done today in software applications (Figure 3.12).

However, it should be pointed out that in live performances, mixing demands continue to involve the need for hardware consoles. Audio equipment manufacturers are still in the process of innovative design research and development to produce ever newer evolutions of equipment (Figure 3.13).

Panning is the procedure of right–left positioning of the sound elements involved, either as each instrument is placed in the orchestra or as the spatial placement of an effect, for instance by adding a delay on an instrument that will widen the aural impression to create a stereo image. Mixing for surround multichannel reproduction is also a procedure related to panning as it can be an extension of it. Recently, panning has been enriched with more sophisticated tools that take into account spatial and spectral parameters beyond the traditional left–right leveling technique that have become familiar.

A *mixdown* often involves final adjustments to panning and other key decisions. Eargle (2003) suggests that even the final mixdown should be evaluated again the next day, when "fresh ears" can be brought to bear on the sound build. At all stages of editing and mixing, metering tools, such as VU meters, waveforms and spectrograms, are helpful. They provide indications that may lead to identification of problems and the prevention of errors.

Of course, in order for recordings to be listened to as intended by producers, good listening environments are essential and are achieved through good room acoustics and by the optimum layout of reference loudspeakers for the given room. In contrast to the sense of vision, hearing is perceived from 360 ° around us, and it should be reproduced accordingly so as to help the listener create the images that are carried along with every moment in the sound mix. Original placement or positional changes of the sources should be clearly accepted by auditory processes to lead to the stimulation of other senses (Alten, 2013).

Although there are objective criteria concerning the final aural outcome, nevertheless the perceptions of certain characteristics of sound, which are instrumentally measurable, such as "acoustical warmth or intimacy," may lay on the previous experiences of the listener and can be dealt with by

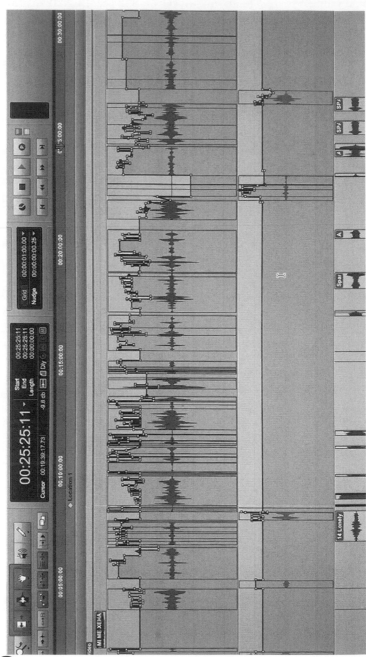

(a)

(b)

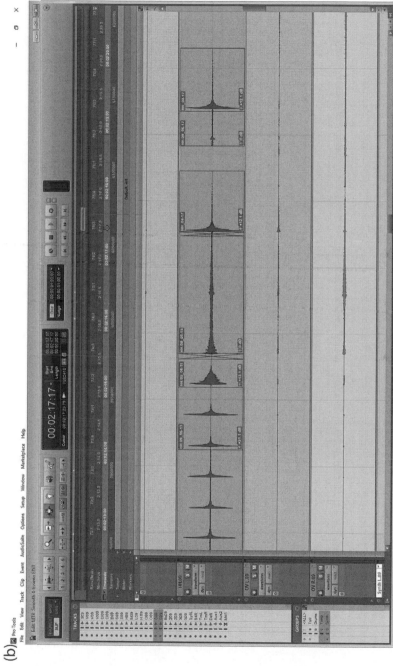

Figure 3.11 Mixing Techniques: (a) creating an envelope and (b) partitioning.

Source: Provided courtesy of Avid Technology, Inc.

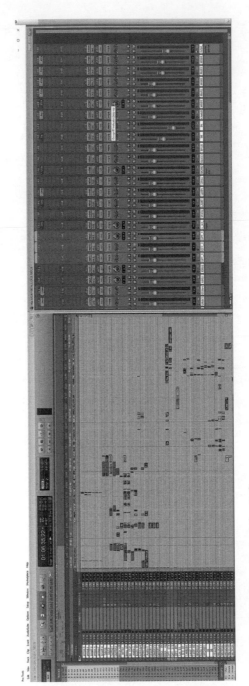

Figure 3.12 Console as Part of the Software,

Source: Provided courtesy of Avid Technology, Inc.

Figure 3.13 Mixing Console Used in Live Performances: The console is enhanced with displays of virtual instruments, along with hardware faders and knobs used as controllers of software inside computerized processing units.

Photo credit: Pete Brown

means of psychoacoustics (Eargle, 2003). As Beranek (1996) states, *intimacy* can be defined as a subjective impression of sound perception due to the specific physics of a space that result from the time that passes until the first reflection, following the direct sound, reaches the listener. Acoustical *warmth* is connected to the clear audibility of bass sounds when all the orchestra is playing. They can both be measured giving indicators of their perceived extent, and their usability rests on listeners having the "feeling" that they are in the original venue where the recording took place.

A significant point that should not be forgotten is that the final production should sound good on any playback equipment even on those of low cost, and this is what reference mixing is about. The final outcome should be listened to on multiple pairs of monitors to make sure the different elements keep their consistency. Also, the mix should be cross-referenced in stereo and mono versions.

The placement of loudspeakers is among the critical considerations when creating the listening environment, along with the speakers' technical characteristics, such as frequency response, sensitivity and linearity. Far-field and near-field monitoring, as well as the use of headphones, is the way by which sound engineers can listen while they are working at the different stages of sound processing. Far-field monitors are large and, as their name implies, are placed farther than near-field monitors relative to the listening position. Near-field monitors are placed nearer, often at a 1 m distance and at the height of the ears; they are smaller and less affected by the room's acoustics (Viers, 2008; Alten, 2013). Headphones are another very helpful device that might help in the reception of certain sound features, for instance in balancing—even better than the loudspeakers.

3.7. Conclusion

This chapter introduces all the audio editing processes until the last step in sound production, i.e. the mastering, where the final adjustments are made in order for the final formatting and packaging are completed to meet the criteria of the designated medium (audio CD, film, etc.). Mastering should not involve any more individual track corrections. The primary tools applied to the final mix are equalization, dynamics control, stereo and/or harmonic enhancement, ambient reverb etc. (Eargle, 2003).

The production should result in a controlled relationship between all the involved sound signals (vocals, speech, music, ambient sounds), without altering their value, their internal balance and rhythm.

The ultimate goal is to attain the highest fidelity while recreating the original sound field, integrating its spatial, three-dimensional nature. This chapter has presented the creative and audio engineering processes that are involved in the end-to-end sound design chain, to achieve maximum realism or impact during audience reception. Overall, "sound holds more than half of the picture," and careful and experienced treatment is necessary to fulfill the production needs.

Acknowledgments

The authors would like to thank their colleagues Prof. George Papaniko-laou, Dr. Kostas Kontos, Dr. Christos Goussios, and Mr. Andreas Georgal-lis, sound engineer, for providing valuable visual material in the figures for this chapter.

Figures 3.5, 3.6, 3.7(a) and (b), 3.10(a) and (b), 3.11(a) and (b) and 3.12 are © 12/3/2018 Avid Technology, Inc. All rights reserved. AVID, ▲▼◢▶ are either registered trademarks or trademarks of Avid Technology, Inc. in the United States, Canada, European Union and/or other countries.

Figure 3.8 De-noising interface with the available modules shown on the right is provided courtesy of iZotope, Inc.

Figure 3.9 Working environment of software for mastering processing applying multiband dynamics is provided courtesy of Magix (Sequoia 14).

Note

1. https://en.wikipedia.org/wiki/Digital_audio_workstation

References

Alten, S. R. (2013). *Audio in media*. Boston and Wadsworth: Cengage Learning.

Benson, K.B. (1988). *Audio engineering handbook*. New York: McGraw-Hill.

Beranek, L. and Martin, D. W. (1996). *Concert & opera halls: How they sound*. New York: Springer.

De Man, B., Mora-Mcginity, M., Fazekas, G. and Reiss, J. D. (2014). The open multitrack testbed. In: *137th AES convention*. Los Angeles: eBrief 165. Available at: www.aes. org/e-lib/browse.cfm?elib=17400 [Accessed March 2019].

Dimoulas, C. A. (2015). *Multimedia authoring and management technologies: Non-linear storytelling in the new digital media* (in Greek). Athens: Association of Greek Academic Libraries. Available at: https://repository.kallipos.gr/handle/11419/4343 [Accessed March 2019].

Dimoulas, C. A. (2019). Multimedia. In: D. Merskin and J. G. Golson, eds., *The Sage international encyclopedia of mass media and society*. Thousand Oaks, CA: Sage.

Dimoulas, C. A., Kalliris, G. M., Chatzara, E. G., Tsipas, N. K. and Papanikolaou, G. V. (2014). Audiovisual production, restoration-archiving and content management methods to preserve local tradition and folkloric heritage. *Journal of Cultural Heritage*, 15(3), pp. 234–241.

Dimoulas, C. A., Kalliris, G. M. and Papanikolaou, G. (2008). Soundfield microphone simulation and surround/3D sound design techniques for 3D-graphics and digital-characters movie production. In: *4th Greek national conference of Hellenic institute of acoustics (HELINA) 'Akoustiki 2008'*, Xanthi, Greece: HELINA (Hellenic Institute of Acoustics).

Dimoulas, C. A. and Symeonidis, A. L. (2015). Syncing shared multimedia through audio-visual bimodal segmentation. *IEEE MultiMedia*, 22(3), pp. 26–42.

Dimoulas, C. A., Veglis, A. and Kalliris, G. (2014). Application of mobile cloud-based technologies. In: J. Rodrigues, K. Lin and J. Lloret, eds., *Mobile networks and cloud computing convergence for progressive services and applications*. Hershey, PA: Information Science Reference, pp. 320–343. doi:10.4018/978-1-4666-4781-7.ch017.

Dimoulas, C. A., Veglis, A. A. and Kalliris, G. (2015). Audiovisual hypermedia in the semantic Web. In: M. Khosrow-Pour, ed., *Encyclopedia of information science and technology*. 3rd ed. Hershey, PA: Information Science Reference, pp. 7594–7604. doi:10.4018/978-1-4666-5888-2.ch748.

Dimoulas, C. A., Veglis, A. A. and Kalliris, G. (2018). Semantically enhanced authoring of shared media. In: M. Khosrow-Pour, ed., *Encyclopedia of information science and technology*. 4th ed. Hershey, PA: IGI Global, pp. 6476–6487.

Dutilleux, P. and Zölzer, U. (2002). Filters. In: U. Zölzer, ed., *DAFX—digital audio effects*. Hoboken, NJ: John Wiley & Sons, pp. 31–62.

Eargle, J. M. (1992). *Handbook of recording engineering*, 2nd ed. New York: VanNostrand.

Eargle, J. M. (2003). *Handbook of recording engineering*, 4th ed. Los Angeles: Springer.

Floros, A., Tatlas, N. A. and Potirakis, S. (2011). Sonic perceptual crossings: A tic-tac-toe audio game. In: *Proceedings of the 6th audio mostly conference: A conference on interaction with sound*. ACM, September, pp. 88–94.

Flueckiger, B. (2009). Sound effects strategies for sound effects in film. In: G. Harper, ed., *Sound and music in film and visual media: A critical overview*. London and New York: Continuum International Publishing Group, pp. 151–179.

Hanson, H. (2007). Sound affects. Post-production sound, soundscapes and sound design in Hollywood's studio era. *Music, Sound, and the Moving Image*, I(1) (Spring), pp. 27–49.

Hollman, T. (1988). Postproduction systems and editing. In: K. B. Benson, ed., *Audio engineering handbook*. New York: McGraw-Hill, pp. 14.1–14.44.

Jekosch, U. (2005). Assigning meaning to sounds—Semiotics in the context of product-sound design. In: *Communication acoustics*. Berlin and Heidelberg: Springer, pp. 193–221.

Langford, S. (2014). *Digital audio editing correcting and enhancing audio in Pro Tools, Logic Pro, Cubase, and Studio One*. Burlington, VT: Focal Press.

Papanikolaou, G. and Kalliris, G. (1991). Processing and editing of DAT recordings. In: *4th symposium on sound engineering & mastering*, Technical University of Gdansk.

Papanikolaou, G. and Kalliris, G. (1992). Computer aided digital audio production in random access media. In: *Proceedings of the international computer music conference*, Delphi, Greece.

Raimbault, M. and Dubois, D. (2005). Urban soundscapes: Experiences and knowledge. *Cities*, 22, pp. 339–350.

Reiss, J. D. (2011). Intelligent systems for mixing multi-channel audio. In: *17th international conference on digital signal processing*, Corfu, Greece, pp. 1–6, doi:10.1109/ICDSP.2011.6004988.

Truax, B. (2012). Sound, listening and place: The aesthetic dilemma. *Organised Sound*, 17(3), pp. 1–9.

Viers, R. (2008). *The sound effects bible: How to create and record Hollywood style sound effects*. Studio City, CA: Michael Wiese Productions.

Wyatt, H. and Amyes, T. (2005). *Audio post-production for television and film*. 3rd ed. Burlington, VT: Focal Press.

Zettl, H. (2011). *Sight sound motion: Applied media aesthetics*, 6th ed. Boston: Wadsworth Cengage Learning.

4

Audio Effects in Sound Design

Brecht De Man

4.1. Introduction

Audio effect is a term used to refer to nearly any signal processing device or algorithm that can be applied to a stream of audio, usually in real time. Used subtly, audio effects make scenes sound convincing, immersive and professional. Used in more extreme ways, entirely new sounds can come about, up to a point where it is as much a synthesis tool as a production tool. The job of their operator—whether human or algorithmic—is to do this while respecting the physical constraints of the system and the medium, and the eardrums and neighbors of the user. To this end, they use the many transformations at their disposal to convey realism, hyperrealism, perception of the character and imagination of the storyteller. This chapter provides an overview of such audio effects and a source of inspiration toward achieving each of these goals.

Audio effects are encountered by a wide range of agents in the sound design process, and this text seeks to cater to each of them. Between the sound recordist, Foley artist, sound designer, mix engineer, producer, SFX library author, game audio developer and synth programmer, among others, they can be anything from purely technical to completely creative tools.

There are several aspects to audio effects, including the concept, their use, their interfaces and their implementation. The first two are the subject of this chapter: what are the important audio effects to be aware of, and how should or could they be applied in your sound design process? The interface of an audio effect can go from a series of knobs or faders—be it on a screen or on hardware—to canvases and curve drawing. While the method of controlling can be very important, especially in Foley-like setups where the effect is played live like a musical instrument, it is too specific to each device or plugin to cover here. Finally, the implementation is another area where essentially identical effects can differ enormously, but this is beyond the scope of this work.

Digital storage and processing equipment has all but replaced tape and valves, and much of the postproduction happens "in-the-box," on a single computer with the appropriate software rather than a few racks of hardware. Of all audio fields, the most analogue aficionados can be found in popular music production, where the sound of music technology from a particular era can be a sought-after aesthetic. That said, there are several possible reasons to use analogue equipment or workflow, such as quick or intuitive access to controls, creative use of an analogue device's coloration and distortion or simply old habits or nostalgia. Implementations and interfaces of digital tools are still heavily influenced by their analogue counterparts, so knowledge of one side is not wasted on the other side, and no further distinction is made in this text.

4.2.   Effect Types

4.2.1.   Balance

The simplest but perhaps the most important of all processing steps, level balance is as easy to explain as it is difficult to get right. To account for variation in the source content, desired emphasis and supposed distance from sources, levels are usually automated to a large extent, by recording fader movements in real time ("fader riding") or drawing the envelope in the digital audio workstation.

Doing little more than amplifying or attenuating a signal by a given, frequency-independent factor, the effect (if it can indeed be considered an audio effect) of a fader is straightforward and predictable. However, the actual loudness of a source is determined by many more factors, all of which need to be considered when choosing the appropriate level. For instance, the perception of loudness is heavily influenced by the frequency content of a source (further adjusted with equalization), its short-term level variations (affected by dynamic range processing) and any other level changes the other effect units may introduce. So read on!

4.2.2.   Panning

The history of stereo starts in the cinema, when Alan Blumlein wanted to address the mismatch between an actor's location and the origin of the sound. With different levels and propagation delays, as would, for instance, be recorded by a stereo microphone pair, the sound could effectively follow the actor on the screen. This relies on the phenomenon of a phantom source, that is, the illusion of a signal emanating from a position depending on the relative levels and time delays of that same signal coming out of the speakers simultaneously. For instance, a source will appear right in between the speakers when its signal comes out of both speakers simultaneously and at the same level, provided the speakers are equidistant with respect to the listener. When coming out of one speaker only, the sound will appear coincident with the speaker location, regardless of the listener's position.

The positioning of sources, or *panning* (for "panorama"), can emerge naturally from a microphone array that captures off-center sources at a different level (for instance, a spaced pair or coincident directional pair) or time (a spaced pair). Of course, the position of the speakers or indeed the effect of headphones will have a profound effect on the perceived localization.

The artificial panning of a monaural signal—i.e. not as recorded by a stereo microphone pair usually happens through amplitude panning, where the position of a sound source is conveyed through a level difference between the left and the right channels. The resulting level difference between one's left and right ear then creates a virtual position for an off-center object that would have the same level of difference between the ears. On headphones, the left and right signals are fed to the left and right ears separately; on speakers, the left signal still reaches the right ear and vice versa, albeit delayed, attenuated and filtered by the obstructing head.

An alternative or complementary method is *phase panning*, or delay panning, where the left and right signals are shifted in time with regard to one another. This then creates the illusion of a position from which the sound would reach both ears at a different time, from immediately left or right (the time difference roughly equals the distance between both ears divided by the speed of sound) to right before or behind the listener (no difference at all, as the source is equally far from both ears).

As the equidistance from the speakers cannot always be guaranteed—consider for instance a theater or even living room—a center speaker is usually added in 2+ speaker arrangements.

For surround and 3D audio, increasingly important in anything from film over game audio to virtual (VR) and augmented reality (AR), panning is extended to a larger number of channels. Here, too, the localization of sources can either be embedded in the recording by using the appropriate microphone array or panned artificially.

The human ear is remarkably good at resolving the angle from which a sound emanates, at least in the horizontal plane. Our resolution of the elevation angle is significantly less precise. However, it may still be worth providing some amount of differentiation in height, which is why some speaker setups will include central top speakers ("voice of God") and otherwise heightened and sometimes lowered speakers, with most speakers still in the horizontal, "ear-height" plane.

4.2.3.   EQ and Filtering

The term *equalization* (EQ for short) refers to the act of electronically making the total frequency response of a system "flat," or equal for all frequencies. This is accomplished through filters, which emphasize or attenuate some frequency bands more than others. This was an important concern for early radio and audio systems, where components of the signal path were severely compromised in this regard. So too were analogue storage media like tape, where preemphasis of high frequencies at recording and de-emphasis at playback could be employed to mitigate noise problems. Nowadays unintended coloration of audio by recording and playback equipment is rarely an issue.

However, equalization has many more uses, both technical and creative, than simply "making equal." First of all, it allows for amplifying or attenuating those frequency components whose relative strength makes the source signal sound unbalanced or annoying. This may be due to a problem inherent to the source or may have to do with the recording method or distance. The interplay of sources is an important consideration as well, especially if there are many that should be audible and intelligible. Sources that coexist in the same frequency range(s) tend to mask each other, so reducing the level of a band of frequencies makes room for the competing source to peek through in that band. For speech, the intelligibility can be enhanced by emphasizing those frequencies that are most important for making out what is said, i.e. those corresponding to the consonants or those masked by other sources. Much like a color filter for visual media, an audio signal filter can subtly or dramatically affect the color of many sources simply by altering the relative strengths of the frequencies. This opens up a wide range of creative possibilities, all with a ubiquitous and relatively simple processor.

The layout of an equalizer will typically take one of two forms. In the case of a graphic equalizer, a number of adjacent frequency bands can be adjusted in level, using faders that effectively visualize the amplification factor as a function of frequency—hence the name. The spectrum can be divided in octaves (each band's center frequency equals double the previous band's center frequency), third octaves (each octave is further divided in three bands) or two-third octaves. A graphic EQ with as much as 31 third-octave-wide bands is very common.

The parametric equalizer, in contrast, is much more versatile as it offers access to not only gain but also to center frequency and bandwidth controls for each filter stage—of which there could be several. In the digital realm, there is no reason why the resulting frequency response cannot be visualized, thereby mostly negating any user experience advantage of a graphic equalizer.

There are about as many possible EQ or filter implementations as there are filter types, but for sound design purposes it suffices to be aware of the following handful.

- A *peaking filter* is the most important component of any equalizer, changing the level of a particular region (defined by its center frequency and the bandwidth or quality factor "Q") with a specified gain factor.
- A *low shelf filter* changes the level of all content below a specified frequency by a specified gain factor. A *high shelf filter* is the equivalent at the higher end of the spectrum.
- A *low pass filter* (LPF), sometimes dubbed high cut filter, removes all content above a certain frequency, leaving anything below unaffected. A *high pass filter* (HPF), also known as low cut filter, does the opposite.
- A *band pass filter*, then, removes anything outside of a specified band. A band stop or (for narrow bands) notch filter does the opposite, blocking the sound in this band.

It is important to note that any filter comes with important trade-offs, so decisions have to be made regarding the bandwidth and steepness of each of the aforementioned EQ stages. Without going into the mathematical intricacies, it may be helpful to remember "you can't have it all." For instance, one has to accept some signal dispersion in the time domain for added steepness in the frequency domain and vice versa. Ultimately, the trained ear will be the judge of what is the most appropriate setting, aided by the intuition that comes with a long experience of toying with different parameters and implementations, on a variety of source content.

4.2.4. Dynamic Range Processing

At the most basic, dynamic processors replace a constantly varying level fader, obviating the need for drawing automation curves (or indeed moving them manually). But they are far more powerful than that, enabling transformations that are barely possible with even the most elaborate level automation.

The first and best known effect in this category is *dynamic range compression*. As the name suggests, it reduces the dynamic range, typically by attenuating any signal that has a level above a certain threshold (downward compression) and only rarely by amplifying low-level sounds (upward compression). In the former case, it leaves any sound with levels below the threshold unaffected and scales those that exceed the threshold by a certain ratio. With a threshold at −6 dB and a ratio of 2:1, for instance, an input level of −4 dB will result in an output level of −4 dB − (−4 dB −(−6 dB))/2, or −5 dB. In other words, as the input level exceeds the threshold by 2 dB, the output level exceeds it by only 1 dB (2:1). If the ratio were 4:1, the output level would only be 0.5 dB above the threshold, and so on. To make the transition around the threshold less abrupt, a knee parameter determines the width of a transition region where some smooth curve connects the linear pre- and post-threshold regions. As a significant portion of the signal will be reduced in level, the overall perceived loudness (and objective average level) will be lower after being processed by a downward dynamic range compressor. For this reason, a makeup gain parameter is often included—this simply applies a positive gain to the entire signal to make up for this loss in level. However, the user should be aware that this results in a net increase in noise: the quiet parts of the signal may be amplified considerably, including any background noise, even if the louder bits are not.

If this level transformation were immediate, then any movement of the signal would be affected so that the waveform would change in shape and timbre. As this results in distortion and artifacts, which is usually undesired, the level changes are artificially slowed down over time. The time constants attack and release time—also referred to as the ballistics of the compressor—specify the rate at which a compressor approaches the full attenuation, and how quickly it reverts back to unity gain, respectively. In other words, the previous input signal of −4 dB will result in the same −4 dB output at first and be turned down exponentially to (almost) −5 dB. As the input signal suddenly drops below the −6 dB threshold again, the 1 dB initial attenuation will gradually revert to 0 dB. Careful setting of these time constants is paramount, as overly short or long ballistics will result in obnoxious effects such as *pumping* (audible reverting to unity level), *breathing* (audible swelling of the background noise as the attenuation reverts to unity level), letting loud transients through (due to an overly slow attack time), and distortion (when the attack time is so short that it acts on a wavelength level, modifying the shape of the waveform—especially a concern at low frequencies).

Compression's logical counterpart is expansion, where quiet sounds are made even more quiet (downward expansion) or loud sounds louder (upward expansion). This is useful as a subtle way to reduce the level of background noise, which would be especially audible and bothersome in quiet intervals. Because of their temporal behavior, compression cannot be reverted simply by applying expansion, as both will introduce a lag around the transients. An extreme yet more popular form of expansion is *gating*, or the attenuation and even muting of any signal below a certain threshold. Here, too, time constants ensure that nasty glitches are minimized and only intervals of sufficient length are turned into silence.

The aforementioned dynamic processors are adaptive to the input signal, that is, the gain applied depends on the recent history of levels up to a certain point. In the case of a "lookahead" function, it can even adapt to the level of the signal in the near future. However, there is no reason why it cannot react to other sources. In this case, the input to the signal processing path is different from the input to the signal sensing path, or *side-chain*, so that the signal is reduced in level when a different source is particularly active. Most commonly associated with radio DJs speaking over a song's intro or ending, this *ducking* effect can be used in clever and creative ways to ensure audibility between competing sources without turning one of them down permanently.

Some processors are somewhere between the spectral (EQ and other filters) and dynamic processors. The side-chain functionality can be augmented by placing processors in the side-chain path, including filters. A popular example of this is the *de-esser*, where a band of high frequencies is emphasized (and/or low frequencies cut) in the side-chain so that the compressor attenuates the frequencies associated with sharp "S" sounds in speech. Another clever combination of filters and compression stages is the multiband compressor. Here, the signal is first split into different bands through filtering, and then each component is compressed separately before being summed together again. The effect of this is that only a particular band of a signal, not the signal as a whole, gets attenuated when that band exceeds a certain threshold. As a result, the output will be closer to a desired *target spectrum*, with overly loud frequency components being turned down and others left alone. A related processor is the *dynamic EQ*, where a certain frequency is boosted or cut only when the signal hits a certain threshold value and leaves it alone otherwise.

4.2.5. Reverberation

An aspect of each nonsynthetic, nonlaboratory sound, reverberation is immensely critical to get right when designing sounds and environments. As such, it is automatically a part of most recordings and something to be considered at the recording stage. Indeed, in an anechoic room, the sound of a gun firing is a very unimpressive "pop"—if a very loud one. So a large portion of the sound we have come to associate with a gun, an explosion, footsteps, and any other kind of especially impulsive sonic events is that of the reverberation it causes.

You cannot get any more "real" than the original, acoustical reverberation, and given a suitable recording position, all parameters—reflections, reverberation time, direct-to-reverberant ratio—will have appropriate values. However, in many cases, the embedded reverberation is not sufficient in amount or quality, and artificial reverberation must be added. In this case, the original reverberation should be significantly lower in level, or it will still be heard as well. This is one reason why recording studios tend to be relatively "dry," that is, not very reverberant. In a situation where many sounds are close-miked, recorded after the fact, taken from sample libraries or synthesized, reverberation can help tie them together by providing a common acoustic, as though they were recorded in the same space.

The two main types of reverberation effects are algorithmic and convolution reverberators. An *algorithmic reverb* has the benefit of being highly customizable, in addition to being less computationally expensive. As such, its parameters can be automated to smoothly vary between settings. A *convolution reverb* applies a particular impulse response to the source audio, simulating a given acoustic (at least between a given source and receiver location) more or less perfectly. This is essential in re-recording (dubbing) dialogue or sound effects, where studio-recorded sound needs to sound like it was recorded in the original space. This can in theory be accomplished using an impulse response measurement and a convolution reverb.

Such impulse responses need not be limited to acoustic spaces—they may also represent a tube, a resonating string, or really any sound whose spectral and temporal characteristics one wishes to mix with the original signal. A common creative use of physics-defying reverberation is the so-called reverse reverb, where the sound builds up prior to the direct sound rather than decaying after it. The buildup is an ideal dramatic device that gives away clues of the sound that follows (e.g. a voice) or that can be followed by an equally dramatic dry sound or silence.

Closely related to the reverberation is the *delay effect*, also referred to as echo. In this case, a delayed version of the signal is summed to the original at a lower level. The output of this is often fed back to the input of the effect, resulting in a train of delayed versions, each a fixed amount lower than the previous. The delayed version can also be filtered or processed in other ways, so that the echoes become increasingly muffled, thin or of some other quality. While less natural than the much more complex reverberation of a space, such an echo can mimic sound bouncing back off a wall or emanating from a canyon, and a group of them come close to approximating the early reflections of room boundaries—without the wash of sound at the end of the reverberation tail, which can have the undesired effect of muddying the mix.

4.2.6. Modulation Effects

Modulation effects are those that get their signature sound from the variation (modulation) of one or more parameters. This can follow a pattern generated by a low-frequency oscillator (LFO) or a random number-based generator (sample-and-hold, sample-and-glide); depend on the input signal's amplitude or some other feature (ADSR, envelope follower); or be controlled manually by the user. All of these modulation approaches can be found in sound synthesis, too, where any parameter can also be modulated to achieve a particular sonic effect. In simpler terms, a modulation effect makes the incoming sound go "weee-ooow," "wub-wub-wub-wub," and, of course, "wah-wah."

While any effect parameter can be modulated periodically or in response to some trigger, the most common types of modulation effects vary level, delay or a filter's characteristic frequency. Unless otherwise noted, typical implementations will have an LFO as their control signal.

In addition to the wonderfully wacky effects that extreme settings induce, subtle quantities of these effects are known to add life, movement and body to their input. Furthermore, they are often used to "spread" or "widen" a sound across multiple channels. In this case, it is essential to induce some difference between the left and right (and other) channels that can be accomplished by, for instance, having a 90° phase shift between the respective control signals.

4.2.6.1. Modulating Level

The easiest modulation effect (to create and to imagine) is the cyclical changing of a source's level, known as the *tremolo* effect. Starting at full volume, the gain can drop to absolute silence or to some intermediate value. It does little more than moving the fader according to some predetermined pattern (the waveform) at a given rate, so you don't have to. This is equivalent to multiplying some periodic curve, between 0 and 1, with the original signal.

Change the level of the different channels in different ways, and this effect becomes an autopanner. In particular, if the left channel is at minimum level when the right is at its maximum, and vice versa, the source moves from one speaker to the other—or between your ears in the case of headphones. More elaborate effects are possible in multispeaker systems and binaural sound, from bees buzzing around your head to circling helicopters.

Most modulation effects are characterized by a low-frequency oscillator. In this context, "low" typically means "below audible frequencies," or less than 20 Hz. If the level of a source is modulated at higher frequencies than this, the input signal becomes a metallic version of itself. This weird and unsettling effect is an example of *ring modulation* and perhaps most associated with the voices of Doctor Who's Daleks and other robots. There is no limit to which signals you can multiply with the source, but the multiplier of choice is almost always a sinusoid.

4.2.6.2. Modulating Delay

Another very satisfying effect parameter to modulate is the delay, in the sense of "time elapsed before the signal/segment/sample is played." By itself, a static delay is inaudible as it doesn't affect the timbre, level or tempo. But it can make an important difference when interfering signals are being played back concurrently, in addition to, of course, making the sound arrive slightly later. Smoothly modulating the delay of a signal, however, is in fact equivalent to modulating pitch. This effect can be witnessed when manually tweaking the delay time of most delay effect units. A vibrato effect, almost equivalent to rapidly moving a string back and forth on a violin or guitar, is a continuous "vibration" of the pitch. But even more interesting effects occur when mixing this modulated signal with the original, feeding the processed output back to the input or processing different channels differently.

A *flanger* creates the illusion of doubling a sound, in the sense that a very similar yet slightly different source is added. It achieves this by simultaneously playing two identical signals, where one is delayed by a varying time interval. The characteristic "swoosh," especially at low modulation speeds, can be reminiscent of jet engines or really any noisy sound where the source or receiver moves with respect to a reflective surface (a plane over a runway, a fountain in a swimming pool). The time between the arrival of the original and the reflected sound at the receiver's ears then changes gradually, which results in cancellations and amplifications of continually varying frequencies. The output is often fed back to the input for more extreme frequency variations.

The *chorus effect* takes this "doubling" a step further by mixing in more than one extra version, all with differing delays. This then essentially creates a choir, or indeed chorus, of sounds that are different only in the "natural" pitch and timing variations that a highly trained vocal group would exhibit.

In addition to upping the perceived number of sources, flanger and chorus effects are also very effective (pun intended) in increasing their perceived width and coverage. It suffices to apply a somewhat different effect to the different channels, usually by means of a phase shift in the modulator.

4.2.6.3. Modulating Filters

Another method to introduce cancellations of certain frequencies is the *phaser effect*, which filters a signal so that its phase is shifted to different degrees at different (and continuously changing) frequencies. When summed with the original signal, the frequencies where the filtered signal is out of phase are cancelled. As opposed to the flanger, where these notches are equally spaced, a phaser doesn't correspond with any physical process and sounds quite other-worldly or synthetic—which, of course, is a very valid target in sound design.

Other filters are also suitable for continuous modulation of the characteristic frequency. In the case of a low pass filter, band pass filter, or boosting peaking filter, the resulting effect is classified as a "wah-wah" (second "wah" optional), due to its onomatopoeic resemblance to the corresponding vocalization. The frequency sweeping is traditionally controlled manually rather than by an LFO and is best known as a foot pedal for guitarists. It also lends itself very well to being triggered by the signal envelope, where the loud portions are typically clear and bright and quieter bits progressively more muffled. In this case, it is referred to as an auto wah or envelope follower.

4.2.7. Vocoder

Most commonly associated with monotonous robot voices, a vocoder (from "voice coder") is at the basis of a wide variety of effects. It can take the phase characteristics of a signal and change its pitch or apply them to a synthetic sound, making a synthesizer, guitar or other noise "talk." At its core is an electronic or digital circuit that filters some input, the "carrier," according to some "modulator." Nonelectronic implementations of this concept exist as well, from placing a speaker on the throat and mouthing words (making the train in Disney's Dumbo say "All aboard!" through its steam whistle) to feeding the sound through a tube in the mouth (the so-called "talk box" popular with guitarists).

4.2.8. Pitch Shifting

Pitch shifting is an essential tool for creating new sounds. A straightforward way is to play a signal back at a different speed—half speed will result in any pitch being an octave down and double speed in an octave up. However, this obviously means the speed changes too, which may not always be desired. Furthermore, as digital audio has content up to half the sampling rate; a half-speed signal will only have content up to a quarter of the sampling rate and so on. Therefore, if at all possible, the original sound should be recorded at a sufficiently high sample rate, or the resulting sound will be "dull." Using more sophisticated techniques (such as a phase vocoder), a signal's speed and pitch can be altered independently. For spoken and sung audio, simulated vocal tract characteristics can be tuned to great extents from male to female and from monster to chipmunk.

4.2.9. Distortion

Essentially a defect of a processing device or storage medium, *distortion* can be a sought-after effect with interesting sonic properties. Many forms of distortion exist, though the most common is amplitude distortion (or harmonic distortion), where the output amplitude is a nonlinear function of the input amplitude. In contrast with dynamic range compression, this input–output curve is applied immediately, so that it does affect the waveform and can result in a significant change in timbre. While the curve can take any shape, they are often approximately linear around zero (quiet signals pass through virtually unaffected) and increasingly horizontal at high positive and/or negative amplitudes. Depending on this shape, a certain pattern of harmonics will appear, which are frequency components at integer multiples of the existing frequencies ($1f$, $2f$, $3f$, . . .). A pure tone at 100 Hz may therefore give rise to a train of tones at 200 Hz, 300 Hz and so on. Because of a distortion effect's nonlinear nature, the same tone at a higher level may result in a different and usually stronger harmonics pattern. If multiple tones are present (which is usually the case), a different form of distortion, called intermodulation distortion (IMD), arises as well. In this case, the distortion effect will create frequency components not just at those frequencies corresponding to multiples of either of the original components but also at sums and differences of these multiples. These are no longer *harmonic* with the initial signal, meaning the frequencies do not form musical intervals such as octaves and fifths with the input.

As one can imagine, due to the limitless number of input-output curves, a wide range of "colors," is possible with distortion. Since it is such a creative effect, there are no best practices to study—anything goes. However, it can be useful to observe a few mathematical properties:

- If the input–output curve is odd ($f(X) = Y$ if $f(-X) = -Y$), only odd harmonics are generated.
- If the input–output curve is even ($f(X) = Y$ if $f(-X) = Y$), only even harmonics are generated.
- If the input–output curve can be expressed as a polynomial function of order N (where N is a natural number), harmonics will only appear until the Nth harmonic.

4.3. The Effect of Effects

4.3.1. Realism

The first goal for the application of this long list of signal processors is to create or reconstruct the sound of the environment at hand, including all sound sources and any effects from sound propagation and obstacles. As the soundscape rarely results from a single recording by a microphone array but rather from a mix of a great many close-miked, re-recorded or synthesized sounds, these transformations will often have to be applied "by hand," simulating what the audience would hear if they were present at the fictional scene.

This starts with observing the laws of physics. A source's location will dictate its distance to the receiver position and therefore its amplitude (balance), propagation time (delay), angle (panning) and amount of air absorption (EQ-ing out the high frequencies). It will affect reverberation as well: a closer source will lead to more direct-versus-reverberant sound, and a longer predelay, i.e. the gap between the arrival of the direct sound and the wash of reverb. With the exception of the diffuse tail—a reasonable assumption—the impulse response of the reverberation will be different, too, for every source-receiver pair.

Acoustics further help position a sound behind a wall or below the floor, for instance when simulating the "party next door" effect. The muffled, bass-heavy music and crowd noise is the effect of a low pass filter with a slope of 6 dB per octave (12 dB in case of a double wall), with the resonance frequency and damping factor determined by the mass and elasticity of the structure.

Calculations of the exact resonance frequencies, absorption coefficients and so on are seldom needed, but it doesn't hurt to seek inspiration in physical models and toy with the parameters. Ultimately, your own ears (and your client's) will judge what's "realistic," even though it isn't scientifically accurate.

For many sounds—from dinosaurs to alien spaceships—there is no real-world reference to approximate, making a sound designer's job both easy (there is no wrong answer) and incredibly hard (there is no right answer)! But the audience's expectation, especially based on titles from the same series or genre, may be an important guideline.

4.3.2. Hyperrealism

After learning the rules, you can start breaking the rules. Diligently observing the laws of physics may lead to ugly sounding spaces, unimpressive fights, or dull soundscapes. Audiences have grown accustomed to beautiful and rich sonic environments, exaggerated explosions and punches that sound more like bass drums. So simply recording (or recreating) what goes on usually results in a disappointing and unexciting ensemble, even if it is life-like in theory.

At the same time, the most violent, ear-piercing and overwhelming sounds have to be toned down to serve the limitations of the playback system and comfort of the listener. The intelligibility of dialogue, from faraway whispers to discordant shouting, is a fundamental requirement that trumps any quest for realism. Of course, sometimes the intent is to make something difficult to hear or impossible to understand or to convey negative emotions by making an event sound harsh and almost painful. As with everything, moderation is key, and the desired effect, targeted venue, medium and (in an interactive situation) user preferences will dictate the appropriate exaggeration or restraint of the sonics. Multichannel audio allows for the placement of a greater number of sources without masking the most important elements.

4.3.3.   Perception and Interaction

When we experience the real world, we are not only subject to the physical processes that make and shape sounds but also to how sound waves interact with our body and how our ears and brain decode them. By being present, we affect the environment, and because of this we expect things to sound a certain way that cannot be informed by a mere application of the laws of physics in the void. In this sense we are not usually invisible observers, who do not take up no space or block sound—pursuing this would simply sound odd. Instead, we are usually simulating what we would hear if we were at the scene, even if the first person isn't part of the narrative, so we are not embodying a character.

For instance, our head and ears are obstacles to wind, resulting in audible turbulence around us. Sources sound different depending on what direction they come from, due to the shape of our ears and reflection off and diffraction around our head. And we are used to being a certain distance from the ground, with the corresponding reflections.

Within our ears, a number of phenomena affect the way different sounds are perceived. For instance, the ear effectively acts as a compressor, damping vibrations at different stages when sensing high levels. As a result, the real-world dynamic range from threshold of hearing to threshold of pain is significantly reduced in perceptual terms, where all sound is scaled depending on the loudest element of the current (or very recent) soundscape. A very direct example of exploiting these properties can be illustrated in the design of a nearby explosion. At realistic levels, while preserving the intelligibility of dialogue, such an event would both shatter eardrums and exhaust any medium's or system's dynamic range. The perceived loudness can be conveyed in part by rapidly reducing the level, just as our built-in compressor would, and leaving the level there for a while to indicate (temporary) hearing loss. The classic "tinnitus" tone and lack of other audio following a loud and traumatic event is another effective approach.

An equal-loudness contour, showing which amplitude levels are perceived as being equally loud at different frequencies, looks different for different perceived "loudnesses." For instance, high-amplitude sounds at different frequencies are much closer in perceived loudness than low-amplitude sounds, where, for instance, low frequencies require a much higher level to be perceived equally loud as a high-frequency sound. This offers another opportunity to suggest loudness without excessive increase in level: exaggerating especially the low end makes a source appear louder than it actually is.

Another interesting property of human hearing is the ability to single out a particular sound from a mix of inharmonious noises, like a conversation partner in a noisy room—a phenomenon aptly referred to as the cocktail party effect. Just like the choice of visual focus is made for us by committing to a framing and focal depth, so too the focus of our ears can be guided by "zooming in" on what's supposedly important. This is a valid excuse for making dialogue much more intelligible than it should be in real life: the sound design dictates what the audience should listen to, and the added emphasis will feel natural.

A different category of interaction effects comprises the distortions and artifacts associated with audio equipment. While separate from human hearing, we have grown accustomed to the way certain things sound through microphones and recording devices. Ironically, the perception of realism may be increased when the objective quality of the recording is poor: a windy beach is best represented through almost deafening, low-frequency turbulence noise, with dialogue shouted over it. Again, loudness can be suggested without exhausting headroom by introducing the distortion of a recording device when the recorded sound exceeds the device's dynamic range.

4.3.4. Imagination

When all the rules have been applied, realism is maximally pursued and everything is clearly audible, it is time to mix things up and do the opposite. Sound is a very powerful narrative tool, and any deviation from what's expected can induce a clear emotion. The audio effects applied to (nonmusical) audio can make a world of difference with regard to the story it tells. The mood of the scene can be set, the mental state of the protagonist underscored, a future event announced.

As an example, absentmindedness of any sort usually goes hand in hand with a low level (up to complete silence), high frequencies being filtered out, or high levels of reverberation. Similarly, upon abruptly waking up, the sound may suddenly become loud and dry, with a full frequency spectrum. A shift to the next scene—even when the visuals are yet to follow—can equally be clarified by first making the soundscape quieter, more reverberated and filtered, before coming in "in full force." Crazier effects may be chosen to provide a first-person experience of various kinds of drugs, dreams, hallucinations and memories, including but not limited to delay effects, modulation effects, pitch shifting and dynamic panning. Elements that are meant to appear synthetic, virtual or magical are also best processed heavily to distinguish them from the real world.

Audio effects can further help make the distinction of a narrator from everything else, usually by simulating "proximity": a close and dry microphone recording with ample low end.

Further Reading

Reiss, J. D. and Andrew McPherson, A. (2015). *Audio effects: Theory, implementation and application*. Boca Raton, FL: CRC Press.
Case, A. (2012). *Sound FX*. New York: Taylor & Francis.
Zölzer, U. (2002). *DAFX: Digital audio effects*. New York: John Wiley & Sons.

The Mix Stems

Voice, Effects, Music, Buses

Neil Hillman

5.1. Introduction

Since the advent of the talkies, if there has been one awkward and unavoidable question that has left the lips of every line producer and postproduction supervisor, it is, "How long will it take you to mix?" And for over 90 years, generations of supervising sound editors and re-recording mixers have found this as open and beguiling a question as ever challenged humankind; its deceptive simplicity being as succinct, paradoxical and imponderable as any Japanese *koan*.

It's also not particularly helpful that there is, probably, only one "universal truth" that dubbing mixers or re-recording mixers unanimously agree on, and that is however much time is allocated for delivering a final mix, the finished soundtrack would always—*always*—benefit from more time being made available for mixing.

Therefore, any attempts to definitively answer this conundrum tend to be based on an idiosyncratic, rule-of-thumb duality that goes something along the lines of either, "I'd budget for somewhere between 1 to 5 minutes of running time per hour, we may get lucky" (i.e. "I'm not prepared to work any extra hours unpaid on this film, and I have the confidence to tell you so") or "Well, we may get lucky and push through 10 minutes of running time each hour, if the tracklay is good" (i.e. "I desperately need the work or the credit on my curriculum vitae").

One anecdotal qualifier to this comes from my own working lifetime as a sound editor and mixer who has been engaged on a wide variety of film, TV and advertising projects—and I can say with some confidence that the time taken for any mix will almost always precisely match the time imposed by the schedule, or, more pragmatically, the time that the client has actually paid for.

Every mixer has his or her own way of approaching and structuring a mix, and today's practitioners need to be mindful not only of the audio quality and artistic integrity of the final mix (to my mind the crowning glory of postproduction) but also to the financial significance of efficiently creating separate *mix stems* that will go on to form the component parts of subsequently syndicated formats, international re-versions or foreign language masters. Indeed, it can be argued that a modern mixer's skill base needs to include not only highly proficient hands and ears for *mixing* but also considerable expertise in the area of digital *systems management*, arranging an efficient workflow for deriving the alternative mixes that modern theatrical, broadcast and online platforms also demand and ideally in as least commercial studio time as is physically possible. By this measure are reputations made and repeat business generated.

These challenges, however, are not limited to audio postproduction. Contemporary outside broadcast (OB) facilities, particularly those employed on premier sporting events, also have multiple-mix requirements that can require the delivery of Dolby Atmos streams for 4K/UHD[1] pictures, 5.1 surround sound to accompany HD pictures, and Lo/Ro[2] Stereo downmix audio streams for SD television transmissions. What is more, at an outside broadcast, they need to be produced live and simultaneously.

For mixers sitting in the hot seat, there is another kind of double jeopardy to overcome. It's not only the time they actually have to complete their task that is foisted upon them; the actual raw materials they will work with are also imposed on them from upstream "editorial" colleagues (e.g. the output of the dialogue, music and effects editors.) In the case of broadcast television, these are often likely to have been roughly assembled and leveled not by a sound editor but by the picture editor, to the satisfaction of the program's director and producer, before delivery to audio post.

Precisely because of this audio aggregation process, made more easily possible by the increased functionality of picture editing software, there is a danger that the mixer, or specifically the act of mixing, while rightly recognized by a few enlightened souls to be the culmination of the audio postproduction process, gets regarded by others in the production chain as a primarily technical exercise (albeit a vague and complex one) that has to be carried out at the end of the line by technicians, satisfying a bunch of technical stuff, the impact of which on any temporarily mixed material is rarely, if ever fully understood. Inevitably, at some point in all mixers' working life, they will endure the fallout from a revelation that eventually befalls all directors—a phenomenon that occurs when they critically listen, with the picture editor, on edit room speakers. Then the mixer must prepare to be told in no uncertain terms that the picture editor's mix is louder, that the dialogue miraculously possesses more clarity, and that the music definitely has more presence. As a mixer, I have received my fair share of these "picture-mixes"; where audio levels have set up home beyond the meter end-stops, yet get described in misty-eyed terms by the director as being altogether "more visceral" than any of my "legal-level"[3] mix. (Unfortunately, the interpersonal skills required at this point are somewhat outside the scope of this chapter, but don't worry, they develop naturally over time spent in the mixer's chair.)

Each mixer then has his or her own philosophical approach to crystallizing the output of the art and delivering the primary final product (along with whatever secondary, submaster mixes are requested by the production company or distributor), and this can manifest itself in different ways: in contrasting physical preferences, for instance—such as a favorite console and layout—or a bespoke "catch-all" digital audio workstation (DAW) template. Yet all of these diverse methods are designed to achieve the same conclusion—a final mix that meets the satisfaction of the director, while meeting the technical requirements of any "deliverables" document presented by a television broadcaster or film distributor.

To my mind, this eschewing of a one-size-fits-all mentality is one of the richest aspects of belonging to what might reasonably be described as the mixer "community"—a fluid and informal, global, autonomous collective of mutual support and collaboration among practitioners, manufacturers (and gradually) progressive academic institutions, an arrangement that somehow manages to organically disseminate new knowledge and suggest developments to current practices better than first inspection might suggest. Indeed, it is perfectly acceptable and to be encouraged that mixers develop their own mixing methods rather than slavishly following an instruction manual, but where might one new to the art choose to start?

Perhaps seven key areas to the mixing and stem-creation process bear closer examination, as well as informing and affecting the decision-making process of most mixing professionals:

1. *The Information Architecture*—This concerns the segregation of audio clips on the timeline (or microphone sources in the case of, for example, live sport mixing); the organization of the tracks (or mixer channels in the case of, for example, live sport mixing); the grouping of tracks (or microphone sources in the case of, for example, live sport mixing) according to the general logic of voice, effects and music.
2. *The Gain Structure*—The more sounds that are "layered," the higher the overall levels become, so the relative levels of "sounds" need to be organized to prevent clipping, distortion or failure to meet a broadcast-compliant peak level and or cinema loudness level.
3. *Monitoring Paths*—Multiple monitoring is required for checking mono/stereo/multichannel/multidimensional mixes.

4. *Grouping*—Group faders are a convenient way of controlling dense submixes, such as those that occur in the creation of dialogue, music and effects stems.

5. *The EQ Process*—The use of frequency equalization (EQ) can be valuable in blending the audio spectrum or enabling the "nesting" of separate sounds, so that they sit together better in the mix (not to be considered as "sound design" in the context of this chapter).

6. *Output Formats*—For live sport sound originating at an outside broadcast, for instance, the output signal is often "wrapped" or encoded into a Dolby E[4] file before sending to the television transmission suite. In audio postproduction, specific deliverables are required not only by distributors and broadcasters but by colleagues who will work on re-versions later in the process. In these instances, the delivery medium can vary from storage on a rugged, portable computer drive to a networked Pro Tools session or as items such as. OMF,[5].AAF,[6] .bwav/.wav[7] and poly WAV[8] files being passed down the line.

7. *The Impact of Further Picture Edits During the Mixing Process*—This is a postproduction issue rather than a live mixing concern and brings into question how "fixed" a mix and any automation is to the so-called locked-cut[9] received from the picture editor.

To find a sense of common ground and to illustrate how these seven factors might influence practitioners, bringing about different approaches both to the mix and to the mix-stem creation process, I [NH] put questions to a panel of three highly experienced mixers, who currently work on different areas within the sound-for-moving-pictures sector:

1. *Pip Norton [PN]*—Pip is a feature film re-recording mixer, presently working in the UK on the mixing stages of the world-famous Pinewood and Shepperton Studios group. Pip originally trained in film at the BBC before moving to AIR Lyndhurst to take a position as a staff mixer, enabling her to spend more time working on feature films. She has extensive experience of mixing theatrical release movies, as well mixing international versions of blockbusters for major distributors such as Paramount and Sony Pictures. Pip and I have mixed two feature films together: *Here and Now* (2014) at AIR Lyndhurst and *Finding Fatimah* (2017) at Pinewood.

2. *Alan Sallabank [AS]*—Alan started his career in Bristol at the BBC Natural History Unit, followed by a spell at Aardman Animation, before moving to London. Gaining experience as a mixer at several premier facility houses, he now runs his own audio postproduction company, 8dB Sound. Alan has worked as a mixer on many genres, which range from TV documentaries and drama to independent film

and studio features. Since 2009, he has mixed every season of the hugely popular prime-time drama *Doc Martin* (2004–present), which is syndicated worldwide, and we have collaborated in the past by virtue of my contributing ADR for several episodes.

3. *Ian Rosam [IR]*—Ian started his career at BBC Cardiff before moving to the commercial ITV network, interspersed with mixing front-of-house audio for bands touring throughout Europe. He has supervised the audio arrangements for Sky Television's Premiership football service from inception in 1992, taking the audio output from mono to today's UHD, Dolby Atmos experience. He has been a planning consultant and audio quality control supervisor for every summer and winter Olympic Games, FIFA World Cup and UEFA Euro Championship since 2002, and he continues to be an operational, hands-on mixer. We have worked together on very many Sky Premiership football games.

5.2. Question 1: The Information Architecture

[NH]—How do you segregate the clips, organize the tracks and then group them according to the general logic of voice, effects and music; and how does this influence the bussing structure you use as your usual template for creating mix stems?

[PN]—A drama mix for me will consist of three stems: Dialogue, Effects (FX) and Music. A dialogue stem will typically be subdivided into dialogue tracks, loop group and dialogue reverb returns, each with a dedicated auxiliary feed (aux) fader. The FX stem will consist of separate mono FX, stereo FX, mono atmospheres (atmos), stereo atmos, FX reverb returns, Foley and Foley reverb. The Music stem can actually turn out to be the simplest, depending on how the music tracks have been delivered to me. As I work with Pro Tools, I will describe how I set the mix out on that platform.

Starting at the top of the edit page will be my main character tracks. If two mics (e.g. lavalier and boom) are available I will have them on adjacent tracks. From there, the order will be loop tracks, a minimum of two mono pre-reverb returns, a minimum of two reverb returns, mono FX tracks, stereo FX tracks, mono backgrounds, stereo backgrounds, multichannel FX, a minimum of two reverb returns for them, Foley footsteps, Foley moves, Foley spot effects, two Foley pre-reverb returns, two Foley reverb returns, stereo music tracks, multichannel music tracks and then the music reverbs.

On the desk I will have separate layers of dialogue, FX and music; and I will then also have an extra layer of aux faders. From these stems, I can make all the deliverables. Obviously, they will all need a bit of tweaking, but at this stage I will mostly be working with the aux and stem faders.

[NH]—How do multichannel and multidimensional audio affect these arrangements [e.g. 5.1, 7.1, 9.1 and Dolby Atmos]? What arrangements are required to be provided for in your mix structure to accommodate downmixes specified in distributor or broadcaster deliverables?

[AS]—I always work backwards from what I know (or hope) will be the "largest" delivery format. For modern primetime TV drama, especially those with a co-production partner, you can safely assume that there will be requirement for a 5.1 mix, plus other deliverables. Depending on the distributor, they may also want a main stereo and even in some cases a mono delivery; but this is not always the case.

The most detailed deliverable requirement I've come across was for a large U.S. production house. They want to be able to completely manipulate the mix after it had been delivered, so that they could change things like the score, background music, sound FX and of course the language spoken by the actors. They broke it down like this:

Location Foreground Dialogue

Foreground ADR

Loop Group Dialogue/Background Walla

'Native' Language Dialogue—any dialogue included in the original version but not in the same language

Music Score (vocal free)

Music Score (vocal free) Undipped

Music Vocals (if featured)

Music 'Source' (incidental/background)

Full length undipped versions of all Music Score cues (delivered as separate nonsynchronous files)

Full-length untreated versions of all 'Source' Music cues (delivered as separate nonsynchronous files)

Location Sound FX

Foreground laid-in Sound FX/Sound Design

Background Sound FX

Foley Footsteps

Foley Spot FX

As the production was in 5.1, you can imagine how many tracks and how much hard drive space this particular delivery took. [. . .] The Producers then wanted hard copies burned to multiple DVD-Rs (if anyone still remembers those?)

The U.S. Production Company also sent a representative over to the UK to supervise me mixing the Music and Effects (M&E) stems, although they hadn't sent one for the original final mix destined for the

UK. But this is understandable—the market for the re-versioned and repackaged mixes far outweighed that for the original version, so they needed to be sure in their own minds that these markets would receive the best possible product.

[PN]—My M&E stems will mostly be made from the Music and FX stems, with extra Foley and some sync FX appearing from the dialogue tracks. I generally also send my dialogue and FX stems to a "music free" print track. This can prove really useful for trails, recaps or if there are any music changes. If the primary output is a Theatrical mix, then rather than use an offline process to make the R128-compliant TV mix, I rebalance the stems. My mix template includes postfader inserts on each of the stem aux faders to feed the downmix plugins, which allows me to simultaneously print the stereo downmixes.

[IR]—My console layout on the Calrec Apollo is arranged to take account of delivering either in HD (stereo and 5.1 audio) or UHD (stereo, 5.1 and Dolby Atmos). The mixing desk inputs are arranged over two fader layers (A and B) and are shown along with the Group, Master and Auxiliary routing in Table 5.1.

Table 5.1 Ian Rosam's Calrec Apollo Mixing Console Channel, Group, Master and Auxiliary Allocations for a Sky Premiership Football, UHD Outside Broadcast.

Channel Number	Source
1	Soundfield 5.1
2	Soundfield 4.0 height (slaved to channel 1)
3	Fill mic left
4	Fill mic right
5	Stereo gantry mic
6A	Dug out reporter mic, live insert into game if required
6B	Spare tunnel mic if camera channel audio fails
7	Camera A Interview RE50
8	Camera A FX on-cam mic
9	Camera B Interview RE50
10	Camera B FX on-cam mic
11	Camera C Interview RE50
12	Camera C FX on-cam mic
13	Camera D Interview RE50

(Continued)

Table 5.1 (Continued)

Channel Number	Source
14	Camera D FX on-cam mic
15	Camera E Interview RE50
16	Camera E FX on-cam mic
17A	Steadicam 1 FX
17B	Camera F Interview RE50
18A	Steadicam 2 FX
18B	Camera F FX on-cam mic
19A	Stadium P.A. to 5.1 when required
19B	Stadium P.A. to Atmos (slaved to channel 1)
20	Pitch mic 1
21	Pitch mic 2
22	Pitch mic 3
23	Pitch mic 4
24	Pitch mic 5
25	Pitch mic 6
26	Pitch mic 7
27	Pitch mic 8
28	Pitch mic 9
29	Pitch mic 10
30	Pitch mic 11
31	Pitch mic 12
32	Pitch mic 13
33	Spare Commentators mic via SQN from gantry
34	1st Commentator
35	2nd Commentator
36	3rd Commentator
37A	Spare 3rd Commentator
37B	Remote referee from reverse video circuit 1 or 2
38	Commentators group to Main 1 & 3
39	Group 1 master FX mix to Main1 & Main 3, master to slave Master 5 & Master 6
40A	Group 2 master FX MIX to M4
40B	Group 2 master direct out to M2

Channel Number	Source
41A	Line A tracks 1 & 2
41B	Line A tracks 3 & 4
42A	Line B tracks 1 & 2
42B	Line B tracks 3 & 4
43A	Line C tracks 1 & 2
43B	Line C tracks 3 & 4
44A	Line D tracks 1 & 2
44B	Line D tracks 3 & 4
45A	EVS Z tracks 1 & 2
45B	EVS Z tracks 3 & 4
46	"Spot-on" music
47	"Spot-on" FX
48	
49	
50	
51A	Reverse video circuit 1 tracks 1 & 2
51B	Reverse video circuit 1 tracks 3 & 4
52A	Reverse video circuit 2 tracks 1 & 2
52B	Reverse video circuit 2 tracks 3 & 4
53	GFX sound FX (when reqd.)
54A	Whoosh audio for replay wipe
54B	Reverb for whoosh
55	Mix from Presentation truck for 3rd commentator
56A	Mix from Presentation truck
56B	Mix from Sky

Groups	Function
Group 1	FX to M1, M3 and master for slave faders for M5 & M6
Group 2	FX to M2 & M4
Group 3	
Group 4	High levels via SoundField upmixer
Group 5	Commentators and dug out mic

(*Continued*)

Table 5.1   (Continued)

Groups	Function
Group 6	Presentation (if required)
Group 7	Reporter (if required)
Group 8	Ball kicks and on-camera FX
Group 9	
Group 10	In-vision to M1 & M3
Group 11	In-vision to M4
Main Outputs	*Function*
Main 1	5.1 HD main mix AES 1 dirty vision
Main 2	5.1 HD FX mix AES 2 dirty vision
Main 3	5.1 HD clean mix (no GFX FX) AES 1 clean vision
Main 4	5.1 HD International Sound AES 2 clean vision
Main 5	Atmos FX 5.1 bed
Main 6	Atmos FX Height 4.0
Auxiliary Outputs	*Function*
Aux 1–16	Various feeds to various destinations: e.g. prehear for Production, Reporter, Commentators

5.3.   Question 2: The Gain Structure

[NH]—Fairly obviously, the more sounds that are "layered," the higher the overall levels become, so the relative level of the "sounds" need to be organized to prevent clipping, distortion or a failure to meet either a broadcast-compliant peak level or a theatrical loudness-level. How important is it at this stage to establish the levels of the quietest sounds (e.g. room tones) and the loudest sounds (e.g. explosions) relative to the "average" sounds (e.g. dialogue)?

[PN]—The dialogue will generally be delivered from editorial to the mixing stage balanced, but not mixed. This means that whilst it should not need 20 dBs worth of gain either way, it probably will require me to make realistic fader moves to achieve the optimum balance. The reason for completing this process before the final mix is that if you do need to boost a clip by 20 dB or so, then it may well need some form of de-noising, and this is certainly best done prior to the final mix. But if any processing *is* done ahead of the mix, I would expect to have the unprocessed clip available too, but muted within the delivered track-lay of the Pro Tools project.

Establishing the levels between the quietest/average/loudest sounds also relates to the monitoring levels you choose at this stage. If I am mixing for a theatrical release, I will be mixing at 85 dB; for TV it will be lower, possibly 79 dB, but that will depend on the room. And, of course, having set your monitoring level, do not touch it!

[AS]—The concept of "gain management" has become far less of an issue since the advent of floating point digital mixing and acquisition technology and 32-/64-bit processing.

Instead, I prefer to call this stage "dynamic range management." ' Even though we now have far higher peak levels that we can hit in broadcast for instance [−1 dBFS measured on a digital True Peak meter versus −10 dB as read on an old BBC-style PPM], *dialogue dynamic range* in particular still needs careful management. Dialogue generally drives the story, so it absolutely needs to be the solid corner-stone of the mix.

After working on a large U.S. drama series, I was told that the way they viewed the mix was this: the dialogue tells the plot, the music sets the emotion and the effects place the other two in the correct context. The important thing to remember is that whilst the Director may consider this to be their *magnum opus*, the producers and distributors are very much viewing it as a product rather than a work of art, for which maximum compatibility means maximised sales.

[NH]—At this stage of setting up the mix, what are your thoughts on the provisioning or placing of dynamics (compression, limiting on individual sources, overall limiting), as well as any course/fine EQ-ing of individual signals, which can affect the overall gain structure?

[PN]—For me, a mix always begins with the dialogue pass. I generally have a compressor in every dialogue channel, but it is either barely working, or it's bypassed, ready if needed for a particular scene. I will also have a compressor on the dialogue stem, and the level that this is set at will depend on the mix. As a safety measure, I have a limiter inserted in all stems. In addition, all dialogue channels will have active filters and EQs. There will generally also be gates, expanders, a C1 [a Pro Tools side-chain compressor] and a de-esser inserted in the channel—ready, just in case any of them are needed. Everybody has their own idea on compression and whether to trigger it pre- or postfade. I tend to use as little compression as I can get away with, but the above signal path gives me access to all options. Each way has its pros and cons, with different combinations working better on different scenes. I do find if more compression is needed, it is better to have several compressors each doing their own little bit, rather than simply having one that is working very hard.

Once I have completed the dialogue premix, I tend to mix in the Foley, matching the reverbs and perspectives of the dialogue. Next are the background tracks, then the FX tracks and then finally the music. But all of these are balanced around the dialogue. Once all the premixing is completed, the aux and stem faders can be used to help give the mix some shape.

[AS]—I use a combination of in-line channel dynamic range management, including gates, expanders, de-essers and compressors, along with bus compression, and then master mix and stem True Peak limiting. Of course, the other good way to manage dynamic range is to use your faders. [. . .] The danger of the dialogue's dynamic range being excessive is that intelligibility falls and those viewers watching in less than

Table 5.2 Ian Rosam's Calrec Apollo Dynamics Settings for Sky Premiership Football, HD and UHD Outside Broadcasts.

Source	Parameter 1	Parameter 2	Parameter 3
Commentators	Group compressed	3:1 compression ratio, −18dBFS threshold	Attack 1 ms, Release 100 ms
Presentation	Group compressed	3:1 compression ratio, −18dBFS threshold	Attack 1 ms, Release 100 ms
Reporter	Group compressed	3:1 compression ratio, −18dBFS threshold	Attack 1 ms, Release 100 ms
Interview mics.	Individual channel EQ and compression	3:1 compression ratio, −18dBFS threshold	Fed to Group 10 and 11 for delay and routing
SoundField mic.	Multichannel EQ and compression	3:1 compression ratio, −18dBFS threshold	Slow attack 50 ms, long release 400 ms
Group 8 mono FX	Group compressed	3:1 compression ratio, −18dBFS threshold	Slow attack 50 ms, long release 400 ms ideal for ball kicks
Main Outputs	Limiter 50:1	−8dBFS threshold	Attack 1 ms, Release 100 ms

ideal conditions get turned off by the mix and eventually switch off or change channel.

[IR]—The consideration of the dynamics [is] enormously important on something like football, where there is a huge sound pressure level from the crowd, but a much lower level for things like the sound of the ball being kicked on the pitch; then there is the reporter at the touchline and the commentators to bed in against that crowd. Table 5.2 shows my usual compressor/limiter arrangements for each element of the football mix.

5.4. Question 3: Monitoring Paths

[NH]—Following on from the IA and your gain structure arrangements, how do you establish monitoring paths for mono/stereo/multichannel/ multidimensional mixes; and what is your "main monitor"? Do you switch across to monitor the down mixes?

[PN]—I tend to monitor my print tracks. Whilst premixing, these tracks would be on input, then I switch to playback monitoring when printing. I have my monitoring set so that I can separately monitor each stem, which means I can track down any strange noises! It is also important to be able to cut or solo individual speakers. These days, I do not monitor stereo as I mix. I do, however, monitor the stereo when I am mixing the broadcast-compliant R128 version; and again, I monitor that through the print tracks.

[AS]—For me, the best method of bringing order to the track management, the dynamic range control and achieving an overall balance, means I set up my main monitoring configuration first; then the stems required to accommodate the delivery requirements, followed by the fold-down mix paths and then, lastly, the subgroups that the different categories of tracks need to feed into.

The practical way of achieving this depends very much on the workflow dictated by your equipment infrastructure—which DAW you are using, for instance, or if you are also using 'external' DSP mix technology, such as an AMS Neve DFC. If you are mixing "in the box,"[10] which obviously you can within platforms like Pro Tools, there are shortcuts which can help you manage the path count; and similarly, on the DFC, its assignable configuration can make life a lot easier.

Currently, the standard delivery for TV drama, regardless of local distribution requirements, is 5.1; with a LoRo ITU specification nonmatrix encoded stereo fold-down, which both need to meet the loudness and peak level requirements (usually the R128 standard) of the distributor. TV has incredibly stringent delivery specifications

that require adhering to. On feature films, you are more likely to be
restricted by the busing abilities of the mixing venue. There is no point
agreeing to deliver in 7.1 if the mix facility is only 5.1, for example.
[. . .] So if your workflow involves having to go to another venue to
final mix, this also has to be taken into consideration."

[NH]—Ian, when you're mixing live sport, what are you looking for in the
integrity of what you are hearing in the downmix?

[IR]—The Dolby Atmos for Sky's 4K UHD service, the 5.1 for their HD
service and the stereo mixes all get checked by me at random points
during the program, simply by switching across the various monitoring feeds available on my [usually Calrec Apollo] console; but sometimes during presentation parts of show—the buildup to a match, or
the half-time and postmatch analysis segments, for instance—I will
mix with the stereo monitors as my reference. During play, I generally
switch between monitoring the 5.1 and the atmos mixes and less frequently check the stereo mix. It's also worth pointing out that the 5.1
mix is not a downmix from the atmos stream, which in essence is its
own entity. (The commentators and the stadium public address (PA)
system are fed as audio objects into the Atmos 7.2.4 stream from direct
outputs of the mixing console.) Downmixing only occurs to derive the
stereo from the HD service's 5.1 stream. Just like drama colleagues
who work hard to establish clarity in their dialogue tracks, my prime
concern is always that the Commentators are clear and unimpeded by
any other sound, across all three of the outputted mix formats: stereo,
5.1 and Atmos.

5.5. Question 4: Grouping

[NH]—How do you arrange your "group faders" and what do they consist
of? For postproduction colleagues: do you—how do you submix to
allow the convenience of group faders to control "dialogue, music and
effects"?

[PN]—I tend not to use Pro Tools groups for mixing; but as I described
earlier, I do, however, make good use of the aux faders and route other
faders to them.

[AS]—Here is how I set up the groups in the TV drama *Doc Martin*:
    Location Foreground Dialogue
    Foreground ADR
    Loop Group Dialogue/Background Walla
    "Native" Language Dialogue—any dialogue included in the origi-
nal version but not in the same language
    Music Score (vocal free)

Music Vocals (if featured)
Music "Source" (incidental/background)
Location Sound FX
Foreground Laid Sound FX
Sound Design
Background Sound FX/"Atmospheres"
Foley Footsteps
Foley Spot FX

I keep this as an admittedly wide spread of sources not particularly for overall balance control, but, instead, in case the delivery expands beyond the usual requirements for the mix, plus an M&E and a DME, or any other requested changes to the delivered stems. If that does happen, it simply requires a change in the routing of the outputs of my groups.

[IR]—I like to keep the group faders out of the immediate way on the console, leaving me a clear and fast access to the full "spread" of commentators and FX faders. Console surface "real estate" is at a premium on a live show mixed in a truck, so what gets placed on what layer of the desk is obviously an important consideration for me.

5.6. Question 5: The EQ Process

[NH—How do you go about EQ-ing individual clips, or blending the audio spectrum of separate sounds, so that they sit together better in the mix? (Not to be confused with any "sound design" manipulation that might take place before the final mixing stage.)

[PN]—I work hard to avoid muddy mixes and my default filters on each dialogue track are set at 80 Hz and 15k Hz. These may vary, depending on the tracks, but it is my starting position. I like to use lots of spot FX to build up the soundscape, and I am not averse to EQ-ing these to stop a buildup of mush. For example, when I'm building a bird track, I would have individually placed bird calls, with any inherent traffic or wind noise EQ'd down. Another aspect is mixing music under dialogue; here, my motto is a simple one: "Use it or lose it." If the music is fighting the dialogue, I will EQ a dip in the music from 2 kHz to 5 kHz to help the speech consonants come through.

[AS]—I think that this is entirely what post-production mixing is all about—the telling of a story, first and foremost—and the art of mixing lies in manipulating the dialogue, the effects and the music in such a way that it more effectively tells the story. This is why it's so helpful when a score is written around the dialogue and the composer has a good idea of what sound effects are taking place. Most of the time, the

Table 5.3 Ian Rosam's Calrec Apollo Microphone EQ Settings for Sky Premiership Football, HD and UHD Outside Broadcast.

Source	Parameter 1	Parameter 2	Parameter 3
Pitch FX mics (× 13)	120-Hz high pass filter	4-kHz bell curve, Q = 1, +5 dB	8-kHz bell curve, Q = 1, +5 dB
Commentator lip mics	No EQ used/ minimal EQ used	Small adjustments to help "sit" in mix	
Reporter and interview mics	No EQ used/ minimal EQ used	Small adjustments to help "sit" in mix	
SoundField mic	Minimal EQ used		

"nesting" I do is generally on background atmospheres tracks, to help them bed in with the production sound and to carefully reduce any frequency ranges in the music that may be clashing with dialogue. Well recorded, edited and mixed Foley should be able to run all the way through the mix and should bed in perfectly with the location sound.

It is worth mentioning, by the way, that when deliverables specifically request a "fully filled" M&E mix (which is subtly but significantly different to an "M&E" mix), it is exactly that—it is a soundtrack that is fully filled with music and effects without a consideration of how the dialogue finds a space to be heard. So, a separate mix pass will be required because there is no way to generate a properly "fully filled" M&E automatically as a substem or "mix-minus" without compromising either the full mix or the "fully filled" M&E.

[IR]—As far as my choice of EQ is concerned, this has been arrived at from years of "tweaking," but it now remains pretty much settled on the settings shown in Table 5.3, with a generous high pass filter used in conjunction with two distinct "lifts" centered around 4 kHz and 8 kHz for the effects mics.

5.7.   Question 6: Output Formats/Deliverables

[NH]—In postproduction, what deliverables are required by distributors/ broadcasters? What provisions need to be made for e.g. foreign dubs of movies or overseas syndication for TV drama? What is the delivery medium—is the soundtrack and any submaster mixes now universally delivered to a drive? What files do you provide—e.g. a Pro Tools session, .aaf, .omf, .wav, .bwav, poly WAV files?

[PN]—For theatrical mixes I usually deliver the main mix, with a stereo fold-down, and then dialogue, music and FX stems [often referred to as the DME mix], with a stereo fold-down for each individual element. There would also be an M&E mix for international use, and I personally like to generate a music-free mix as I've found it useful in the past for re-versioning. I would then need a day to tweak the main mix, using the stems, to make the broadcast-compliant R128 mix. These mixes would all end up being delivered as .wav files on a portable drive. For TV mixes, I would be mixing to meet R128 levels on the main mix straightaway, but everything else of the stem deliverables would remain the same. The Pro Tools session itself is of course saved as part of the project's assets.

[AS]—TV drama delivery is now pretty much like film delivery—predominantly file based. We supply the mixes and stems as .wav files (noninterleaved in the case of 5.1 mixes) at 48-kHz, 24-bit resolution to the picture department, who then include the requested mixes and stems in a Prores video file, which can accommodate up to 14 audio channels.

As the saying goes, "Time is money" when mixing, and therefore there is a compromise that needs to be arrived at for the vexed question of what constitutes the best, or most efficient, use of studio time whilst still ensuring a good level of quality control. I create all my LoRo folddowns in real time, using the Avid Downmixer mainly, or occasionally I'll use either the Neyrinck SoundCode or the Nugen Downmixer plugins. This is so the resulting folddown mix can be monitored as it goes through, and its dynamic range can be managed to ensure it is a good representation of the fuller 5.1 mix.

[IR]—We have a standard configuration for our various mix outputs that is recognized internationally, whether it is at an English Sky Premiership game, a UEFA Champions League match in Europe or a FIFA World Cup tournament. Table 5.4 shows the standard audio outputs for football in HD and Dolby Atmos.

5.8. Question 7: Further Picture Edits

[NH]—How "fixed" is your mix/automation to the "locked-cut" received from editorial? Is it commonplace to receive recut pictures even when at the mixing stage these days? Does your mixing strategy allow you to easily adjust/accommodate in the event of a last-minute recut? Do you factor this possibility in until the very last pass yourself? Do you avoid automation until the last pass or use it from the very start of mixing? Might tie neatly back into your answer to question 2 on

Table 5.4   Ian Rosam's HD and UHD Outputs for a Sky Premiership Football Outside Broadcast.

HD Output	Content
AES 1	Main Mix
AES 2	FX Mix
AES 3	Dolby E Main Mix
AES 4	Dolby E FX Mix
UHD Output	*Content*
AES 1	Main Mix
AES 2	FX Mix
AES 3	Dolby E Main Mix
AES 4	Dolby E FX Mix
AES 5	Dolby E Atmos feed
AES 6	Dolby E Atmos feed

gain structure and the adjusting of clip levels at an early stage. How do you "chase" edits made by editorial with your existing mix? Can you give an example of being presented with this problem and your solution?

[PN]—Each facility has different mixing desks; these days they can all be used to control Pro Tools, as well as being a desk in their own right. Although I love using the DFC console, especially for its EQ when mixing dialogue, I now tend to use the desk to enable me to actually mix *within* Pro Tools. That way I can switch studios and make changes at any stage. I use automation from the start, never making hard pre-mixes, e.g. printing. wavs of the premix elements and then playing that. wav back in the mix. Mixing within Pro Tools means I only have to cut one session rather than separate tracks and automation sessions. I use Virtual Katy or a cut list to edit, depending on the number of changes. I then go over each edit to make sure it works.

[AS]—A wise, long established re-recording mixer who had spent his entire working life in feature films once said to me, "There's no such thing as a final mix—there's just the last approved temp mix." Nowadays in TV audio postproduction, we can't—and don't—treat anything as "locked" until it is transmitted. Until that point, events worldwide that you have no control over can have an impact on your piece. Maybe an actor featured passes away or, worse, gets embroiled in a damaging scandal.

Shortly after the death of Princess Diana, I had to remix several episodes of a drama that was due for imminent transmission, as they'd featured a black Mercedes limousine being driven at speed into an underpass, whilst being pursued by scooters. After the catastrophic tsunami in Thailand, I had to remix a piece that featured a tsunami. . . .

There is also the problem of international music clearance. Quite often the same music cannot be used the world over, and the picture edit has been cut to a specific music track. In one extreme case, I had a character singing along to the Queen song "Don't Stop Me Now," but it couldn't be used in the overseas mix. So that required not only a recut of that particular scene but also the ones around it.

I don't know how I'd cope with any of these things if I were operating today with what was known in the late 1990s as event based automation—where automating multiple events was locked to the timeline. [Believe me—it was a pain to make changes—NH.] For some time now, automation—the manipulation of the individual clip's levels, its EQ treatment, reverbs and so on—is tied to the individual clips themselves. Move a clip to any other point on the timeline or another track, and, very conveniently, the automation can follow it.

This means that if a recut comes in after the initial locked pictures have been mixed to, sound software like Conformalizer is able to reconform an entire Pro Tools session—transposing the mix data, plugin settings, automation and track information, along with the video track—to fit the newly cut version. It can work from a change list supplied by the cutting room, it can compare two EDLs to generate its own change list, or it can compare two different video files to work out for itself where the changes are. It's really incredible technology. However, it comes at a relatively hefty price, and I am also not in favour of letting the cutting room become too complacent!

[NH]—So, any final thoughts about setting out "the mix"?

[AS]—Yes! Here is my one "small-but-major" gripe: dialogue divergence in 5.1. [. . .] Please can it stop? To my mind, it makes a mockery out of the entire reason for having a center speaker in the first place. Fairly obviously, the whole point of the center channel is to have a central anchor point, which serves to lock the soundtrack to the center of the image, no matter how far off axis or how far to the left or right of the "sweet spot" you are listening. However, when you use divergence on center signals, spreading dialogue into the front left and right channels, the sound those listeners hear when they sit outside the sweet spot will simply appear to be "off center." You might as well just rely on a phantom center and get rid of the center channel altogether!

You will have gathered that, in general, the poor use of the center channel is an irritation of mine, and, in particular, I think that there's no point in specifying that the center channel must be used for dialogue only. That just means that the FX and music have to rely on a phantom center, which in turn means that the rest of the soundtrack is disjointed from the dialogue, as well as making it much harder to bed-in any ADR. Plus, again, it relies on the listener being in the sweet spot to hear the intended mix balance.

In a nutshell—if a sound comes from the center of the screen, then feed it to the center channel, if it is becoming obstructive to the dialogue by being in the same channel—then we need to use our mixing skills to do something about it; after all, that's what we're being paid for as mixers!

5.9. Summary

It has been suggested, with some *chutzpah* here, that the final mix represents the triumphant culmination of any moving picture production, and while fellow mixers might consider this to be a reasonable point of view, any celebration of the art form that is this kind of mixing also needs to be contextualized with what is actually going on by this stage of the production process. Just to arrive at the final mix may have been years in the gestation for a feature film producer (who, by the way, it is worth remembering is the person ultimately paying your bill). The director's time frame is no less demanding on television drama or documentaries; their output is commonly the result of many months in the preparation, setting up, shooting and picture editing.

Therefore, it's fair to say that pressures are brought to bear on a re-recording mixer: on the one hand, there will be *external* pressure from production,whose imperative by this stage is often (but not always) simply to get the final mix "over the line" and delivered. On the other hand, there is another kind of self-induced, *internal* pressure felt by a mixer that often remains unspoken on the mixing stage. It derives not from constantly ensuring that the final mix satisfies the director (always there as an external pressure) and in turn pleases the producer. Nor is it the "given" that the mix will meet the specified technical and quality standards for cinema, broadcast or online (and probably all three). Nor is it the concern that the submaster mix stems, those precious audio jewels that will be reset later by other mixers skilled in handling such gems, will indeed turn out to be imperceptible duplicates of the original masterpiece.

Instead, what can exist for a re-recording mixer is an internal pressure arising from wanting to deliver a soundtrack that not only satisfies all of these needs but also brings about an additional aesthetic integrity to the narrative aims of the pictures, one where the sound serves not only to adumbrate, complement or reinforce the image in the physical realm of the screen, but where a somewhat more metaphysical artistic "truth" is revealed—first to the mixer but in any case becoming apparent only at the final mix stage. Recognizable as the sometimes discomforting, often intangible hairs-on-the-back-of-your-neck-standing-up moments, these are initiated—and initially experienced—by those re-recording mixers that possess awareness and sensitivity.

This heightened state is fragile, as little as a 1-dB adjustment either way can take it to one side or the other of the optimum balance, and this "magic in the mix" is not there all of the time along the timeline. But it can definitely be there some of the time—and this holds true whether a mixer is working alone, in-the-box mixing a micro-budget or broadcast project, or playing a part in a movie mixing team by looking after the dialogue and music or the sound effects, in a big theater, on a large console, controlling three Pro Tools rigs. Get to know them well enough, and a mixer might recount his or her own moments of *satori*, those moments of comprehension and understanding that miraculously and spontaneously become somehow unlocked by the act of mixing sound.

One final thought: for mixing and for creating the mix stems themselves, it pays to keep it simple. Particularly when working in the box, it can be easy to cascade plugin after plug in after plug in. The fact that the audio engine might tolerate this while maintaining the system's overall responsiveness, or that maximizing the RAM count has made this possible, or that you have a processor that can take this in its stride, or that you've invested in a virtual rack that can run eight plugins on one insert point—none of this means that you should. Experienced mixers have a tendency to use fewer tools rather than more. They keep the main mix clean, and they remember that someone, somewhere down the line, is going to work on their submaster stems and that any processing that has forever colored the original sound is likely to cause colleagues an issue.

Listen and be honest as you appraise what it is that these effects and dynamic processing are actually giving you, compared to the artifacts that they are adding. Easy mistakes to make as an inexperienced but over-enthusiastic mixer include the overuse of dialogue noise suppression or adding copious reverb to match ADR to production dialogue. The adage of "less is more" definitely applies here; so think carefully before developing a Gollum-esque obsession for the use of those "precious" plugins at every opportunity.

That *Lord of the Rings* reference cross-fades to a conversation I had on this topic with Michael Hedges,[11] a multiple Academy Award–winning Re-recording mixer, as we sat together at the console of Mixing Theatre 2 at Park Road post in Wellington, New Zealand. This is the home of Peter Jackson's prodigious Weta Studios, and it is where Michael mixes multimillion-dollar epics for the likes of Peter Jackson and Steven Spielberg. We had just listened to a collection of movie excerpts, all beautifully mixed by Michael, to appraise and highlight the capabilities of one of the most sophisticated mixing theaters in the world (and the place where he gets to ply his trade on a daily basis). Amid as large a Pro Tools rig (640 channels of playback), as impressive a Euphonix console (662 inputs), and as comprehensive a speaker array (the theater is THX Certified with Dolby Atmos monitoring) you could ever wish to see, Michael made a profound point:

> When people rely heavily on multiple plugins, particularly when they work in the box, I, you, the audience, can actually hear the processing affecting the soundtrack. [. . .] As re-recording mixers, our aim is to create a soundscape that the audience doesn't even realize they are being affected by; and for me, silence is one of our greatest tools.

Which got me thinking: if Michael Hedges can resist using every tool in his toolbox when he mixes, then we probably should too.

Notes

1. 4K / UHD—Although often advertised as one-and-the-same at 4,096 by 2,160 (an aspect ratio of 1.9:1), cinema 4-K resolution is greater than that used for ultra-high-definition (UHD) television at 3,840 by 2,160 (a smaller aspect ratio of 1.78:1)
2. LoRo (Left-only/Right-only) refers to a stereo downmix from a 5.1 signal and is also known as an ITU Downmix. It is fully stereo and mono compatible and the Left=only channel is created by adding Left, plus Centre (at 3 dB), plus Left surround (at 3 dB). The Right-only channel is created by adding Right, plus Centre (at 3 dB), plus Right surround (at 3 dB). The 5.1 LFE channel is ignored in LoRo.
3. Broadcasters worldwide require an adherence to a strict audio quality control standard for average loudness and peak signal levels, defined in the UK and Europe by EBU R-128 and in the United States by ATSC A/85.
4. Dolby E is an audio encoding / decoding technology that enables broadcasting and post-production facilities to carry up to eight channels of sound over a two-channel (stereo) infrastructure.
5. OMF—One of the logistical challenges in video and audio post-production is the sharing of media between different platforms and workstations. The original Open Media Framework (OMF 1) was an early file format (still current in OMF2 version) designed to work cross-platform for the exchanging of video and audio sequence information between different editing systems. In audio post-production, an omf is most often used for transferring audio sequence, track, and clip information from video editing systems to audio applications. Nowadays, the generic term "OMF" usually refers to an "OMF2" file.

6. AAF—The Advanced Authoring Format (AAF) is similar to the OMF in as much as it is a professional file interchange format designed for film and video post-production. It was created by the Advanced Media Workflow Association to allow (amongst other things) larger file sizes and an increased metadata capability.

7. .bwav /.wav—A broadcast wave (.bwav)—sometimes also referred to as a broadcast wave file (.bwf)—is different to a standard pulse code modulated (PCM) wave file (.wav) by virtue of it containing metadata in the file header that can place the start point sample-accurate on an editing system timeline (it records the start point as the number of samples before "midnight" [i.e. 24:00:00:00] with reference to an audio timeline.) However, software or systems unable to read this file header information simply read the audio data within the file.

8. Poly WAV files are multichannel .bwf files that contain extra metadata for e.g. identifying channels.

9. The "locked cut" will receive no further picture editing, i.e. the pictures are now locked to the sound.

10. *In-the-box mixing* is when the audio clips and tracks of a project are contained and processed within the computer program operated by a digital audio workstation (DAW) with the software of the DAW rather than hardware, facilitating the adjustment of e.g. volume, dynamics and reverberation. Conversely, *out–of-the-box mixing* is the more traditional practice of bringing each audio track to the channel input of an external mixing console for signal processing. DAWs using a control surface (also known as a controller), for e.g. tactile transport control, level adjustment or automation of certain audio parameters, are still considered to be providing "in-the-box mixing.

11. Michael Hedges has won two Academy Awards for sound mixing (for *The Lord of the Rings: The Return of the King* and *King Kong*) and has received two Academy Award nominations (for *The Hobbit: The Desolation of Smaug* and *The Lord of the Rings: The Two Towers*). A native New Zealander, he is the Senior Re-recording Mixer at Park Road post, Wellington, New Zealand.

6

Mixing and Mastering

Paul Geluso

6.1. Introduction to Mixing and Critical Listening

Mixing sound is a performative practice that requires intuition and technical expertise. Being prepared, organized and working with intention can speed the process along, thus leaving more time to be creatively engaged. Ultimately, the goal of mixing is to provide the listener an audio program that best translates the artistic intention of the work.

For the beginner as well as the seasoned professional, critically listening to well-mixed soundtracks can help a mix engineer develop a personal concept of what a good mix is. Whereas the lay listener enjoys the entertaining and informational aspects of the program, the critical listener can perceive and discriminate individual physical properties of sound, including pitch, amplitude, time and timbre, as well as analyze aesthetic qualities of the mix made by its creators.

Aesthetic Qualities of a Sound Mix

Balance
Spectral content
Dynamics range
Sound source location
Sound source dimension
Sense of space and envelopment

The aesthetic qualities of a mix define its distinct sound. For example, the producers of psychedelic rock in the late 1960s put the vocals much lower in the mix than their rock and roll predecessors. This subtle balancing effect creates a sense of disorientation. Later in the mid-1970s, the disco era ushered in an up-front, punchy and bright bass drum sound to punctuate the pulse of the music. These mixing decisions were made by innovative producer/engineer teams that helped defined the genre of recorded music, in part by the way it was mixed.

Foremost, mixes are crafted to create a sonic illusion that supports the artistic intent of the work's creators. For example, the high and low frequencies of a voice are often pushed electronically in a mix to be audible above amplified instruments in a way that is beyond what is humanly possible acoustically. The vocal sound is easy to perceive this way in the mix and thus appears to sound very *natural* to the listener—when, in fact, it is not. On film soundtracks, on- and off-camera sounds such as voices, car door slams and footsteps are often presented much richer and fuller than they are in real life, but they do not appear unrealistic because they support the story being told. Ultimately, it is the mix engineer's responsibility to create a final balance, equalization, dynamic range and a sense of direction, emersion and space in a way that works both technically and artistically.

6.2. Mix Session Preparation

6.2.1. The OMF

In the postproduction world, the mix session is typically handed off to the mix engineer as an open media framework file known as an OMF. The OMF can then be imported into the mix engineer's compatible DAW to pick up where the previous film editor or audio engineer left off. If the audio mix supports an image, a reference movie containing the rough mix should accompany the OMF file. Depending on the organizational skills exercised by the previous editor, the session may require a thorough reorganization before any serious mixing can begin. If the current engineer is working with the same software as the previous engineer, an active plugin inventory should be made to determine whether plugins need to be purchased or rented to restore the mix. Preparing printed multitracks, with and without effects burned in, provides a safeguard when sessions are changing hands.

6.2.2. Session Organization

In the previous chapter, the concept of mix stems is discussed in depth. Stems are not only useful for mixing but are often requested by clients in addition to the full mix at the time of delivery. In postproduction, mono and stereo stems can be used to create alternate mixes, multilanguage mixes and upmix content to surround and 3D formats.

In the audio postproduction industry, stems are usually created for the following three classifications of sounds: dialogue, effects and music. Music and effect tracks are referred to as M&E tracks. The effect tracks can be broken down further into subgroups for sound effects, ambiance, Foley etc. Once the number of stems has been determined, busses must be created to keep the stemmed audio separated during the mixing process. The full mix is essentially the summed output of the stems. No audio should go directly to the full mix bypassing the stem busses.

For music sessions, stem busses serve as a way to group sounds for level control and effect processing. For example, a pop mix can have stems for:

- Drums.
- Percussion.
- Bass.
- Keyboards.
- Guitars.
- Lead vocal.
- Backing vocals.

When a composer is delivering music tracks to a filmmaker, stems are often requested in addition to the full mix so that sounds within the final mix can be adjusted in postproduction to best serve the likes of the director. For example, they may want to turn up the strings during an emotional spot or turn down a soloist to make way for the dialogue at the final film mix session.

In any case, once the number of stems has been established and the busses for them are created, it's time to conform the existing session to the new bussing plan. All of the audio regions must be monitored to make sure they are labeled correctly, grouped and routed to the proper buss. Additional tracks may be required. For example, a documentary film mix template session may start off with a modest number of tracks. The template tracks can then be populated with the appropriate sound regions, adding additional tracks as needed. (Figure 6.1.)

When preparing for a music mix, it is necessary to meticulously listen to all of the tracks individually while checking for technical and musical issues including poorly executed edits and unwanted noises. Tracks should be organized in a logical order and grouped in a logical way. For example, an engineer may start with drums followed by percussion, bass, keyboards, guitars and finally vocals and backing vocals. A working balance and panning plan should be established early on so that sounds can be monitored in place in the rough mix.

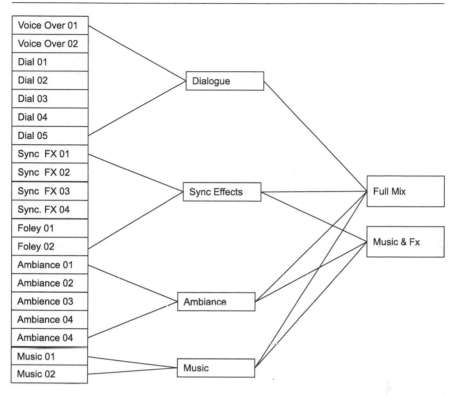

Figure 6.1 Track Organization for a Film Mix (left), Subgroups (center), and Mix Outputs (right).

6.3.   Monitoring the Mix

6.3.1.   Speaker Placement and Calibration

The mixing environment can have a serious effect on how we perceive balances, timbres and imaging. No two rooms, headphones or speakers sound identical. For example, the perceived low-frequency response of a sound system can change radically as a listener changes location due to the room's geometry, dimensions and acoustic treatments. In addition, our perception of balance, spectral content and dynamics can shift as the monitoring level changes. The mix engineer should be confident that their mix will translate well in a variety of listening environments. Proper speaker placement and monitoring level calibration are vital.

For stereo monitoring, speakers should be placed equidistant from the listener, about 30° to 45° off center. Most professional control rooms have a main stereo speaker spread from 5 to 10 feet, depending on the dimensions of the room, size of the console, and the distance of the desired mix position to the baseline of the speakers. The ideal monitoring location is called the sweet spot. This position should be free of strong reflections from nearby walls, ceilings or other flat surfaces. If the speakers are near a large boundary, such as a wall, ceiling, or floor, the bass response may be enhanced. Most active speakers have an equalization feature built in to compensate for the effect of the speaker placement and room acoustics. Speaker equalization settings can be guided by the use of calibration software and measurement microphones, although the final adjustments can be done to the liking of the engineer by ear. Averaging the response from a few locations within the listening area will guard against overcompensation due to a sudden spike or dip in the frequency response caused by strong reflections or a room's resonance. In any case, the final speaker calibration is typically checked by listening to multiple known reference recordings on the system. If the perceived bass, midrange or treble jumps out or is lacking on one of your trusted reference tracks, further calibration and experimentation with speaker location may still be needed.

Calibrating the playback level with a reference signal is done so that consistent mixing levels can be achieved. When the dialogue target level is between −18 and −12 dBFS (typical for film mixes), it should sound very full but not too loud through the system. Broadcast sound and film mixing studios typically use a pink noise reference level at 79 to 85 dB SPL C-weighted for calibration. For smaller mix rooms, an SPL reference of 70 to 75 dB may provide a more comfortable mixing volume.

To calibrate a speaker system, generate a pink noise test signal at −20 dBFS from the DAW session and adjust the sound level at the listening position, one speaker at a time, so that the SPL meter (set to C-weighting) reads the desired reference level for each speaker. Once completed, the position of the control room volume knob should be noted so that the same monitoring level can be easily achieved at a later date. If a master volume setting does not work for all of the speakers, individual speaker levels will require adjustment at the speaker amplifiers. Once a comfortable monitoring level is found and noted, it should remain consistent through the mixing session. To better understand the psychoacoustic phenomenon of how our perception of frequency and dynamic range is dependent on loudness, a study of published equal-loudness contour curves will be informative.

6.3.2. Subwoofers

Subwoofers can be used to reproduce a dedicated low-frequency effect (LFE) signal or to support the lower frequencies on systems with main speakers that are not full range. Placement for the subwoofer should be

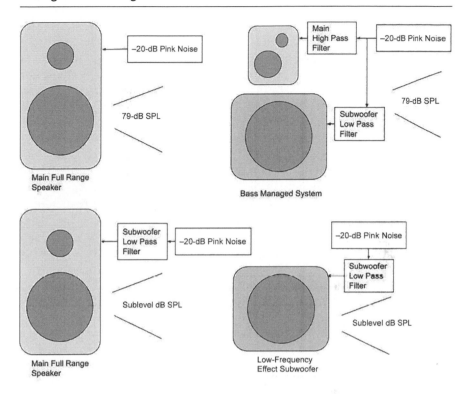

Figure 6.2 Speaker Calibrations: Full-range speaker calibration (top left), bass managed system calibration (top right), determining the equivalent subwoofer level for calibration using a full-range speaker (bottom left), and low-frequency effect subwoofer calibration (bottom right).

near the front center of the stereo or surround system but not exactly on center to avoid reinforcing room modal frequencies.

In theory, the subwoofer should be calibrated to deliver the equivalent low frequency from one full-range speaker in the system. Therefore, the cutoff frequency of the subwoofer must be determined. If it is not published, the frequency response of the subwoofer must be measured. While sweeping a sine wave though the subwoofer upward from 20 Hz, monitor an SPL (sound pressure level) meter to see at what frequency the level begins to rapidly drop off. A studio-grade subwoofer typically has a cutoff frequency between 40 and 80 Hz. A subwoofer's cutoff frequency may be as high as 250 Hz for consumer-grade systems with satellite main speakers using a bass management system. Since the low-frequency response in a room will vary by location, it is best to perform subwoofer calibration tests at a few locations around the room to find an average setting.

The next step is to determine the sound pressure level that the subwoofer should be putting out. Using one of the main full-range speakers, send a pink noise test signal passing through a second-order low pass filter set to the cutoff frequency of the subwoofer. For example, if the subwoofer low pass filter is 60 Hz, you will see the SPL level in the room drop off by about 8 dB. This reduced SPL level is what the subwoofer alone should produce. If the main system was calibrated at 79 dB SPL, and the subwoofer signal is low pass filtered at 60 Hz, the subwoofer alone will put out closer to 71 dB SPL with the filtered reference signal sent to it. Obviously, this will vary somewhat depending on your system and the bass response of your room. Using test signals with an SPL meter will certainly get you close to a good calibration point, but I find it is necessary to make the final adjustment by ear using commercial music and film mix references.

6.3.3.   Bass Managed Systems

Bass managed systems use a crossover network to route the lower frequencies to the subwoofer and the upper frequencies to the main speakers. In these systems, it is not unusual to find a range of crossover frequencies from 80 to 250 Hz depending on the size of the main speakers. Using a bass managed system, the main speaker and the subwoofer work together in concert to produce a full-range sound (see Figure 6.2). The subwoofer level should be set so that the transition from sub-bass region to the low-frequency region of the main speakers is smooth. The subwoofer level should be adjusted so that a swept sine wave test signal to a speaker paired with a subwoofer should be smooth without sudden dips or surges from 20 HZ to 1 kHz.

6.3.4.   Surround Sound

Surround sound systems are based on an array of speakers that surround the listener. In addition to the left and right speakers found in two-channel stereo systems, a dedicated center speaker is added that creates a three-channel stereo system in front of the listener with left and right speakers placed +/−30° of center. In addition to the center channel, left surround and right surround speakers are placed to the side-rear of the room. All of the speakers are placed on an equal radius arc measured from the listening position. The Audio Engineering Society (AES) recommendation for 5.1 surround speakers is to place them +/−110° off center. In 7.1 surround systems, a pair of side surround channels are added at 90° off center, and the surround channels are placed farther to the rear at +/−135° (see Figure 6.3).

Optimal surround speaker placement is not always physically possible due to the shape and dimensions of the room. Even so, sound from the center channel will remain perceptually locked in the center even if the

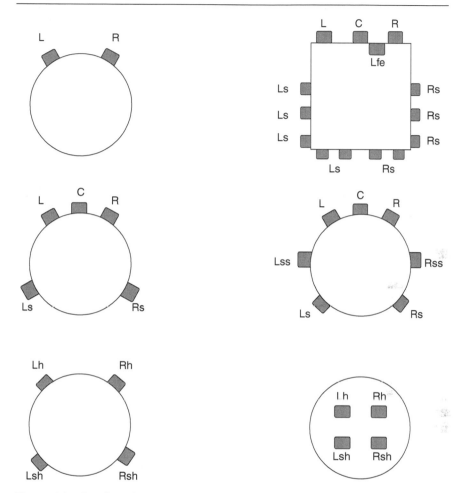

Figure 6.3 Speaker Placements: stereo (top left), 5.1 theater surround (top right), 5.0 surround (middle left), 7.0 surround (middle right), height layer (bottom left), overhead height layer (bottom right).

seating area is large. In theaters, multiple surround speakers are typically ganged and delayed to provide an immersive sound experience without taking precedence over the center and stereo speakers located behind the screen.

6.3.5. Height Channels

Height channels add a vertical dimension to sound reproduction. The basic AURO 3D recommendation is to add an upper quad speaker layer at 45°

directly above the main layer of speakers. The left height speaker is placed above the left speaker, the right height speaker is placed above the right speaker, the left surround height speaker is placed above the left speaker, and the right surround height speaker is placed above the right surround speaker. A basic Dolby Atmos speaker configuration includes height channels mounted to the ceiling overhead, often in two rows.

6.3.6. Headphones

Headphones, for better or worse, provide a listening experience that is unaffected by the acoustics of the room and the acoustic shadow cast by the human head and body. In effect, the room is a complex filter causing temporal and spectral shifts due to reflections from boundaries and objects in the room. Using speakers, the perceived low-frequency response is drastically affected by the room's geometry and location of the listener. Since headphone monitoring is unaffected by the room, low frequencies can be monitored accurately in a professional set of headphones that can reproduce frequencies to 20 Hz. Technical issues such as digital clicks, poor edits and system noise can be easily detected using headphones. But monitoring through headphones is not without its own set of issues. A headphone listener may experience an uneven high-frequency response that is dependent on how the headphones are seated on the head. Additionally, headphones cannot reproduce the same physical sensation of bass transmitted through the body that we are accustomed to in a room with live sound. Headphone monitoring also lacks the acoustic crosstalk that occurs between our ears and the effect of head shadowing, the outer ear, and our bodies. In headphones, the mix may appear unusually wide at times, while at the same time, some sounds are perceived *in the head*. Speaker simulation on headphone processors are now readily available as plugins for most DAWs. They provide an effective way to check the mix in headphones when speakers are not available.

6.3.7. Object-Based Mixing

Channel based mixes, such as stereo, 5.1 surround, 7.1 surround, 9.1 surround with height etc., have a 1:1 correlation between the number of mix output channels and the number of discrete speakers channels assigned for playback. For example, a stereo mix is created specifically for two-speaker systems, whereas a 5.0 mix is created specifically for a five-speaker system. Object-based systems do not commit to the number of playback speaker channels until the time of playback. This is accomplished by encoding directional information with discrete channels of audio during the mixing process. On playback, a renderer is used to direct sounds based on the specific playback speaker configuration. For example, a renderer can create a multichannel output to support a 22.2 speaker configuration even though the program was mixed on a 14.1 speaker system. Also, an object-based mix can be easily translated to a version for headphone monitoring.

6.3.8. Ambisonic Systems

Working with ambisonic systems, as few as four channels of audio can store a fully immersive sound mix. These four channels, known collectively as B-Format, hold directional and phase information based on three directional axes (X, Y, Z) and an omnidirectional component (W). Location accuracy can be improved greatly using higher-order ambisonic systems (HOAs). For example, a 2nd-order 3-D system uses nine channels, whereas a 3rd-order 3-D system uses 16 channels to significantly increase location accuracy. Mixing mono, stereo, 5.1 surround etc. in ambisonics requires an ambisonic encoder to convert these channel-based signals into B format or HOA. On playback, an ambisonics decoder is required to monitor through speakers or headphones. Since the playback channel count and orientation are not fixed, a head tracking system can be used to provide the listener with an interactive virtual acoustic environment over headphones.

6.4. Metering

6.4.1. Analog Metering

Analog VU (volume unit) meters measure the average amplitude of an audio signal. A typical analog VU meter is scaled from −20 to +3 VU. Typically, 0 VU is used as the target level with only momentary peaks going into the red zone above it. The meter's VU scale is in decibels (dBs) and indicates the ratio of the signal level to a reference signal level. For example, 0 dB indicates a 1:1 ratio. Professional audio equipment is typically calibrated so that 0 VU = 4 dBu whereas 0 dBu = 0.775 V at 600 Ω. If the amplitude of a signal is doubled, the meter will indicate a 6 dB boost, whereas if the amplitude is halved, the meter will drop 6 dB. Incidentally, the dB scale on large faders follows the same principle; a 6 dB change up or down indicates a doubling or halving of the signal amplitude respectively. Since an analog VU meter indicates an average, it does not indicate the instantaneous peak level. Therefore, most analog meters have a peak light to provide warning that a momentary overload has occurred. Peak lights on analog devices are typically calibrated to lite up around 24 dBu, thus leaving about 20 dBs of safety above 0 VU (assuming +4 dBu calibration).

6.4.2. Digital Metering

Digital metering can be set up to indicate peaks or averages relative to the full scale. Most DAW meters default to peak metering where 0 VU = 0 dBFS (decibels full scale), thus leaving no built-in headroom above 0, as analog meters do. Above 0 dBFS, clipping will occur, so the engineer must leave enough headroom to avoid inadvertent audible distortions due to overloads on peaks. On most DAWs, a target mixing level can be set manually in the user preferences if desired. For example, a meter color break level of −12dBFS, −18 dBFS or −20 dBFS can be established as 0 VU for metering purposes (see Figure 6.4).

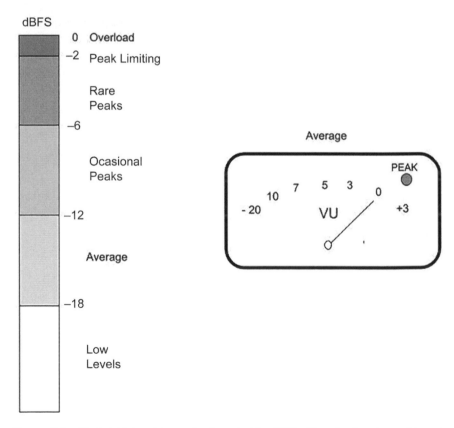

Figure 6.4 Digital Meter: The meter is scaled in dBFS with color breaks and target levels indicated (left) versus analog metering (right) with a target level of 0 VU.

6.4.3. dBFS Target Levels

Target levels should be kept to leave enough headroom for mixing and processing while maintaining a good signal-to-noise ratio. In the digital domain, the sample word size determines the signal's amplitude resolution and thus effects the self-noise of the system known as the *noise floor*. Each bit essentially doubles the resolution of the system, therefore adding about 6 dB of dynamic range per bit. If a mix is made initially at a very low level, the amplitude resolution is not optimal. For example, a signal peaking at −18 dBFS on a 16-bit system is only using about 13 out of its 16 bits, whereas the same signal on a 24-bit system still has about 21-bit resolution. Working at higher bit resolutions, like 24 bits or more, prefader signal levels should average between −18 and −12 dBFS with occasional momentarily peaks above −12 dBFS but very rarely peaking above −6 dBFS. Maintaining these target prefader levels will leave enough headroom for normal mixing, equalization, dynamic and effects processing. To safeguard against bus overloads, a peak-limiting plugin set to about −2 dBFS is effective.

6.4.4. Loudness Metering

Peak and average level metering does not necessarily reflect how loud a mix will be perceived. The *loudness unit* (LU) calculation takes into account the spectral content, silences and peak levels over time to better represent how we perceive loudness. LUFS (loudness unit full scale) metering provides a way to approximate the human perceived dynamic range of an audio program. For example, the established standard target level for film and broadcast is about −24 LUFS. If the mix is averaging −12 LUFS, it is an indication that too much compression and/or peak limiting with makeup gain is being applied thus you are not taking full advantage of the available dynamic range. Furthermore, playback systems with an automatic loudness correction feature enabled will compensate and actually turn on your mix down on playback. Loudness standards will be discussed in more detail later in the chapter.

6.5. Signal Flow

6.5.1. Trim Pots

Trim pots adjust the signal level before any processing takes place on a channel strip. If the input signal is too low or too high, adjusting the trim so that a target level is achieved will help position the channel fader close to unity with enough room to move up or down while making adjustments during the mixing process.

6.5.2. Inserts

Inserts provide a physical send-and-return point for routing signals to and from a console or audio interface for external processing. On digital audio workstations (DAWs), inserts provide a way to add plugins, virtual instruments and external processors into the signal flow. Some typical processors applied via inserts are noise reduction, equalization, dynamics and tuning plugins.

6.5.3. Auxiliary Sends

Auxiliary sends (aux-sends) provide a way to send multiple copies of the signals to multiple busses or outputs. Aux-sends can be assigned for prefader or postfader operation. Signals sent postfader will be affected by the main channel fader, whereas prefader signals will not. Some typical processors applied via aux-sends are reverbs, delays and side-chain dynamics. Auxiliary sends can be used to create submixes as well. For example, aux-sends can be used to create multiple headphone mixes or group signals for effect processing.

6.5.4. Channel Faders

A *channel fader*, also known as track fader on DAWs, controls the level of the audio signal to be sent to the next mixing stage. This is where the fine-tuning of a mix takes place. At the 0 dB position of a fader, also known as unity, the channel fader essentially serves as a pass-through. Be aware that boosting the signal at the channel fader past unity may cause an overload down the line. I highly recommend using a control surface with at least eight faders whenever possible so that you can adjust multiple faders independently at once, thus capturing a multidimensional performance.

6.5.5. Master Faders

Master tracks control the overall signal level of an assigned audio bus (sometimes spelled buss) or output. Since many signals can be summed on a bus, the master bus fader may need to be lowered to prevent overload. If bus levels become too high during the mixing process, there really is no need to lower all of the channel faders at once—it is better to simply reduce the master fader to correct the issue. Most modern DAWs provide ample dynamic range, so lowering the master fader by about 6 dB will not cause an increase in system noise down the line but will free up some headroom. Similarly, the master fader can be used to correct low bus levels. If bus signals are low, a maximizer plugin (a specialized peak limiter with automatic makeup gain) can be used on a master track to boost levels without fear of clipping.

6.5.6. Auxiliary Tracks

Most DAWs have an option to create auxiliary tracks with assignable inputs and outputs. The auxiliary track functions much like a regular track, except it does not have an associated playlist lane to edit and sequence audio regions. Auxiliary tracks can be used to reroute signals, send or receive multiple signals from a physical output or input, create subgroups and provide paths for group or parallel processing.

6.5.7. Clip Gain and Normalization

Clip gain adjusts the amplitude level of an audio region. It can be used to help prepare for a mix by matching the level of consecutive regions or portions of regions by splitting up the region. In this way, using the clip gain feature creates a prefader mix. Good prefader mix levels will mean that less dramatic fader moves will be required during the final mix process.

Normalization is an automatic process through which gain or attenuation is applied to a region or audio so that the peak level reaches a set limit. No compression or expansion is applied during the normalization process. Normalizing regions to 0 dBFS does not leave any headroom for boosting or processing the signal down the line; therefore it is a better practice to normalize regions around −6 dBFS.

6.6. Panning

6.6.1. Mono and Phantom Sources

A *monaural system* can be described as a single audio signal amplified by a single speaker. In a true mono system, the perceived location of the reproduced sound stays with the speaker no matter where the listener is located. If a mono sound signal is duplicated for stereo reproduction, hence a dual-mono stereo signal, and sent to two speakers arranged for stereo monitoring, the location of the sound source will be perceived in between the speakers, thus creating a phantom center image. If the listener moves out of the sweet spot, the phantom center image can become unstable and appear to follow the listener to the closest speaker.

6.6.2. Amplitude Panning

By lowering one side of a dual-mono stereo signal, the virtual sound source can be perceived off center at various locations across the stereo field bounded by the loudspeakers. Panning potentiometers, also known as pan-pots, were added to early stereo mixing consoles to accomplish this task. Mathematical stereo panning laws are used to ensure that there are no perceived sudden boosts or dropouts while moving sounds in the stereo field. For example, when a signal is panned from hard left or hard right to the center, left and right signals are both reduced about −3dB using the sine/cosine panning law so that there is no perceived change in overall level.

6.6.3. Delay Panning

Alternatively, phantom center images can be moved using a short delay in the channel opposite the desired motion. In practice, short delays up to 2.5 ms work well for this application. The effective delay time will vary depending on the spectral and transient nature of the signal. If a short delay is applied to the left channel of a dual-mono signal, the image will be perceived from the right, even though the levels in the left and right channels are still equal. In addition, the sound may now have a perceived sense of space because the delayed signal will simulate an acoustic reflection similar to that in a real room. The drawback to using the stereo short delay panning technique is that a mono downmix may exhibit coloration.

6.6.4. Equalization Panning

Equalization can be used to move sounds in the stereo field as well. For example, by low pass filtering one side of a phantom center between 2000 Hz to 700 Hz, the sound image will move toward but not fully to the opposite side. This creates a spacious effect as well because the low pass filtering emulates human head shadowing that would be felt in a real acoustic space. Further low pass filtering will move the image fully to one side. Like delay panning, the panning effect will vary depending on the spectral and transient nature of the sound.

6.6.5. Combined Panning Techniques

Using a combination of level, delay and equalization panning can create a more natural sounding effect than using just one or two of these techniques alone. A multidimensional panning effect occurs naturally when stereo microphone techniques are used because all sound sources are picked up by both microphones in some way. Even without using formal stereo microphone techniques, recordings made with multiple microphones and multiple musicians playing live in the same room can have a similar effect. That said, most pop and rock music producers prefer total sound isolation while recording to have full control over the sound for mixing, editing and processing during the mixing process.

Reverberation can help create a natural panning effect as well. For example, if we pan a trumpet signal hard left for clarity and location while simultaneously sending it to a stereo reverb or to a mono reverb centered, the left channel will have a dry plus reverberated signal, whereas the right channel will only have only the reverberated signal. The trumpet will be perceived clearly located to the left but with a sense of stereo spread (see Figure 6.5).

6.6.6. Panning Stereo Sources

Stereo recordings can contain complex interchannel phase, spectral and level relationships. This information allows us to perceive multiple directional sounds within a single recording. Panning the left and right signals of a stereo recording toward the center creates a partial dual-mono signal, thus narrowing the stereo image. To pan a stereo recording to favor the

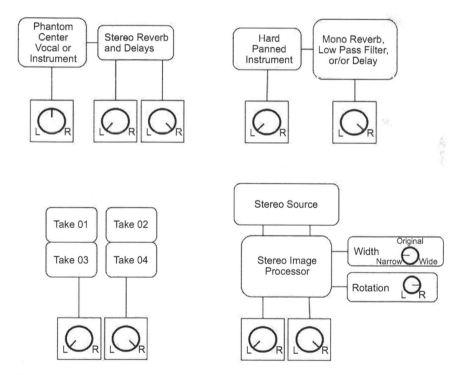

Figure 6.5 Placements: Mono source in the phantom center of the stereo field with a stereo effect (top left), adding an opposite panned mono effect for stereo effect (top right), placement of natural doubled takes for stereo effect (bottom left), narrowing the stereo image before panning a stereo source (bottom right).

left or right, the stereo width needs to be reduced prior to panning. Stereo image plugins that have width and rotation controls provide an effective way to adjust and reposition stereo images within a mix with ease and accuracy (see Figure 6.5). Stereo imaging plugins are especially useful when panning stereo sampled orchestral library instruments.

6.6.7. Mid-Side (MS) Processing

Stereo recordings captured using a middle-side microphone technique can be converted to a two-channel stereo signal using an MS decoding matrix:

Left = Middle + Side
Right = Middle − Side

Conventional stereo signals can be converted to a mid-side (MS) signal, as well using a MS encoding matrix:

Middle = Left + Right
Side = Left − Right

While in the MS domain, the reciprocal stereo signal's image can be narrowed or widened by varying the level of the side signal before converting the MS pair back to stereo again. Equalization and other processors, including delays, reverb and compression, can be applied independently to the middle and/or side signals for effect as well. Many plugins have an MS feature to explore.

6.6.8. Stereo Phase-Correlation Meter

The stereo phase-correlation meter monitors the phase relationship between left and right channels of a stereo program. A correlation approaching +1 indicates that the left channel is near identical to the right channel; thus, the image will be perceived in the phantom center. A correlation approaching 0 means that the left and right channels are not alike, indicting a very wide stereo image and a perhaps a perceptual hole in the middle of the stereo image. A correlation hovering around +0.5 is ideal, indicating that the stereo program has an even distribution of sound across the stereo plane without any severe phase issues.

A correlation approaching −1 means that left and right channels are nearly identical but one side is polarity inverted. In this case, the stereo image may be perceived unnaturally wide, but the mono downmix will suffer from coloration and signal loss.

6.6.9. Mono to Split Stereo

The width of a dual-mono sound recording cannot be narrowed or widened because it contains no stereo directional information, but a mono source can be used to create a pseudo stereo image using a mid-side-based *split* technique. The unprocessed mono signal can function as the middle signal of the mid-side pair. To create the pseudo side signal, virtuality any effect processor can be applied to the original mono signal. An equalizer or filter will create a split equalization stereo effect. A short delay, less than 2 ms, will create a split filter effect as well, whereas a longer delay, more than 20 ms, or a reverberator, will create a wide echo or reverb effect. This can be done manually using individual plugins and an MS matrix or by using a dedicated mono to split stereo plugin.

6.6.10. LCR Panning

The front end of a surround sound system, consisting of left (L), center (C), and right (R) speakers, can be considered an LCR stereo system. Sounds placed in the mono center channel will be perceived in the middle of the stereo field no matter where the listener is located. For movie houses, this keeps the dialogue and other synchronized sounds locked to the center of the screen. On mixing consoles and DAWs equipped with LCR panning, the amount of signal sent to the center channel when panning from hard left to hard right can be controlled. At one extreme, the sound can exist only in the C channel when the pan-pot is up the middle. This is the most useful setting for film dialogue and sync-sound applications. At the other extreme, the LCR pan-pot can function like a two-channel stereo pan-pot with no signal at all going to the center channel, thus creating a phantom center image.

6.6.11. Left Surround and Right Surround Channels

Left surround and right surround channels can be accessed using a surround panner. Without a dedicated surround panner, surround channels can be created using aux-sends, busses, direct outs etc., as long as there are enough physical outputs on the system to send signals to all of the speakers. Some consoles equipped for surround have a front-to-rear pan-pot in addition to an LCR stereo pan-pot and an LFE (low-frequency effect) subwoofer send. Some analog and digital consoles are outfitted with a joystick for quad and surround panning. DAW surround panners typically provide a virtual XY surface with a movable dot or zone that represents the position and spread of the sound source in relation to the speaker channels.

It is common for film mixers to use surround channels to support effects in motion, ambiance and momentary effects. It is important that the surround content does not feel oddly disconnected from the left-center-right content. For example, ambiances can be placed in the left and right channels as well as in the surround channels so that the effect feels more immersive rather than stuck in the back and sides of the room.

Music in surround can be mixed from a concertgoer's perspective where the surround channels contain mostly ambiance—or from a "you're-in-the-band" perspective where the surround channels contain an abundance of direct sounds like the front channels. In any case, the challenge lies in preserving the impact that people have become familiar with listening to stereo while making use of the panoramic soundstage possibilities.

6.6.12.   Low-Frequency Effects (LFE) Channel

The send to the subwoofer is typically controlled by a separate low-frequency effect (LFE) rotary pot or slider on the surround panner module. The LFE should only fire momentarily for maximum effect—otherwise, the listener may tune the low-end effect out or become fatigued or sickened by continuous bass conduction felt directly through the body. A sub-bass focused mix pass may be required to automate what sounds should and should not be sent to the LFE channel.

6.6.13.   Height Channels

Adding height channels allows the vertical spread and orientation of sounds to be explored. They can be used to create truly immersive and enveloping sound environments and realistic fly-over effects. Height sound can greatly enhance the sense of depth and space in a mix and also provide important spectral information to support sound sources located in the main surround layer.

6.6.14.   Vocals and Dialogue

For music mixes, lead vocals are typically mixed up the middle because they carry vital artistic information. Imagine if the left channel goes out on a system, and the lead vocal vanishes. Vocals need to be placed in the mix so that they are easy to hear; otherwise the listener may become fatigued trying to understand them. Vocal effects can be spread across the stereo or surround sound field by using multichannel delays, reverberation and electronic or natural doubling effects. Doubling a vocal or instrumental part, by recording multiple performances of the same musical part and panning the individual takes across the stereo or surround channels, can create an immersive effect. For example, this technique can be used to create a subtle chorus of backup vocals or electric guitar parts. The doubling effect can also be achieved during the mixing stage by electronically introducing slightly pitch- and time-modulated copies of a part spread across the stereo or surround channels.

Film dialogue is typically mixed up the middle as well. In surround mixes, the dialogue works best confined to the center channel alone. This is done to keep the dialogue centered on the screen. Attempting to widen the dialogue sound by bleeding the dialogue signal into other channels is not recommended because it may blur the location and color the sound of the dialogue depending on where the listener is seated.

6.6.15. Music Beds in Surround

If the music has been recorded and/or mixed in surround, the original surround channel assignments can be used in the final mix. Music can exist in all channels, but it should be reduced in center channel to make way for the dialogue. If the music source is in stereo, some of the left and right channels can be bled slightly into the center channel to keep the music centered in the theater. It is not a good practice to send the identical music or effect signal to the surrounds and to the front stereo pair without processing the surround signals somehow. The surround signals can by widened using a stereo image plugin, by adding reverb, or by reducing the high end, thus creating a more effective immersive effect (see Figure 6.6).

6.6.16. Individual Instruments

Hard panning instruments all the way left or right was common in the early days of stereo record production. By the early 1970s, it was more common for rhythm section instruments to be placed in the center of or spread across the stereo field, thus leaving only candidates like percussion and backup vocals to be hard-panned. The location of instruments in classical music recordings typically follows the seating arrangement used when it was performed. Jazz recordings can follow the live performance staging as well, but many are produced in studios where the performers are placed in a circle facing each other, behind gobos, or in isolation booths. In any case, the production team must decide what panning and spatial effects best serves the music. One mixing strategy is to offset a sound panned hard left with a sound panned hard right. To create a sense of room depth for hard-panned sound sources, the opposite channel can contain a reverberated, delayed, or equalized version of the sound to create a sense of stereo spread (see Figure 6.6).

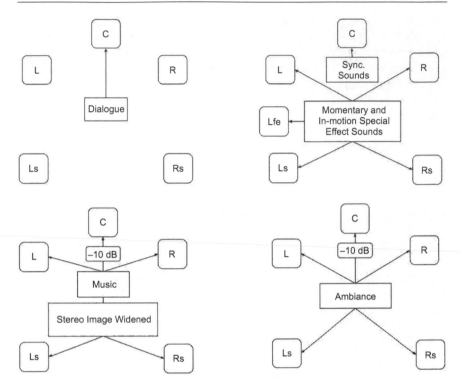

Figure 6.6 Surround Panning Strategies: For film dialogue (top left), stereo music upmixed to 5.0 (bottom left), placement of sounds synchronized with the moving image and other momentary or in-motion effect sounds (top right), and the placement of purely ambient sounds (bottom right).

Stereo image processors can be used to exaggerate the sense of envelopment by enhancing the stereo width of a stereo sound source or stereo effect. This approach is commonly used in electronic pop and dance music on stereo synth pads and effects to compensate for mono drum, bass and vocal sounds that are typically panned up the middle. This technique can also be used to upmix stereo sources for surround mixes (see Figure 6.6).

6.6.17. Creating Spaciousness and Envelopment

Multichannel delays and long reverbs can be applied to create the sense of an external space, as in a church, concert hall, parking garage etc. Similarly, the sense of envelopment can be created by placing a sound in multiple speaker channels with a slightly different timbre, level, phase and/or modulating effect in each. Dense, short reverberation is a good example of an effect that creates an enveloping effect through complex continuous time and tamberal shifts. Multimono modulating effects such as choruses, flangers and phasers are effective as well.

6.7.   Equalization

6.7.1.   Frequency Bands

If too much bass or low midrange is present, a sound can appear *unfocused* or *boxy*. If too much high end is present, a sound can feel *shrill*. Many terms to describe sound, including "dark," "bright," "fuzzy," "dull" etc., are borrowed from our other senses. Although analogous terms have their place, at some point they no longer communicate what you really mean and can cause confusion in the control room. Using the correct technical terms to describe specific properties of sound requires understanding the relationship between the physical properties of sound and how we actually perceive sound. For example, the *fundamental frequency* determines the musical pitch of a sound. *Octaves* refer to the doubling or halving of a given frequency. *Harmonics* are a complex series of frequencies numerically related to the fundamental frequencies. *Formants* are the characteristic harmonics that give an instrument or voice its unique and identifiable timbre. Physical properties of a sound change over time as well. Critically listening to each portion of a sound in time—its attack, decay, sustain and release—is vital while making any equalization adjustments.

The theoretical audible frequency range for humans is from 20 Hz to 20 kHz. Due to the nonlinear nature of the human perception of sound, the center of the audible range can be considered 1–2 KHz. For identifying bands quickly, the audible spectrum can be broken into bands: lows, mids and highs. These bands can be broken down further to correspond to aesthetic attributes associated with the spectral content of a sound.

The sub-bass region, below 50 Hz, is where sounds are felt physically perhaps more than heard as intelligible sounds when isolated without their upper band counterparts. Sub-bass can be routed to a subwoofer for support or isolated for a special effect. Many sound systems cannot reproduce this band; therefore, before adjusting the sub-bass, make sure that your system can reproduce it accurately. Large full-range speakers or systems equipped with subwoofers are needed to perceive this range properly. When monitoring through speakers, the perception of this band can be severely affected by the room's acoustics and your location in the room. Headphone monitoring is not affected by room acoustics and can reproduce sub-bass frequencies at your ears but cannot reproduce the physical

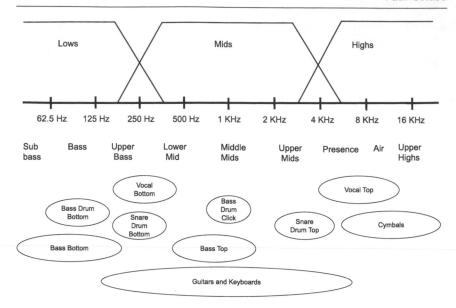

Figure 6.7 Low, Midrange and High-Frequency Equalization Bands: Separated by cutoff frequency (top), common aesthetic descriptors (middle), and an example of a characteristic band equalization approach for mixing voice and musical instruments (bottom).

feeling of bass we experience directly through sound pressure felt on the rest of the body.

At the other end of the spectrum is the upper high-frequency range, from about 12 to 20 KHz. In this range, isolated sounds may be perceived as a sensation rather than as an intelligible sound as well. Analog tape systems tend to gently attenuate this band. High-resolution digital systems, for better or worse, can faithfully record and reproduce the upper high-frequency band accurately granted that the sample rate is high enough. A gradual analog-like attenuation in this band can in fact *warm* up the sound. Some sound systems with single full-range speakers, like TVs, computer speakers and portable radios, cannot reproduce this band accurately, although most speakers equipped with tweeters can. Most headphones can reproduce this band as well, but be aware that the seating of the headphones each time you put them on may affect your perception of some high frequencies.

The region just above the sub-bass region, from about 50 to 200 Hz, is where an abundance of lower fundamental and bass frequencies exist. Like the sub-bass region, this region needs to be balanced properly with other bands to not overpower, or muddy up, the mix. The bass region needs to be present just enough so that instruments and voices sound natural but not overpowering. Unfortunately, when monitoring through speakers, the perception of this band can be severely affected by the room's acoustics and your location in the room as well. Therefore, it is a good practice to listen from multiple positions in the room or monitor with professional headphones before making any vital changes to this region.

The whole midrange band, from about 250 Hz to 4 KHz, contains the frequency bands that just about all sound systems, large or small, can reproduce. Not enough midrange may make a program sound "thin" and even unintelligible, whereas too much energy in this band can create a "low-fidelity" effect. I regularly check my mixes with this band, isolated by filtering out frequencies below 250 Hz and above 4 KHz. If the mix works without its extreme highs and lows, it will hold up under just about any playback condition. Further, mix adjustments made to correct issues heard during the isolated middle band test usually improve the full-frequency mix as well.

An abundance of energy in the low midrange region, from about 250 Hz to 1 kHz, can creates a "boomy" or "boxy" sound. Alternatively, a lack of energy in this range can make a sound "brittle" or lacking some punch. Often, close-miked instruments will have an abundance of energy in this band that needs to be reduced.

The upper midrange frequency band, from about 1 to 4 KHz, is where our hearing is the most sensitive. A boost in this area can certainly boost the *presence* of an instrument and increase speech intelligibility of a voice, but an overabundance of energy in this range can cause short- and long-term discomfort to the listener when monitored at higher volumes, or the "nails on a chalkboard" effect resulting in a listener's fatigue. Ducking this band can bring a warmth back to a "brittle" sound.

A boost in the high-frequency range from about 6 to 10 kHz, also known as the *air band*, can help open up and unveil a sound in the mix. The top end of the human voice and cymbals are typically active in this range. A complementary boost in the bass band may be required to compensate for an aggressive boost made in the air band on a voice or voice-like instrument like a saxophone. Also, too much high end can bring out resonant vocal sibilance, like Ss, Ts and Zs sounds. To compensate for unwanted sibilances, a de-esser can be applied. De-essers are commonly used on vocal processing chains and are discussed later in this chapter.

6.7.2.   Filters

Filters are designed to remove a portion of the spectral content of a sound signal. They are qualified by the frequency band they affect, cutoff frequency and the filter slope. The slope is defined by the filter's order number and describes the steepness of the filter response curve.

1st order = 6 dB per octave
2nd order = 12 dB per octave
3rd order = 18 dB per octave
4th order = 24 dB per octave

Filters that preserve the lower portion of the frequency spectrum are called *low pass filters*. Similarly, *high pass filters* preserve the high-frequency portion of the frequency spectrum. *Band pass*, or band cut filters, are essentially paired low pass and high pass filters. They are defined by their center frequency and bandwidth. Typically, filters are used to correct a technical issue like a high-pitched noise, resonance or low rumble—rather than modify or enhance the timbre of a sound for artistic reasons.

6.7.3.   Equalizers

Equalizers were originally designed to compensate for spectral capture and reproduction deficiencies in early recording and broadcast equipment. Today, the use of the equalizer has been expanded to include more creative and artistic applications. Equalizers are defined by their center frequency (Fc), bandwidth (Bw) and the amount they can cut or boost. Fc is described in hertz (Hz), the cut or boost gain in decibels (dBs), and its bandwidth by quality (Q). The equalizer's Q is described by this equation:

$$Q = Fc/Bw$$

where:
Q = quality
Fc = center cutoff frequency in Hz
Bw = bandwidth = (Fc high − Fc low) in Hz

As the bandwidth of a filter increases, the Q goes down. Low-Q filters are considered more *musical* sounding because they do not radically alter the spectral balance within the sound. Medium- or high-Q filters can perform more precise adjustments within a sound to correct a technical issue or to bring out a very specific part of the sound for effect.

When applying equalization, I recommend starting with a low Q around 0.5 and a low cut/boost amount of +2 dB. Then apply more EQ as needed, simultaneously increasing the Q and the cut/boost amount. Some programmable equalizers and EQ plugins provide this function automatically. Many utility multiband digital EQs default to a higher Q setting than necessary. Some specialized or boutique EQ plugins provide more *musical* presets with lower default Qs and more sensitive user controls.

6.7.4. Characteristic Frequency Band Approach to Mixing

The audible frequency spectrum, from 20 Hz to 20 kHz, can be considered the *width* of a sound canvas on which multiple sounds can be organized and coexist. Active frequency bands from one sound may fully or partially overlap with active frequency bands of another sound, thus causing one sound to be partially or completely masked by another. The strategic and creative use of equalization to feature characteristic spectral bands of a given sound can certainly bring clarity to a mix. The following is a list of characteristic frequency bands for a select group of musical sound sources from a specific mix session. Obviously, an analysis like this needs to be done by ear and case by case because not all instruments and vocalists sound the same. Nonetheless, some general observations can be made that will apply to most mixes.

Bass drum—bottom at 60–80 Hz, top at 1–2 kHz
Snare drum—bottom at 200 Hz, top end at 3–4 kHz
Hi hats   bottom at 700 Hz, presence at 5–10 kHz
Cymbals—bottom at 200 Hz, presence at 7–12 kHz
Rack toms—bottom at 240 Hz, top at 5 kHz
Floor toms—bottom at 80–100 Hz, top at 5 kHz
Bass guitar—bottom at 37–160 Hz, top at 700 Hz–1.5 kHz
Electric guitar—bottom at 240 Hz to the top at 2.5 kHz
Acoustic guitar—bottom at 80–120 Hz to the top at 3–6 kHz
Piano—low end at 60–120 Hz to the top at 2.5–5 kHz
Vocals—low end at 220 Hz, presence at 5 to 10 kHz

6.7.5. Voice EQ

The low- and middle-frequency bands of a mix can become very cluttered, whereas the high-frequency range is typically open. Taking advantage of this situation, the voice can be given more presence in the high-frequency region, from around 6 K to 10 KHz, to bring it out front in a mix. This way, the voice can be present without masking or being masked by other sounds. To compensate for any unnatural brightness, the low end of the voice, typically 200 to 250 Hz can be boosted as well—especially for female voices. Dialogue sound recorded with miniature body microphones often requires an aggressive cut in the low midrange area and a boost in the highs to sound natural. If the vocal sound is overtly resonant in the high end, a de-esser may be needed to reduce the sibilance. De-essers belong to a family of frequency dependent dynamic processors that will be discussed later.

6.7.6. Drums EQ

Often, the low end of a bass drum is centered around 80 Hz, and the click of the attack sound is centered around 1 kHz. Close-miked bass drums recordings are one of the few instrumental sound sources that may require extreme equalization. Often, they require a severe low- to middle-frequency cut at around 300 Hz to get rid of the *boxy* sound. A clear attack sound around 1 kHz is vital to make the bass drum sound appear present. Ironically, too much energy in the low frequencies below 80 Hz can muddy the sound, overengage compressors, and make the bass drum appear less focused. Since there are not many instruments with subbass, the low end of the bass drum may appear too weak when monitored in solo mode. When placed back in the mix, the low end will still be still be present, whereas the upper bands may be masked by other sounds, thus sounding overall muddy. Therefore, the bass drum should sound a bit brighter than needed in solo mode. The same holds true for the bass guitar, bass synths, etc. The snare drum's distinctive low end is typically closer to 200 Hz, and its high end, centered around 3 kHz. The high end that defines its attack is more broadband, or noise-like, than most other drums. The toms will each have their own unique characteristic low- and high-frequency bands depending on how they are tuned.

6.7.7. Bass Instrument EQ

The fundamental frequency area of the string bass has a wide range, unlike the bass drum that is fixed. The characteristic low end of the bass surrounds the low end of the bass drum. The sub-bass area of the bass sound provides a certain warmth to the overall mix, whereas the upper harmonics help the listener determine pitch and rhythm. Applying distortion or a harmonic exciter to the bass will help the pitch and rhythmic information in the bass performance cut though the mix. The sub-bass band provides the physical feeling that the bass is present. The sub-bass and upper bands should be addressed separately because they serve different functions. Using a multiband dynamic processor, as discussed later in this chapter, upper bass and sub-bass frequencies can be dynamically controlled independently so that they are balanced and present but not overbearing. As the saying goes, "If you want to hear more bass (instrument), turn down the bass (frequencies)—and turn it up!"

6.7.8. Guitar EQ

The electric guitar sound lives in the midrange of the mix. The famous low-wattage Marshall and Vox amplifier sounds both have an abundance of midrange in their tone. Careful consideration to the musical arrangement during the mix process needs to take place so that distorted guitars are powerful and present in the appropriate places. Electric and acoustic guitar timbres vary greatly. In some guitar-centric music, the electric guitars are heavily distorted, and the bass part mimics the roots of the guitar chords so that together they form a warm and powerful composite sound. In funk music, the guitar parts stand more alone and are typically percussive in nature, characterized by cutoff chords that leave room for the bass to do its own thing. Acoustic guitar sounds captured by close miking may exhibit an unwanted resonant frequency in the low midrange around 250 Hz that requires attenuation through careful equalization. The high-mid attack sound of acoustic guitar attack can be considered a percussive effect in some musical contexts, not unlike a tambourine or hi-hat.

6.7.9. Keyboard Instrument EQ

Keyboard sounds can span the entire spectral range but typically are featured in the midrange like guitars. In complex arrangements, the low end from the keyboards is often reduced to make room for the bass, or it can be featured if there is no other bass instrument. Similarly, the high end can be reduced to make room for the vocals and cymbals. Out of all the musical instruments, keyboards may require the least amount of equalization, especially electronic keyboard instruments. I typically leave them flat unless there is a severe issue. Acoustic pianos sometimes require an equalization tilt toward the high end to reduce excess bass captured by microphones placed inside the piano.

6.7.10. Acoustic Instrument EQ

In general, orchestral instruments, such a strings, winds, brass and tuned percussion, do not require severe corrective or characteristic equalization. Although close miking can sometimes produce unnatural timbres, dynamics ranges, or resonances in the recording that need to be addressed. If a certain narrow band of frequencies jump out and become distracting, you can identify the musical pitch that is problematic, determine its fundamental frequency and calculate the harmonics to see what frequencies may need to be precisely reduced. If a certain band of frequencies become overly present only momentarily, a multiband dynamic processor (discussed later) can correct the issue.

6.8. Dynamic Processors: Compression and Expansion

6.8.1. Dynamic Range

The *dynamic range* of a program describes the difference between the softest and the loudest portion of audio. The dynamic range of an analog or digital system is bound by its self-noise and the level at which the system will distort. For example, professional tape machines may have a dynamic range of about 75 dB. A 16-bit digital system may have a maximum dynamic range of 96 dB, whereas a 24-bit system may have wider range up to 140 dB. The dynamic range of human hearing is theoretically 120 dB, from 0 dB SPL (threshold of hearing) to 120 dB SPL (threshold of pain). Sound levels above 105 dB can become uncomfortable and even harmful after constant exposure. As far as perceiving sounds near the lower threshold of our hearing range, we rarely experience total silence. The target noise floor of a studio listening environment is about 20–30 dB SPL. The average ambient level in a home or work environment may be 40–60 dB SPL. This leaves us humans with a realistic everyday dynamic range of about 55 dB.

If the program's dynamic range is greater than the listening environment can support, quiet parts may be inaudible. Consider driving in a car with the windows down. The wind noise will create a very high noise floor. The car audio system has a limit to how loud it can go without distorting. The noise in the car will clearly limit the perceived dynamic range of the listener. Obviously, it is more difficult to listen to music produced with a wide dynamic range under such conditions, like classical music or a film soundtrack, whereas pop music typically has a narrow dynamic range and is thus easier to hear in noisy environments. To limit the dynamic range of audio signals, compressors and limiters may be used. Once the dynamic range is limited, the overall level can then be raised.

6.8.2. Leveling Amplifiers

Leveling amplifiers are known for their simple but effective design and functionality. The circuitry typically contains an input amplifier, a control section, a variable output amplifier and a multifunction VU meter. The attenuation of the signal is automatically controlled by the circuitry that is monitoring the input signal. The higher the input level gets, the more gain reduction is automatically applied—thus leveling the output level.

For signals that fall below the set threshold, no gain reduction is applied. Unlike modern compressors, the leveling amplifier ratio and/or threshold is fixed. On some models, an input gain control is provided to drive the input signal past the threshold, thus activating the gain reduction. Other models simply have gain and peak reduction controls to adjust the output level and the amount of gain reduction. Also, an option for compression or limiting is typically available. Some VU meters can be switched between monitoring the gain reduction and the output level. The sound of vintage analog leveling amplifiers are still in high demand today. Many vintage models have been re-issued and emulated in plugin versions.

6.8.3. Compressors

Modern compressors are based on the leveling amplifier concept but provide a comprehensive set of controls for the user:

- *Input Gain*—controls the signal level before compression
- *Threshold*—sets the level at which the gain reduction will be engaged
- *Makeup Gain*—sets the output level after compression
- *Ratio*—sets the steepness of the gain reduction curve that the compression circuit will follow
- *Limit Mode*—sets the compression ratio upward of 10:1
- *Attack Time*—the delay before the gain reduction starts after the input threshold is crossed
- *Release Time*—the time that the compressor remains engages after the input has dropped below the threshold
- *Key Input*—allows an external signal to trigger the compression
- *Knee*—a progressively steeper compression curve is applied
- *Optical Mode*—uses an optical system to detect the input signal level or the simulation of one

6.8.4. Limiters

Limiters are essentially compressors with very high compression ratios. They provide a way to control inadvertent amplitude peaks. Analog limiters need about 2 dB of headroom to contain momentary peaks in signal amplitude. Digital limiters, such as maximizer plugins, can work within a fraction of a decibel. In general, they are not designed to be *musical* sounding devices like compressors. Instead, they are optimized to prevent signal overload. They typically have very fast attack and recovery times to conceal their effect. If the peak limiter is firing constantly, the program may sound unnatural and even distorted. The limiter provides a guard against unexpected signal clipping and distortion and allows the overall level of the program to be safely raised.

6.8.5. Expanders and Gates

When *expanding* a signal, the dynamic range is increased; softer sounds become softer, and louder sounds become louder. *Gates* are downward-only expanders with thresholds. When the input signal level drops below the threshold, the expander attenuates the output level following the expansion curve. Gates are commonly used on snare drum, tom-tom and bass drum channels so that the signal is essentially off when the drum is not is being hit. Like compressors, the threshold, ratio, attack time and the release time can be adjusted. In addition, most gates have a range control that sets a second threshold to stop the expansion, thus returning to a 1:1 ratio again, but at lower level to prevent the gate from completely cutting off the signal.

6.8.6. Dynamic Noise Reduction

Gating can be used as a form of noise reduction by cutting out unwanted background sounds when the signal goes low. Multiband gating is a more complex technique used by stand-alone noise reduction systems and noise reduction plugins. Noisy frequency bands, like hiss that lives around the 10-kHz range, can be attenuated when there is a not enough energy in that band to mask the noise. When a loud sound is present in a noisy band, there is no need for noise reduction because the noise is masked by the intended sound. The noise reduction plugin can learn the noisy band and target the noise very specifically. If too much noise reduction is applied, intelligibility may suffer, and a chirping artifact may be audible. Noise should only be reduced to the point where it ceases to be distracting. For example, some audible hiss may not distract the listener, but reducing the hiss too much may remove an important part of the intended sound and cause audible artifacts. Extreme caution should be used while applying noise reduction. If the noise can be reduced or removed, that is good, but not at the cost of the timbre of the original sound.

6.8.7. Key Input

Using a *key input* allows an external signal to remotely control the gain reduction. For example, the keyboard part in an electronic music track can be reduced momentarily when the bass drum is attacked, thus creating a synchronized rhythmic effect. For a podcast or a documentary film mix, the music can be compressed with a key input from the dialogue track so that the music automatically ducks to make way for the dialogue in the mix.

6.8.8. Multiband Compression

Multiband dynamic processors have discrete compression and expansion settings for multiple frequency bands. For example, a four-band multiband dynamic processor can divide the input signal into four signals: lows, low-mids, high-mids and highs. A compressor and expander may be assigned to each band operating independently. Whereas conventional equalizers may cut or boost a fixed amount, the dynamic equalizer listens to the input and dynamically varies the amount of equalization applied over time. For example, if there is a harsh burst of energy in the high midrange in a piano recording only during loud passages, cutting the high-mids permanently may create an unwanted dull sound during softer passages. By using a multiband compressor, attenuation of the high-mids will occur only during the louder passages, leaving the quieter passages unchanged.

6.8.9. Dialogue Compression

Dialogue for a film mix typically averages between −18 dBFS to −12 dBFS, peaking only momentarily above −10 dBFS. A compression ratio of 3:1 with a knee and set to an average of −3 dB gain reduction, with only momentary instances of −6 dB of gain reduction, may help the mix engineer maintain a target level. If the gain reduction becomes too extreme, the dialogue clarity may suffer. In this case, the channel trim or clip gain needs to be adjusted word by word so that the compressor does not overfire.

6.8.10. Vocal and Soloist Compression

In rock and pop productions, the singer is typically closed miked, and the performer may have a wide dynamic range. The vocal part typically requires a steady presence in the mix so that words or parts of words do not get lost. To steady the vocal level, using a fair amount of compression may be necessary. Since the softer parts may be less dynamic than the loud parts, a knee function is often used to contain loud bursts but not over-compress the midlevel dynamics of the performance. Following a knee compression curve progressively increases the compression ratio as the input level rises, thus smoothly transitioning from compression toward a limiting function and back again. If a knee compression function is not available, two compressors in series can be used with the second one set to a higher threshold and a more aggressive compression curve.

6.8.11. Musical Instrument Dynamic Processing

Close-miked instruments usually benefit from compression during the mixing process. The microphone choice and placement can greatly affect the way the attack, sustain and release portions of the sound are captured. The unique timbre of a sound is defined not only by its spectral content but also by its dynamic envelope—how the sound changes over time. Close miking instruments can result in an exaggerated dynamic range. To restore the instrument to its natural sound, the way we would hear it acoustically in the room, some compression must be applied to the signal. If the compressor attack time is fast, the transient portion of sound can be diminished. The release time must be set so that the compressor recovers in time for the next note. Applying compression can lengthen the perceived sustain portion as well. For example, a bass guitar recording can benefit from a 5:1 ratio compression with 3–5 dB of gain reduction to enhance its sustain. A snare drum sound can be enhanced using an upward expander followed by a compressor and gate. A close-miked piano, cello or guitar may benefit from light dynamic processing to restore its natural dynamic range.

6.9.   Mix Balance

6.9.1. Balancing Overview

Balancing the level of individual sound sources against one another is perhaps the most important job of the mix engineer. When layering multiple sound sources, practical and aesthetic qualities need to be addressed, and creative solutions need to be found. Mixing may involve prioritizing certain sounds over others because the listener can only pay attention to a few sounds at once before fatigue sets in. If important sounds, such as a voice or a melodic instrument, are obstructed by other less important sounds in the mix, the listener can become fatigued trying to comprehend what the artist means to express and may tune out or turn it off.

One mixing strategy is to prioritize just a few sounds at a time during certain sections of the program. This priority can change dynamically as the program progresses. This is indeed the art of mixing. For popular music, the priority may fall on the vocal and the sounds that carry the pulse of the music like the backbeat of the drum part. In dance music, it is the downbeat of the bass drum that is king. On top of these fundamental elements of a mix, other sounds may come and go from the foreground to the background and back again. The bass and the melodic and harmonic instrumentation may need to be audible at all times as well but not prioritized when the vocal is present. If the percussion becomes buried to far down, the listener may become fatigued or disinterested, trying to lock onto the pulse of the music. For example, some Latin music productions have the percussion mixed hotter than just about anything else in the mix. If multiple layers of sound effects or instrumentation are intended to be heard, each sound should be indeed somewhat audible, or else it is perhaps better to remove it altogether if there is a struggle for attention. On the other hand, composite sounds can be created in a mix using two or more sounds intentionally to create the impression of a single complex sound. This is a very powerful sound design and musical mixing concept. For example, the sound of a dragon may consist of the sound of a horse's neigh and slowed-down baby crying. Combining a piano and a glockenspiel can create a new interesting composite keyboard sound. In these two cases, the mix is done so that neither sound takes priority.

6.9.2. Voice Levels

Meticulous word-by-word clip gain or volume automation may be necessary to maintain the voice's presence in the mix. Once the dialogue is leveled out and compression, equalization and effects are set, the other music and effects can be mixed in according to taste. In music, most engineers create the music bed first, then set the vocal in it. An alternate approach is to start with the vocal alone sounding full, then add the instrumentation carefully one by one around it. The latter is a method less traveled but still an interesting approach that I was taught by one of Elvis's producers. In any case, be methodical and create an efficient workflow that works for you.

6.9.3. Music and Effects Levels

For mixed-media production, music and effects levels should be full but not compromise the clarity of the dialogue. Documentary film mixes were notorious for very low music and effect levels. Often, music and effect levels are experimented with several times during the postproduction process before a mix is approved. The perceived balance has a habit of perceptually shifting from the mixing session to the test screening although nothing technically has changed other than one's perception. A compromise may need to be made to get to the finish line. For example, in a 5.1 surround mix, when the dialogue is active and music and sound FX are in the background, a gap as much as 10 dB between the average dialogue and the M&E track levels may be needed to keep the dialogue out in front. The key is to follow your intuition, and if it doesn't sound right, it's not right, and the mix needs more work. Again, using a reference mix to get the music and effects levels with respect to the dialogue in place is helpful. Equalization may be the key to final mix adjustments when you are close and cannot seem to get it right.

6.9.4. Mix Automation

Most DAWs, as well as VCA (voltage controlled amplifier) and DCA (digitally controlled amplifier) equipped consoles, have a mix automation system on board. Automation systems can capture a real-time mix pass and allow you to edit existing mix automation. On DAWs, automation can be drawn in as well and is not limited to fader positions alone. Automation features extend to most plugins as well. Automating a compressor, equalizer, delay or reverb can solve a technical problem or be used for a creative effect. Control surfaces with banks of faders give DAW users access to multiple controls at once. Recording an active mix pass constitutes a recorded performance of its own. It is an efficient and effective way to complete a mix, not to mention a fun way to work. Touch sensitivity and flying faders add an additional degree of automation functionality. Automation modes provide different ways to initially print and later update mixes. They are unique to each console or DAW and should be reviewed with the manual in hand while experimenting. The following are some basic automation modes common to most DAWs and flying fader consoles:

- In *read* mode, the faders follow the printed automation and will not record any new automation when a fader is moved.
- In *write* mode, the fader movements are recorded based on the current status of the fader and will overwrite any existing automation data. This is a useful mode for the first mix pass.
- In *touch* mode, the system is only writing in new automation when the fader is actively engaged with the click of a mouse, a movement, or a touch on a touch-sensitive control surface. When the fader is let go, new automation writing stops, and the system reverts to following the previously recorded automation.
- *Latch* mode is similar to touch mode, but the system will keep writing new automation based on the fader position even after the fader is released.

6.9.5. Mixing Checklist

All balances, panning, equalization, noise reduction, effects and dynamic processing should be checked and revised until the mix balance is correct. You will intuitively know what's right based on your history of working on and listening to other mixes. In summary, the following balance inquiries should be addressed while approving a mix:

- What primary sounds are driving the mix, and are they present enough?
- What sounds are secondary, and are they interfering with the primary sounds?
- Do any words or syllables in the vocal or dialogue tracks need to be attenuated or boosted?
- Is the timbre of each source acceptable?
- Is too much or not enough compression/limiting applied?
- Is too much or not enough noise reduction applied?
- Are any sounds jumping out of the mix in a distracting way?
- Are there any dead spots where a sound can be featured or accented momentarily?
- Are all sounds that need to be audible present?
- Can any sounds be lower or removed to leave more room for the critical ones?
- Have the target levels been met?
- Does the mix play well from start to finish without the sensation that an overall volume adjustment is needed?

6.9.6. Midrange Frequency Band Mix Balance

To see how the mix may translate on another playback system, temporarily engage a first-order high pass filter at 250 Hz and a first-order low pass filter at 4 kHz on the master fader channel and see where the mix lands. Note any severe balance problems and address them with the filters removed. Repeat this process until the mix works well with and without the filters on. Experiment with other cutoff frequencies also to see what happens. Surprisingly, the band limit–based fixes rarely involve compromise. Instead, this exercise greatly improves the full frequency mix as well.

6.9.7. Creating the Final Mix

Finalizing a mix can be a time-consuming process involving reviewing, making notes, correcting and re-reviewing. As the mix nears completion, it is vital to listen to longer portions of the mix without stopping before making corrections. The time code of the mix comment should be noted or a DAW timeline marker dropped, while reviewing complete passes. This way, you can find your way back instantly to make corrections. After each correction cycle, the complete mix must be checked from start to finish uninterrupted until the mix is approved.

6.10.   Mastering

6.10.1.   Mastering Overview

Mastering is the last step in the production chain before a copy is created for replication and broadcast. Although there are specialized mastering rooms and expert mastering engineers to support music producers, film and multimedia producers typically master the mix during the final mix session and then screen the final product in a real or simulated end user listening environment. For higher-budget film productions, the audio may be mixed and mastered in a theater-like large mix room with a projected image.

For compilations of various tracks, as found on CDs, DVDs, Blu-rays or albums, the artist name, track name, year, album and the record company should be notated properly. Silences (if any) are set between tracks. All tracks must start and end clean without technical glitches. For CDs, the track start time and track end time are indexed as well. The level of each track is adjusted so that one track flows smoothly into the next. The equalization of each track is checked independently and also against other tracks for consistency. The stereo image of each track should be checked and matched against other tracks for consistency as well. For acoustic music, the amount of reverberation on each track should be matched against other tracks for consistency. A pass listening for any pops, clicks, dropouts or unwanted noises should be made on headphones. Uninterrupted listening sessions going from track to track through speakers should be conducted for final approval. Once printed, the final master needs to be checked again one last time on headphones for any technical issues before it is sent out for replication.

6.10.2.   Mastering Music for CD

For mastering a single or a collection of music recordings, the program should peak as close to 0 dBFS as possible without distorting. Peaking right at the absolute maximum is not advised because a system down the line may indicate an overload or distort. Even 0.1 dB of headroom is better than none and may prevent clipping on certain machines. Since most music playback systems are not calibrated to a loudness standard as commercial movie theaters are, music tracks are typically mastered as hot as possible so that the track does not sound weak next to other commercially released tracks. That said, universal loudness standards are emerging as web streaming and personal listening devices become the most used technology for music consumers.

6.10.3. dBFS and Loudness Standards for Film and Video

As mentioned, the LUFS (loudness unit full scale) meter takes into account, average, peak and full-scale levels, so it may be the only meter you will need at the mastering stage for film and multimedia projects. Nonetheless, for safety, a traditional peak meter should be observed to make sure that the program is free of clipping. In general, masters made for the film and video industry have nominal levels of about −12 dBFS and much greater dynamic range than music programs mastered for CD or streaming services.

Target film and video loudness levels were set initially by the European Broadcast Union. They chose −23 LUFS as the European standard. Shortly after, −24 LUFS became the standard in the United States and Japan. Although these standards exist, broadcasters and film distribution companies may provide a more specific audio specification for the deliverables. This specification may include required file formats for delivery and target average and peak dBFS levels.

6.10.4. Loudness Standards for Music Streaming

In 2015, an Audio Engineering Society technical council recommended target levels for streaming. For maximum sound quality, −16 LUFS as the maximum and −20 LUFS as the minimum were recommended. With the proliferation of streaming services and personal listening devices that use automatic LU-based loudness compensation, engineers may decide to adhere to reasonable loudness levels, and that's a good thing. If they do not, additional gain changes may be applied to their mix on broadcast or streaming unbeknown to them. Hopefully, the adoption of loudness standards and high-resolution steaming will lead to the creation and consumption of higher-quality audio.

6.11. Summary

In this chapter, a philosophy of mixing, the tools of the trade and mixing techniques were discussed. The stereo and surround mixing techniques put forth in this chapter can be applied to sound design, film sound, installation, theater, multimedia and music projects alike. Although analog systems have been mentioned, the focus was on optimizing the mix workflow using computer-based digital audio workstations. Indeed, professional digital tools are available to just about anyone who wants to get involved with mixing sound, although the most important apparatus is our ears and our ability to imagine what the sound should be. A mix concept should be established before any work commences. Once the target is known, the technology is applied to achieve the goal. Mixing is a fun, performative and creative practice that holds a lifetime of challenges.

Designing Sound for 3D Films

Damian Candusso

7.1. Bridging Sound and Vision

With the success of James Cameron's *Avatar* (2009), 3D film production technology highlighted how filmmakers could push creative possibilities into new levels of audience immersion. Digital 3D films set a new benchmark for contemporary cinemagoers, however, to date, the majority of discussions on 3D film production and the viability of the medium have focused on the imagery. With so much attention directed toward the "look" of 3D cinema, the accompanying soundtrack is often overshadowed. When digital 3D film was first introduced, the new format of imagery did not bring with it a new sound format; instead, the preexisting 5.1 format was the default sound format.

The evolution of film sound production and exhibition has expanded from a monophonic reproduction using a single speaker placed within the orchestra pit, to a single speaker placed behind the center of the screen at the front of the cinema, through to contemporary cinema sound formats utilizing many speakers that surround the audience. The soundtrack of today allows an audience to not only hear sounds; through the inclusion and replay of subharmonic frequencies, the audience can now feel sounds. As Sergi (2001, p. 128) notes, contemporary sound systems are not just louder; they are capable of providing intense sound pressures. These systems also provide a 3D auditory experience that no longer envelops just the image on screen but additionally the entire auditorium.

Prior to 1906–1907, factual information and actuality were the dominant content that formed early filmmaking, with cinema more about "showing" than about providing an engaging narrative to engross the audience. Predating the use of narrative in film, Gunning (1986, p. 64) refers to the early conception of the cinema used for showcasing these films as the "cinema of attractions." The same can also be said of many Hollywood blockbusters, where many films are created purely to immerse the audience into the onscreen action by bombarding their senses (Neale, 2003, p. 47). As cinema continues to evolve, the objective of many films is to increase the immersive experience. Kerins argues that through an increase in cuts per minute in film editing, and the breaking of traditional shot and framing structures—possible because of advancements in technology—a new cinema style is reflected that can also be associated with the emergence of the Digital Sound Cinema (DSC). "The DSC style centres on a strategy of immersion in the filmic environment—audiences are, visually and aurally, literally placed in the middle of the action" (Kerins, 2006, p. 44).

Immersive film sound technologies are being marketed as providing 3D sound; however, some would argue that it is almost too difficult to have true 3D sound recreation with a loudspeaker-based cinema sound system. This is echoed in a patent for an object-based audio system using vector-based amplitude panning.

> Typical channel-based audio distribution systems are also unsuited for 3D video applications because they are incapable of rendering sound accurately in three-dimensional space. These systems are limited by the number and position of speakers and by the fact that psychoacoustic principles are generally ignored. As a result, even the most elaborate sound systems create merely a rough simulation of an acoustic space, which does not approximate a true 3D or multi-dimensional presentation.
>
> (Lemieux, Dressler and Jot, 2013)

Object-based audio allows for individual sounds (or objects) to be independently panned in a theater without being locked to a defined speaker channel. For the argument for 3D sound to be true, there must be complete control over the 360° spatial placement of sounds. Acoustics and environment have such an impact on this replication that controlling these variables while using speakers within a cinema space is extremely difficult and ultimately impractical. As every seat in a cinema is in a different location, "it is impossible to predict the position of the listener or of the speakers in any given situation and impossible to compensate for multiple listeners" (Begault, 2000, p. 176). Sound designer David Farmer (2013) and executives from Auro Technologies (Claypool, Van Baelen and Van Daele, n.d., p. 2) also express concern over the limitations of speaker formats.

Prince (2011, p. 187) explains that "cinema marries a compelling presentation of sound and moving images to the depiction of what often are worlds of the imagination. The more perceptually convincing these imaginary worlds can be made to seem, the more virtual and immersive the spaces of story and image become." Although 3D films aim to create a superior immersive experience compared to 2D films, many in the industry are divided on this. In a letter to film critic Roger Ebert, Walter Murch comments on 3D and immersion, stating the importance of suturing an audience through narrative, where the term "suture" refers to the audience "stitching" itself into a film. This can include connecting to characters, identifying with a character's point of view or with worldviews expressed in a film, and then filling in the temporal and spatial gaps between scenes with imagination. Murch explains that:

> 3D films remind the audience that they are in a certain "perspective" relationship to the image. It is almost a Brechtian trick. Whereas if the film story has really gripped an audience they are "in" the picture in a kind of dreamlike "spaceless" space. So, a good story will give you more dimensionality than you can ever cope with.
>
> (Ebert, 2011)

Although the term "suture" can be used to describe this seamless integration between film narrative and spectator as described by Murch for 2D film, this relationship is dependent on the strengths of the narrative. The term "immersion" is also extremely important as it increases the participation and absorption with the narrative through our senses. This is an extension of suture as it represents far more than an emotional connection, but it is also a sensory connection or absorption. Cinematographer Oliver Stapleton shares Murch's view, suggesting that 3D reminds an audience that they are watching a screen, thus preventing emotional involvement. Stapleton suggests that "3D is antithetical to storytelling, where immersion in character is the goal" and that 3D cinema is not a true representation of real life (Dowd, 2012). Although film does not necessarily always aim to reflect reality, it often aims to draw the audience into the narrative and characters.

Film technology companies are continually developing products that increase and reinforce realism. This includes the continual pursuit of increasing the resolution of images during the capture and projection of the film. Contrary to Murch and Stapleton's views, Mendiburu argues that digital 3D is recognized as advancing this progression toward stronger realism, at least through sensory means if not additionally through narrative.

> Because 3D is our natural way of seeing, it brings a feeling of realism to the audience. With 3D, we no longer have to rebuild the volume of objects in the scene we are looking at, because we

get them directly from our visual system. By reducing the effort
involved in the suspension of disbelief, we significantly increase
the immersion experience.

(Mendiburu, 2009, p. 3)

The cinema is one of America's favorite pastimes, and although the theater
of the future may be physically different, "as long as our great filmmakers keep making movies that are emotionally entertaining, audiences will
flock to them" (*Film Journal International*, 2012). Randy Thom argues
that the story should be the focus of 3D filmmaking, not the technology.
Thom notes that his biggest concern regarding industry conversation about
3D sound "is that once again it is focusing on technology rather than art"
(Thom, 2012). This is echoed by sound designer Erik Aadahl, who points
out that going to the movies is about connecting with characters and being
sutured into the story. Aadahl states that "on the most basic level you want
to create an experience where the viewer is pulled into the moment and
experiences everything that the character is experiencing" (Pennington
and Giardina, 2013, p. 183).

Pennington and Giardina (2013, p. 39) advise that "filmmakers must
maintain focus and courage because such resistance can't be allowed to
thwart the potential of a technology that enables a richer and more compelling motion picture art form." Technology is providing many new opportunities for contemporary filmmakers. Although a good narrative has the
ability to suture the audience, technical abnormalities create dislocation
between the audience and the film. The balance between technology, narrative and working methodologies determines the level of dislocation
between the soundtrack and the 3D imagery.

7.2. Industry

With firsthand experience working on the new resurgence of digital 3D films,
the medium has provided many challenges, notably how the soundtrack is
replayed alongside the 3D imagery. There is a distinct and inherent difference in the cohesive relationship of the image and the soundtrack of
3D films compared to 2D films. An audience without knowledge of film
sound is unable to notice or articulate many of the details provided by
complex contemporary soundtracks. It is not through ignorance that they
are unaware of such subtleties of the soundtrack; instead, it is from inexpe-
rience in the intricate details of soundtrack presentation and not having the
privilege to distinguish subtle sonic nuances on a regular basis. However,
just because an audience is unaware of the limitations, it doesn't make it
acceptable to settle on a format or product that could be better.

The introduction of digital 3D films required an upgrade to the contemporary cinema infrastructure. It was necessary to upgrade the projector, and some cinemas upgraded the screen. However, there were no initial changes to the existing Digital Surround (5.1) format. Even the highly awarded James Cameron 3D blockbuster *Avatar* (2009) was theatrically released with a Dolby Digital 5.1 soundtrack. During the following year, a change in sound for 3D did begin to take place with the release of *Toy Story 3* (Unkrich, 2010), introducing audiences to the 7.1 Dolby Surround format. Increasing the speaker channels from six (5.1) to eight (7.1), Dolby Surround 7.1 offers both the creators and the audience an increase of an additional two surround channels as illustrated in Figure 7.1.

Transitioning to digital film has come an opportunity to find the most efficient way to store as much data as possible within the confines of the medium. Since digital cinema was introduced, the resolution of the cinema projection has doubled, increasing from 2K (2048 × 1080) pixels of resolution to 4K (4096 × 2160) and beyond, and we have also seen the frame rate of the images for certain films double. *The Hobbit: An Unexpected Journey* (Jackson, 2012) was the first 3D release to be filmed and projected at the high frame rate (HFR) of 48 frames per second in addition to having a Dolby Atmos release. In a statement to the *Hollywood Reporter*, Jackson states:

> I strive to make movies that allow the audience to participate in the events onscreen, rather than just watch them unfold. Wonderful technology is now available to support this goal: high frame rates, 3D, and now the stunning Dolby Atmos system.
>
> (Giardina, 2012)

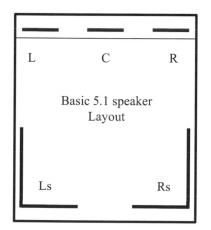 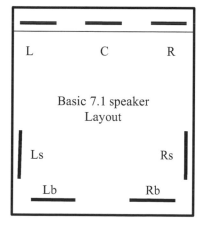

Figure 7.1   Speaker Layouts for 5.1 and 7.1 Formats.

Dolby Atmos provides a new sound exhibition format for 3D as it allows a 9.1 bed plus up to 118 audio objects outputting to 64 speaker channels. Dolby Atmos draws upon working methodologies used in both 5.1 and 7.1 surround sound by using a hybrid "object-based" panning model. In addition to the 9.1 bed, individual objects within the soundtrack can be isolated and independently positioned around the cinema, including overhead in the ceiling with far greater directional accuracy than previous surround systems. The system differs from traditional sound formats where the mixing of sounds, including their speaker channel placement, is rendered during the mixing phase of production. Object-based audio in Dolby Atmos, however, contains metadata that is used to calculate the panning during exhibition, based on each individual cinema configuration.

Although Dolby Atmos has a strong foothold in immersive cinema sound, Auro Technologies now owned by Barco, a name synonymous with cinema projection, provides an alternative immersive sound format. The Auro 11.1 configuration is not object based, instead building upon the existing 5.1 format with an additional height and an additional ceiling layer, providing a three-tiered format. In 2013, Barco, Auro Technologies and audio technology developer DTS made the announcement that they would join forces to support an open standard for immersive object-based cinema sound, recommending DTS's object-based MDA (Multi-Dimensional Audio) as their preferred format. As of 2018, there are 3,600 Dolby Atmos installations (Bowling, 2018) and 450 Auro installations (Manser, 2018) worldwide. Compared to the 150,000-plus digital screens, this highlights the disproportional gap in investment between the image and sound for 3D films.

MDA uses object-based panning within a space, not sounds derived from speaker channels. The reason for this is that it doesn't matter what format the film is mixed in or what format the cinema has installed; the open format algorithms allow the sound data to be decoded and adopted by whatever system is installed within the cinema, thus translating the position of a sound and not tying the sound to a specific channel (Lambert, 2014). In early 2015, DTS unveiled DTS:X, its open-platform, next-generation, immersive, multidimensional audio technology (DTS, 2015). Auro will also transition from their 11.1 format to their new MDA-based platform Auro Max format, in the very near future (Manser, 2018).

The original Digital Cinema Package (DCP) specification allowed for up to 16 independent channels of uncompressed audio. However, in March 2014, the Society of Motion Picture and Television Engineers (SMPTE) expanded the standard to allow for additional tracks due to the emerging immersive audio formats (Society of Motion Picture and Television Engineers®, 2014, p. 6). As the number of audio channels increases, so too does the cost in speaker infrastructure. Although formats with additional channels beyond 7.1 exist, many film sound editors continue to monitor their work in the editing and design phase using the 5.1 or 7.1 format. One of the advantages of the Dolby Atmos format, however, is the ability to use the Dolby Atmos workflow and to have either a software or hardware renderer downmix into the required format for editing and monitoring prior to the mixing stage. This allows increased creativity and reduced costs compared to all spatial panning having to be done on the mixing stage.

Undoubtedly, the biggest influences on Western cinema are the films produced in the United States and in particular Hollywood. With the North American Box Office intake being US$11.4 billion in 2016 (MPAA, 2016, p. 6), the North American film industry is the largest in the world. The introduction of digital 3D films is well-known to have resulted from the desire to increase box office revenue, but with debate over the validity of 3D as a serious format for film narrative, many in the film industry continue to regard the medium as nothing more than a gimmick that demands a premium ticket price.

Contrary to the financial motives of the (re)introduction of 3D film to bring the cinema audience back, several films have proven that strong narrative and filmmaking can be executed through the medium with several 3D films being nominated for Academy Awards® for Best Motion Picture of the Year. A sample of films include:

- *Avatar* (Cameron, 2009)
- *Toy Story 3* (Unkrich, 2010)
- *Hugo* (Scorsese, 2011)
- *Life of Pi* (Lee, 2012)
- *Gravity* (Cuarón, 2013)
- *Mad Max: Fury Road* (Miller, 2015)
- *The Martian* (Scott, 2015)

The film sound industry is unlike many other industries in that there is no single regulating governing body. Instead, it is comprised of many unions, guilds and technical organizations that include:

- Motion Picture Sound Editors (MPSE)
- Cinema Audio Society (CAS)
- The Society of Motion Picture and Television Engineers (SMPTE)
- Audio Engineering Society (AES)
- European Broadcast Union (EBU)
- Moving Picture Expert Group (MPEG)
- Association of Motion Picture Sound (AMPS)

Major studios fund many film productions, and it is often these studios that decide what format the film will be released in. This includes both the imagery and the sound format. Throughout the history of film sound, many of the studios have developed their own proprietary sound technologies. Early cinemas could only exhibit films produced by particular studios as they all used specific formats, necessitating various playback options. Digital film exhibition has become more universal, allowing any cinema to replay any film. The introduction of digital cinema allows the Digital Cinema Package (DCP) to contain all of the relevant formats of the film as digital data. A cinema not equipped to replay a Dolby Atmos format soundtrack, for instance, will have a soundtrack automatically rendered from the immersive version of the mix into a Dolby Digital 5.1 mix. As multiple sound formats require additional time in the mixing stage of the filmmaking process, it is up to the studio to determine whether funding a particular format is worthwhile, especially when it comes to an immersive sound format release.

7.3. Language and Describing Space

There is debate from within industry as to how best approach sound for 3D. This is further complicated with sound for 3D films and 3D sound not being the same. Industry is not so much concentrating on "sound for 3D film," but rather creating "immersive 3D sound." One of the issues contributing to this misconception is a lack of language. Despite the widespread use of surround sound, there is a lack of language able to describe the localization of sounds within the cinema space, particularly along the z-axis.

The volumetric space within a cinema is complicated as both the 3D vision and the surround sound are independently bound to their own conventions and technical limitations. This highlights not only a void in the articulation of spatial sound language but also a void in describing sound that is bound to 3D imagery. Adopting the visual xyz coordinate model and applying it to sound enables an accurate solution. Two-dimensional films do not require accurate sound placement in z-space as the image is on a single plane, unlike 3D films. Bonding sound with 2D imagery along the x-axis is possible, as the placement of onscreen sound is positioned utilizing a combination of the three screen speaker channels. The surround channels primarily support off-screen cues and are not image dependent. The introduction of 3D vision has meant surround channels are no longer used for off-screen sound, but, more importantly, they are now required to support the visuals along the z-axis in z-space.

Increased options in software panning tools have opened up new opportunities for sound designers and re-recording mixers. A hurdle faced working in the new immersive sound formats has been a limitation in tools for sound editors. For some time, vendors have been releasing plugins that allow for binaural panning and panning within a 360° sphere. These were often proprietary or needed ambisonic or binaural decoding to work, and they were therefore unsuitable for cinema application. It has been the introduction of virtual reality that has been the catalyst to push immersive sound tools to everyone. Despite several DAWs allowing multiple channel bussing and outputs, including Cuckos Reaper (Reaper, 2018), it is perhaps the Avid Technologies release of Pro Tools 12.8 with the inclusion of full Dolby Atmos integration (Avid Technology, 2017) that now allows editors to work natively in immersive sound.

Despite enabling industry practitioners to input pan parameters, due to acoustic and speaker format variations between cinemas, this panning is not accurately replicated. During the author's PhD research (Candusso, 2015), it was necessary to clearly define and articulate the positioning of sounds within a 3D space. It became evident that, although a sound can be positioned in a specific location within the 3D space during the editing and mixing process, it is seldom reproduced accurately during exhibition. To identify these differences, the author coined the terms "positional data" and "positional rendering" to differentiate between the location given to a sound during production versus the actual position of the sound during exhibition.

7.4. Acoustic Space

Positional data is the pan location assigned to a sound during the production process.

Limitations to positional data can be a result of the sound team having to work with 2D imagery or not having enough time to pan individual sounds. By not having access to 3D imagery (even if only used for referencing), the sound department is unable to identify the position of objects along the z-axis within z-space. In this context, the z-axis relates to depth and the positioning of sounds off the screen by situating them more in the cinema auditorium (z-space) or, conversely, apparently beyond the screen. As time spent in the final mix is expensive, there is seldom time to adjust panning in the final stages of production. Further compounding the issue, many sound editors and mixers are creating soundtracks for either 5.1 or 7.1 releases, and it is not until after mixing has begun that immersive releases are considered.

The *positional rendering* on playback refers to the ability of a cinema sound system to accurately reproduce a sound to be in exactly the same apparent location as where the sound is located with the positional data. Contemporary 3D films are in either 5.1, 7.1 or immersive sound formats, with the spatial positional rendering accuracy of the soundtrack confined to the configuration of each format. For example, the sound in the left surround channel will be apparently reproduced differently between 5.1, 7.1 and the immersive sound formats. Due to varying speaker placements, there are discrepancies in the spatial positioning of sounds. This highlights some obvious shortcomings in the ability of contemporary speaker systems to provide a homogeneous audiovisual 3D experience.

Two-dimension does not have this problem, as sounds that are placed in the surround speakers are not bonded to the image. For example, a helicopter flying in from behind the audience will use the surround speakers on the helicopter's approach, but it will then play through the front screen speakers once the helicopter is visually on screen. Three-dimension differs as the surround speakers can additionally be used to complement onscreen visuals that are in z-space. Although various techniques and methods are often able to create the illusion of sounds bonded to the imagery within z-space, the spatial precision dissipates due to speaker crosstalk.

Despite 5.1 allowing sound to envelop the viewer, it is difficult to place a sound accurately within z-space. This dislocation is a noticeable concern only with 3D films. Following requests by directors to bring sound into the room similarly to the visuals, Wil Files states, "[I]t's very, very difficult to bring anything into the room, and that's the thing that we often get asked" (2012). Thierry Barbier, 3D producer at AmaK Studio, has observed through his experience "that a 3D movie sound mix is more of a quadrophonic mix, with emphasis on front-to-back effects, rather than a surround mix with stereophonic voices and ambience effects" (Mendiburu, 2009, p. 156).

The immersive sound formats in addition to increased channels, also offer increased frequency response, allowing improvements to the pan resolution. The increased pan resolution of immersive sound formats can also lead to positional inaccuracies being highlighted. The contemporary soundtrack can suit 3D films, depending on the framing of the 3D visuals. This is complicated with 3D, as the soundtrack not only adds to the experience, but in many instances, it also has to support the imagery by bonding to on-screen visuals. David Farmer (2013) echoes the concern about contemporary speaker formats, implying that "sound speakers that aren't customized to each listening position make 3D sound from projection speakers currently near-impossible." Technology vendors also acknowledge the limitation of contemporary speaker formats, with Auro stating that "surround sound combined with overhead speakers alone cannot reproduce the full 3D auditory experience" (Claypool, Van Baelen and Van Daele, n.d., p. 4). In many instances, contemporary sound formats fail to accurately match the position of the image.

While working on the first Australian 3D animated feature film, *Legend of the Guardians: The Owls of Ga'Hoole* (Snyder, 2010), a problem presented itself through the inability to position sound accurately within z-space (Figure 7.2). An object could be panned around the room, from left to right of screen or from the front to the rear of the cinema; however, the sound object could not be located within the center of the room. This meant that sounds could not be attached or positioned accurately with the associated imagery; an issue not encountered on 2-D films. No matter how the various sound effects were designed, and no matter how the objects were panned throughout the 5.1 surround sound space, limitations of the 5.1 format did not allow for the accurate positioning of the sounds along the z-axis.

A specific example is the sound of a sword tip passing slowly in front of the audience members' noses. The sound was dislocated from the

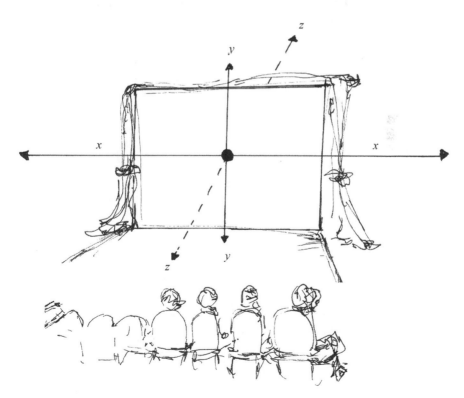

Figure 7.2 z-Axis.

3D imagery showing a blade converged off the screen in strong negative parallax. Parallax identifies where an object appears in 3D space relative to the screen (Figure 7.3). The three parallax states include:

- *Positive Parallax*—Objects appear to be behind the screen, farther away from the viewer.
- *Negative Parallax*—Objects appear to be in front of the screen, closer to the viewer.
- *Zero Parallax System* (ZPS)—Objects appear to be flat with the screen.

The objective was to create the illusion that the sound spatially matched the three-dimensionality and depth of the accompanying imagery. The tip of the blade originated from the neutral parallax (screen plane) before extending into negative parallax as it traveled from the screen into the audience. Problems arose when trying to replicate this with various layers of sounds using the limited parameters of 5.1 panning. Although the base of the blade could be panned as it was anchored on the screen plane,

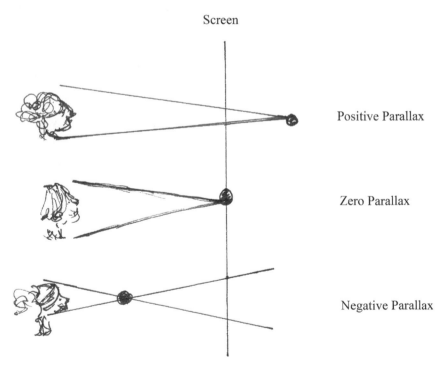

Figure 7.3 Parallax Views.

the tip of the blade in strong negative parallax proved extremely difficult. When the sound was panned from the screen into the surround speakers, it became diffused throughout the theater. To achieve an acceptable outcome, several sound elements were panned independently to create the aural illusion that the overall sound was attached to the blade. This, however, was a huge compromise.

7.5. Layering Sounds for 3D Immersion

Traditionally, sound effects and ambiences are the greatest contributors in occupying the complete speaker array of contemporary surround sound. The atmospheres situate the audience within the onscreen environment, whereas the sound effects are individual sounds that accompany specific actions or are used nondiegetically for narrative purposes. By providing the sense of location as imagined by the onscreen characters, the atmospheres are the primary sound component that contributes to immersing the audience. Kerins describes how, through creating the right multichannel ambient space, the foundation allows audiences to feel they are immersed in the diegetic world (Kerins, 2010, p. 167).

Often recorded on location, the ambiences and atmospheres are generally recorded in at least a two-channel stereo format. Depending on the recordist, atmosphere recordings may be recorded with as many as six tracks (5.1) and more or, in a growing trend, an ambisonic format. Ambisonics is not a new technology, having existed since the 1970s. Using multichannel sound encoding, the format provides the ability to create a planar (2D) or periphonic (3D) sound field. As stated by Rumsey and McCormick (2009, p. 535), the format offers "a complete hierarchical approach to directional sound pickup, storage or transmission and reproduction, which is equally applicable to mono, stereo, horizontal surround sound, or full periphonic reproduction including height information." The four-channel B-Format offers a non-channel-specific format, with any number of decoded sound format possibilities able to be derived including mono, stereo, 5.1 and beyond. This is particularly useful for the recording of atmospheres for films where an ambisonic recording can be decoded directly into the 5.1 format. The advantage of recording in an ambisonic format is the ability to decode to any speaker format. Ambiences and atmospheres are conventionally edited in discrete tracks, either stereo or surround, and then mixed to the relevant surround format used for the film. These sounds play throughout the full range of cinema speakers, providing a constant foundation on which all other sounds are then layered.

Sound effects, on the other hand, are often recorded in mono (or sometimes stereo), then edited and embellished through sound design and panned to follow the onscreen or narrative cues. Through the combination of editing, panning and creative volume mapping of each element, the scene begins to resemble actuality. There is depth and space, and the audience begins to be situated within the scene as these sounds evolve and envelop the cinema. To increase the detail further, these tracks will often have additional sounds multilayered to provide each individual speaker channel greater separation and increased dimensionality. For example, maybe in an exterior city scene, in addition to the general city ambiences, specific background sounds highlight the space. This may include a specific siren playing only through the left rear surround, and a traffic light ticking in the right screen speaker. These sounds are played loud enough not to dominate or draw attention within the scene but merely to distinguish where they are. This separation creates the illusion that the space is wider than if both of these sounds were to be played in monaural together. This methodology translates to immersive technologies. David Acord (2012) from Skywalker Sound was very positive about the effectiveness of creating space through additional speaker channels in the Auro format. He suggests creating a slightly different ambience for the height layer that differs from the lower layer allowing the 11.1 format to work very effectively. This methodology follows the Japanese Broadcast Network a three-layered 22.2 immersive sound format. When describing the use of layers in the 22.2 UHDTV format, it was suggested that specific sounds be assigned to specific layers (Hamasaki et al., 2008, p. 48) as highlighted in Table 7.1.

The use of reverberation is also used widely to provide a sense of space and depth. By adding reverberation to the atmospheres, the space can be opened up or even closed in, creating the illusion of a vast expanse of land or, conversely, the sense of claustrophobia. Reverberation is the lengthening of a sound due to the source sound being reflected and echoed off

Table 7.1   UHDTV Sound Layers.

Upper Layer
- Reverberation and ambience
- Sound localized above, such as loudspeakers hung in gymnasiums and airplanes and at fireworks shows
- Unusual sound such as meaningless sound

Middle Layer
- Forming the basic sound field
- Reproducing envelopment

Lower Layer
- Sounds of water such as the sea, rivers and drops of water
- Sound on the ground in scenes with bird's-eye views

surfaces. In sound production, this process is applied by hardware or software. Sonnenschein (2002, p. 84) advises that when considering reverberation to make a space feel larger, caution should be taken if it is applied to all elements, which may just make it muddy. He also suggests that applying reverberation to a single sound may provide a "vastly more effective expanse."

One of the most contested areas of concern with sound for 3D is in the sound mixing. Since the introduction of surround sound and digital surround sound, new creative opportunities became available. The introduction of mixing and exhibiting in immersive sound formats challenges the notion that any form of surround sound is 3D sound. Immersive formats are being marketed as true 3D sound since they have additional height and/or ceiling speakers when compared to 5.1 or 7.1 surround sound formats. This implies that the traditional surround formats are 2D because there is no height dimensionality as all speakers are located on a single horizontal plane.

If the image is detached from the screen, then the traditional soundtrack emanating from behind the screen has now lost its spatial connection with the image. David Farmer shares this concern. Farmer (2013) argues that an audience is taken away from a film when the sound and image are competing on two separate planes. This is compounded when the dialogue is split across multiple channels. An example is when a character stands over the audience, and the dialogue is split between the ceiling and the center speakers. Farmer recognizes that unless the audience is seated within the "sweet spot," the sound will appear to be biased to whatever speaker is closer to the listener. The *sweet spot* refers to the position at which a listener sits, where the sound quality and sound field (apparent placement of sounds) are at their optimum.

Will Files (2012) suggests that perhaps this is a limitation of the 5.1 format. One of the most common and consistent dislocations that occurs in 3D is the dialogue playing from the center speaker with the character(s) positioned along the z-axis in negative parallax. Knowing that dialogue has traditionally only been mixed to the center channel, it is apparent that despite a film being in 3D, this convention continues. There are some exceptions to films having the dialogues mixed exclusively in the center channel for 5.1 releases, for example, *Gravity* (also with a Dolby Atmos release). Files also suggests that:

> you're much more likely to bring the dialogue off the screen and into the surrounds if you're in 7.1 than if you're in 5.1, because if you bring something off the screen in 7.1, it's only going halfway into the auditorium instead (of) all the way wrapped around the audience.
>
> (2012)

Despite advancements beyond the 5.1 format, it is evident that industry recognizes filmmakers are not taking advantage of immersive sound formats or opportunities. Fairweather (2012, p. 35) argues that sound is an equal partner and can contribute as much meaning as any other cinematic feature, suggesting that the director must work it to his advantage. This is also a concern for Randy Thom, as discussed during his keynote speech at the 2014 *Immersive Sound Conference* in Los Angeles. Thom urged the sound community to encourage directors and screenwriters to "design scenes and moments around immersive sound to take advantage of the sonic cinematic experience" (Giardina, 2014).

A common underlying issue with immersive soundtrack releases is that the decision to release in an immersive format is often considered only in the final stages of production and not throughout the entire production workflow. With an immersive release often not factored into the production process early, these mixes demonstrate many of the shortcomings associated with the 5.1 releases. This is both a time issue and a problem of the immersive formats being unable to have sounds positioned accurately within z-space, despite offering height channels. Directors and producers need to consider immersive sound early in the production process and factor in additional time and funds for immersive versions of their films. Additional time may also need to be factored into the location recordings. Do the dialogues need to be captured differently so that each character can be panned independently? Do the atmospheres need to be recorded in an ambisonic format?

7.6. Does the Contemporary Soundtrack Suit 3D Films?

Since the introduction of digital 3D films, there have been several advancements in sound technologies for cinema, including Dolby Surround 7.1, Dolby Atmos and Auro 11.1. Although the first Auro 11.1 release was Hemingway's 2D film, *Red Tails* (2012), both Dolby technologies were launched with 3D releases including *Toy Story 3* in 7.1 and *Brave* in Dolby Atmos. However, as Andy Nelson (2012) and Ioan Allen state, these formats are not exclusive to 3D (Fuchs, 2012, p. 7). Since the release of *The Hobbit* films in Dolby Atmos, John Neill, head of sound at Park Road Post Production (owned by Peter Jackson), praises the advantages of immersive technologies. Neill (2013) states, "Dolby Atmos, even when used as a 9.1 system, is far better than 7.1, the difference being greater than the leap from 5.1 to 7.1." The release of both Dolby Atmos and Auro 3D has seen the marketing of these immersive formats pitch their respective products as providing 3D sound. Although cinema sound technology has changed, it has not been a direct result of 3D films specifically. Conversely, the same is also true; in many instances, if a film has both a 2D and 3D release, exactly the same soundtrack is used for both versions of the film.

A recurring concern creating the soundtrack for 3D films is that the 3D visuals are rarely seen or played synchronized to the soundtrack until the final mixing days or the final playouts of the film. With the entire sound production never compared with the 3D imagery, it is not surprising that there is often dislocation between the image and the soundtrack. Without working to 3D imagery, the detail and depth of the shots cannot be matched aurally. Although a sound team may be aware that a film will be released in 3D, it is often not until the final mixing stage that the 3D imagery is seen, if at all. Wayne Pashley noted differences between mixing for *The Great Gatsby* (Luhrmann, 2013) and *The Lego Movie* (Lord and Miller, 2014). During *The Great Gatsby*, the sound team had access to 3D imagery that they consistently referenced during the mixing process, but during *The Lego Movie*, the film was not seen in 3D alongside the accompanying soundtrack until the completion of the mixing. This is not uncommon for many 3D films. Brent Burge (2014) states that with 3D films, the problem is that there is no time to experiment with sound. He finds this disappointing as the only real time he sees his work alongside the 3D images is after it is finished.

Sound design elements and the acoustic placement of sounds, especially along the z-axis, become too much of a risk without seeing the imagery. Mixers are cautious of the implications of misaligned sound in exhibition, continuing to work to traditional 2D mixing practices. Keeping the dialogue in the center channel is an example. Restrained panning results in the sounds remaining on the screen plane as opposed to complementing the 3D imagery in z-space. By applying more panning to objects including dialogue and Foley, the relationship between the soundtrack and the 3D image can be improved. Rob Engle, with Sony Pictures Imageworks, backs up this notion, stating that "the typical use of the back channel mostly for ambient sound is outdated. More of the voice tracks should make it to the surround channels to fill up the room space with the actors' lines, along with their images popping out of the screen" (Mendiburu, 2009, p. 156). However, additional panning comes at a cost, as more time is needed to place each sound.

Will Files (2012) states that "in general I think the aesthetic is going more and more towards, in Hollywood at least, bringing things into the room, bringing things off the screen—a lot of directors are wanting to put things in the audience's face and put the audience there with the characters." Erik Aadahl echoes this, explaining that in 3D your eye may be drawn to the foreground, and you may wish to match it sonically (Pennington and Giardina, 2013, p. 183). Industry-leading sound designers Files's and Aadahl's statements suggest that there is a desire from practitioners to bring sound off the screen and into z-space.

To gain full benefit of superior sounding immersive soundtracks, the format must be considered early and incorporated into the entire sound production workflow. If an immersive mix is not considered early on in the sound production process, will it have any added value?

References

Acord, D. (2012). *Personal interview*. Nicasio, CA: Skywalker Sound.

Avid Technology. (2017). *Protocols 12.8 release notes*. Available at: http://avid.force.com/pkb/articles/readme/Pro-Tools-12-8-Release-Notes [Accessed November 2017].

Begault, D. (2000). *3D sound for virtual reality and multimedia*. NASA. Available at: https://ntrs.nasa.gov/archive/nasa/casi.ntrs.nasa.gov/20010044352.pdf [Accessed April 2013].

Bowling, S. (2018). Personal e-mail.

Burge, B. (2014). Personal e-mail.

Cameron, J. (2009). *Avatar*. Motion picture. Twentieth Century Fox Film Corporation.

Candusso, D. (2015). *Dislocations in sound design for 3D films: Sound design and the 3D cinematic experience*. PhD thesis, Australian National University.

Claypool, B., Van Baelen, W. and Van Daele, B. (n.d.). *Auro 11.1 versus object-based sound in 3D*. Technical report. Available at: https://www.barco.com/secureddownloads/cd/MarketingKits/3d-sound/White%20papers/Auro%2011.1_versus_objectbased_sound_in_3D.pdf [Accessed March 2019].

Cuarón, A. (2013). *Gravity*. Motion picture. Warner Bros. Pictures.

Dowd, V. (2012). Has 3D film-making had its day? *BBC News*. Available at: www.bbc.co.uk/news/entertainment-arts-20808920 [Accessed April 2013].

DTS. (n.d.). *Cinema for studios & creators | DTS, DTS professional cinema for studios creatives*. Available at: http://dts.com/professional/cinema-for-studios-creators [Accessed March 2018].

Ebert, R. (2011). Why 3D doesn't work and never will. Case closed. *Robert Ebert's Journal*. Available at: www.rogerebert.com/rogers-journal/why-3d-doesnt-work-and-never-will-case-closed [Accessed April 2013].

Fairweather, E. (2012). Andrey Tarkovsky: The refrain of the sonic fingerprint. In: J. Wierzbicki, ed., *Music, sound and filmmakers: Sonic style in cinema*. New York: Routledge, pp. 32–44.

Farmer, D. (2013). Personal interview. Park Road Post, NZ.

Files, W. (2012). Personal interview. Nicasio, CA: Skywalker Sound.

Film Journal International. (2012). The theatre of the future. *Film Journal International*, 115(6), p. 3.

Fuchs, A. (2012). Soundsational! *Film Journal International*, 115(5), p. 66.

Giardina, C. (2012). 'The Hobbit' to receive Dolby atmos sound mix. *The Hollywood Reporter*. Available at: www.hollywoodreporter.com/news/hobbit-receive-dolby-atmos-sound-382197 [Accessed April 2013].

Giardina, C. (2014). Randy Thom urges directors, screenwriters to 'Create Moments' for immersive sound. *The Hollywood Reporter*. Available at: www.hollywoodreporter.

com/behind-screen/randy-thom-urges-directors-screenwriters-730846 [Accessed September 2014].

Gunning, T. (1986). The cinema of attraction. *Wide Angle*, 3(4).

Hamasaki, K., et al. (2008). A 22.2 multichannel sound system for ultrahigh-definition TV (UHDTV). *SMPTE Motion Imaging Journal*, 117, pp. 40–49. doi:10.5594/J15119.

Hemingway, A. (2012). *Red tails*. Motion picture. Twentieth Century Fox Film Corporation.

Jackson, P. (2012). *The hobbit: An unexpected journey*. Motion picture. Warner Bros.

Kerins, M. (2006). Narration in the cinema of digital sound. *Velvet Light Trap*, 58, pp. 41–54.

Kerins, M. (2010). *Beyond Dolby (stereo): Cinema in the digital sound age*. Bloomington: Indiana University Press.

Lambert, M. (2014). Toward an open-standard surround-sound format. *Motion Picture Editors Guild*. Available at: http://www.content-creators.com/features/Editors Guild-Website/Editor%27sGuild_0314/Images/Editor%27s_Guild_Website_MDA_ Proponents_Demo.pdf [Accessed March 2019].

Lee, A. (2012). *Life of Pi*. Motion picture. Fox 2000 Pictures.

Lemieux, P-A. S., Dressler, R. W. and Jot, J-M. (2013). *Object-based audio system using vector base amplitude panning*, US9197979B2.

Lord, P. and Miller, C. (2014). *The LEGO movie*. Motion picture. Warner Bros.

Luhrmann, B. (2013). *The great Gatsby*. Motion picture. Warner Bros.

Manser, C. (2018). Personal e-mail.

Mendiburu, B. (2009). *3D movie making stereoscopic digital cinema from script to screen*. Amsterdam and Boston: Focal Press, Elsevier.

Miller, G. (2015). *Mad Max: Fury road*. Motion picture. Warner Bros.

MPAA. (2016). *MPAA theatrical market statistics*. Motion picture association of America, p. 31. Available at: www.mpaa.org/wp-content/uploads/2017/03/MPAA-Theatrical-Market-Statistics-2016_Final-1.pdf [Accessed November 2017].

Neale, S. (2003). Hollywood blockbusters: Historical dimensions. In: Stringer, J., ed., *Movie blockbusters*. London and New York: Routledge, pp. 47–60.

Neill, J. (2013). Personal interview. Park Road Post, NZ.

Nelson, A. (2012). Personal interview. Los Angeles: Fox Studios.

Pennington, A. and Giardina, C. (2013). *Exploring 3D: The new grammar of stereoscopic filmmaking*. Burlington, MA: Focal Press.

Prince, S. (2011). *Digital visual effects in cinema: The seduction of reality*. Piscataway, NJ: Rutgers University Press.

REAPER | Audio Production Without Limits. (n.d.). Available at: www.reaper.fm/index. php [Accessed March 2018].

Rumsey, F. and McCormick, T. (2009). *Sound and recording*. Amsterdam and London: Elsevier, Focal Press.

Scorsese, M. (2011). *Hugo*. Motion picture. Paramount Pictures.

Scott, R. (2015). *The martian*. Motion picture. 20th Century Fox.

Sergi, G. (2001). The sonic playground: Hollywood cinema and its listeners. In: M. Stokes and R. Maltby, eds., *Hollywood spectatorship: Changing perceptions of cinema audiences*. London: BFI Pub.

Snyder, Z. (2010). *Legend of the guardians: The owls of Ga'Hoole*. Motion picture. Warner Bros.

Society of Motion Picture and Television Engineers®. (2014). *Report of the study group on immersive audio systems: Cinema B-chain and distribution*. Available at: http://survey.constantcontact.com/survey/a07e95c4mughtlrzbx7/a022dai6xc2djg/questions [Accessed April 2014].

Sonnenschein, D. (2002). *Sound design: The expressive power of music, voice and sound effects in cinema*. Studio City, CA: Michael Wiese Productions.

Thom, R. (2012). *Personal interview*. Nicasio, CA: Skywalker Sound.

Unkrich, L. (2010). *Toy story 3*. Motion picture. Walt Disney Pictures.

Compositional Techniques for Sound Design

Sarah Pickett

8.1. Introduction

A foundation in the principles of music composition can be a great asset to the conceptual sound designer. When the designer is tasked with the job of composing original music used in the production, the importance is obvious. Perhaps less obvious and more complicated is the fact that a composer is an organizer of sound (Cage, 2011, p. 5), and so too is a sound designer an organizer of sound. The line between the disciplines of composing music and composing soundscapes is extraordinarily blurry at times and nonexistent at others. Indeed, the only thing that may separate the two is the title itself. This chapter focuses on applying techniques that are traditionally thought of as residing in the toolbox of the composer and utilizing these tools, both in the process of developing conceptual design ideas and also in the manipulation of content. To that end, an examination of the composition theory treatises of Jean Philippe Rameau, Arnold Schoenberg, James Tenney, Pierre Schaeffer and John Cage yields clues as to how compositional techniques can be applied to the process of conceptual sound design.[1] This is by no means an exhaustive list of music composition theorists, but their popularity and longevity put them high on the list as a good place to start. It bears mentioning that nearly as soon as a treatise is published, it is critiqued and becomes subject to the influence of time as our understanding of the sciences change. Rameau's theories are not all still thought of as accurate to the laws of physics, but his ideas were revolutionary for his time and are somewhat clairvoyant in his assessment of the overtone series and how it relates to a theory of harmony. Although subjective, one could also state that his awareness of theory yielded extraordinary results as far as composition goes. Therefore, rather than think of a treatise as an absolute gathering of facts, this chapter will instead examine them in the context of inspiration.

There was a time when composers were thought of as those who sat down in front of an instrument, staff paper or tablet slate in hand, and wrote down the musical ideas swimming round in their head. This process, shrouded in mystery, set composers apart from the average musician. Certainly the rules of counterpoint and harmony could be learned, but the creative spark, the ability to inhabit character and bring it to life with melody, was a gift. The composers who could communicate with their muse, connecting to the divine, became elevated to godlike status. Indeed, the origins of "the musical genius" are inextricably linked to the origins of musical expression (Lowinsky, 1964, p. 340). In the world of classical music and opera, in addition to writing the music, the composer is also the orchestrator. Decisions that shape the color and texture of the music make the composer the ultimate sonic sculptor. In the genre of musical theater, the composer could simply be the person who comes up with the melodic ideas; the harmony and orchestrations are left to someone else (Suskin, 2009, p. 176). When working on the conceptual sound design for a play, the lines often blur between composer and sound designer. Perhaps we can look to the advice given to a young composer wrestling with a piece of dramatic text, this could also be the starting point of the sound design process. For the text is of the utmost importance and without a thorough understanding of the storytelling, we will lose our way.

> If a young artist asks how to write an effective opera, we can answer only: read the poem, concentrate on it with all the power of your spirit, enter with all the might of your fancy into all phases of the action. You live in its personages; you yourself are the tyrant, the hero, the beloved; you feel the pain and the raptures of love, the shame, the fear, the horror, yes, Death's nameless agony, the transfiguration of blissful joy. You rage, you storm, you hope, you despair; your blood flows through the veins, your pulse beats more violently. In the fire of enthusiasm that inflames your heart, tones, melodies, harmonies ignite, and the poem pours out of your soul in the wonderful language of music.
>
> (Hoffman cited in Lowinsky, 1964, p. 324)

In our present-day context, this may seem a bit "over-the-top," but there is still merit in this approach. Having a connection to the material is key. We may take a more analytical approach and delve into abundant research in order to access the design but must always return to the text; whether it be a poem, a screenplay, a play script or the title of our sound installation piece. Bringing a deep passion to the process and the ability to embody character, "you storm" and what a mighty storm you could create. A mind enlivened by creative thought, connected to the text will be one that brings so much more to the design process. The soundscape will "come alive" and reach beyond what is simply a recording of an event and transcend its individual elements. Imagine a *Tempest* that reaches beyond the sounds of nature and captures the passion of a jealous heart. Inspiration, however, is only part of the picture, and technique comes from practicing the skills of the craft.

Examples of music composition exist on a spectrum. The spectrum is wide and can include "note-for-note" notation for live musicians, sequencing for virtual or electronic instruments, plundering and reconstructive sampling, *musique concrète*, aleatory and algorithmic music or any combination of these. The role of the conceptual sound designer for a theatrical production of a play may include all of these styles of music composition in addition to the job of creating the soundscape of the world of the production. James Tenney (2015, p. 153) writes, "In the light of the changes that have taken place in music since 1900, it is evident that any sound is potentially 'musical'—that is, any sound may function as an element in the musical fabric and in a way that is structurally equivalent to any other sound." Tenney was interested in creating music out of everyday sounds. Although sound designers may not have been Tenney's originally intended audience, we can utilize some of his ideas in the process of creating a conceptual design. Sound designers can utilize musical techniques, infusing notions of rhythm, melody and harmony into everyday sounds to bring a soundscape to life. We will begin with wielding one of the fundamental building blocks—rhythm.

8.2.   Rhythm

Rhythm cannot exist without silence[2]. The duration of a sound is defined by its relationship to the silence that comes before and after it. When something is defined in length by either the silence surrounding it, or that same sound repeated, or the next occurring sound, we get rhythm. This is true whether we are dealing with a note that is written with reference to its intended duration during performance or as an independent sound object within a recorded medium. As the birds sing outside my window, I am intrigued by the spaces in their song. When the refrigerator hum stops in order to begin another cycle, it creates a stillness in the room, a stillness now punctuated by the birdsong. A car drives by.

Rhythm is one of the most powerful tools in the sound designer's toolkit. It is deeply connected with the physical manifestations of emotion within the human body—the beating heart. We have an intrinsic connection with the fluctuation of our own body tempo and how our emotions manifest in this context. Establishing a steady rhythm and then gradually increasing the tempo is a certain way to tap into our flight-or-fight instincts. Conversely, a steady rhythm with variations to keep us interested will give a sense of feelings of elation and celebration. For example, if you were to count a simple pattern of 1 2 3 4, 1 and 2 and 3 and 4 and, 1 2 3 4, 1 and 2 and 3 and 4 and, you could feel the steady four quarter notes and the intensity shift created by the simple subdivision into eighth notes. (figure 8.1)

By returning to our quarter-note figure, we return to what is known. Imagine the same technique applied to a very slow pulsing mechanical drone, as an argument between two characters in a scene continues, the machine's tempo gradually increases; maybe it doubles its oscillating beat pattern to increase the tension; perhaps there is an additional element of some clanking that skips a beat; it is syncopated to give a feeling of uneasiness or something unknown. In addition, the naturally occurring rhythms in the recorded material become another means of expression. By editing found sound objects and applying rhythmical patterns, we can alter what was—into something new. The heating system kicks on with a loud clang, followed by a whirling oscillation and returns to its steady beat, and in our scene this is reflected in the action of the scene—perhaps the characters have come to an understanding.

In addition to creating rhythmic patterns to support a scene, we can extrapolate from the written text of the script some linguistic rhythmical patterns that may help us to develop our conceptual design. Any time signature can be broken down into combinations of 2s and 3s. Take an odd meter time signature like 7/8 where the number "8" tells us that eighth notes get one beat, and the number "7" tells us that there will be seven eighth notes in a measure. To the beginning music student, this may be a difficult time signature to comprehend. By understanding that we can count it simply with 2s and 3s, we can unlock its secrets. We can count the phrasing in the 7/8 time signature in various ways 1 2 3, 1 2, 1 2 or 1 2, 1 2, 1 2 3 or even 1 2, 1 2 3, 1 2. (figure 8.2)

By embracing this notion, it becomes much easier to translate text into rhythms. For example, one of the characters in the scene may exclaim, "You are the best man I know!" This could be interpreted as a 1 2, 1 2 3, 1 2 pattern. Once we've done the translation, then we can utilize it to create soundscapes. These soundscapes can now be interwoven with the written material in the form of transition material, underscoring or used to

Figure 8.1 Quarter Notes Followed by Subdivision into Eighth Notes—Perceived Increase in Tempo and Intensity.

develop a rhythmic motif that recurs in multiple variations. We will return to the power of the motif and variations, but for now let's riff on our earlier example of the bird and car. This 7/8 pattern that was translated from the text could be used to create a sound object sequence that went something like this—"tweet-tweet, car goes by, tweet-tweet." (figure 8.3)

This is one possible technique that embraces the limitations of rhythmic notation. In 1911 Schoenberg wrote:

> We measure time to make it conform to ourselves, to give it boundaries. We can transmit or portray only that which has boundaries [. . .]. The productive mind carries the search also into the province of rhythm and labors to portray what nature, one's nature, reveals as prototypes. Hence, our metrical subdivisions with their primitive imitation of nature, with their simple methods of counting, have long been incapable of satisfying our rhythmic needs. Our imagination disregards the bar line, by displacing accents, by juxtaposing different meters, and the like. Yet, a composer can still not give a performable picture of the rhythms they actually have in mind. Here, too, the future will bring something different.
>
> (Schoenberg, 1978, pp. 204–205)

The future has given the sound designer the digital audio workstation (DAW). With this tool we can create, reproduce, replicate, re-edit, reorder and replace rhythms that exist independently of the need to express ideas in musical notation. Why then the exercise of translating rhythmical sound to notation? Because it is another way to wield the material, another way to approach the sound that may not be immediately evident. It also allows us to play around with rhythmic motif and variation before we return to the land of the DAW. It allows for a *musicalization* of the sonic material, to create something which is both soundscape and musical idea. In

Figure 8.2   Counting Patterns for 7/8.

Figure 8.3   Text to Rhythm Example.

addition, if we are ever working with performers who will recreate our materials in a live context, then we must have a way to communicate these ideas. A great shorthand is the utilization of music notation, whether it be traditional or experimental. It is a common language and one that can be quickly embodied and interpreted.

8.3. Creating Motifs and Melodies

The *Oxford Dictionary of Music* (Kennedy, Kennedy and Rutherford-Johnson, 2013) defines a *motif* (Fr. motif, Eng. motive) as "The shortest intelligible and self-existent melodic or rhythmic figure." Two of the most famous examples of the motif being the first four notes of Beethoven's *Fifth* and the John Williams's two-note motif from the film *Jaws*. We can build melody and phrases out of several motifs put together or multiple variations of the same motif. A helpful methodology in developing a melody is first to define its structural tones and then work from there as your basis. A structural tone is part of the melody that is absolutely necessary in identifying the melody—the tones without which the melody is not clearly recognizable. Imagine that you are having a conversation with a friend and you are referring to a tune. What are the notes that you must sing or play in order for the other person in the conversation to understand which tune you are referring to? Let's take the tune *Frère Jacques* as an example. In this example, the nonstructural tones have *x*-note-heads. (figure 8.4)

Often the structural tones are based on a triad or on the most stable tones of the scale—the 1, 3, 4 and 5. In this case, the triad of G major utilizing the 1, 3 and 5 form the structural tones of the melody. Even if the only notes you sang or played from this tune were the structural tones (the notes with the normal note-heads), it is likely that the tune could be recognized. From here, once you establish your structural tones, you can flesh out your melody with various other notes. These could be approaching notes, passing tones, repeated notes, elongated notes and so on. Variations to your melody can be accomplished by varying the notes between

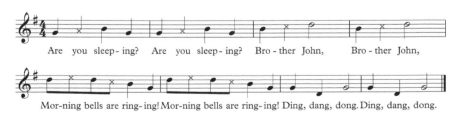

Figure 8.4   Structural Tones of Melody Indicated by Standard Note-Heads.

the structural tones. We can apply this technique to developing variation within a soundscape too.

The next step in transferring this idea to the development of a conceptual sound design is the substitution of the musical note-based motif for a more broadly defined "sounding object motif." This idea is a fusion of Schaeffer's sound object and Tenney's perceived (2015, p. 87) *clang*, which he defines "as a sound or sound-configuration that is perceived as a primary musical unit or aural gestalt." Schaeffer (2017) defines the concept of sound object (*objet sonore*) by what it is not. The sound object is not the source, nor is it the recording material, nor is it defined by a time stamp, nor is it a state of mind—it is both the sound itself as it occurs and the listener's relationship to the acoustic event. This fusion of ideas defines the "sounding object motif," then, as a short figure of sonic objects that together make up a recognizable whole. Nestled in this definition is the implication of the participation of both the creator and the listener.

Before developing a "sounding object motif," it is necessary to define the design objective. Is the design element's primary function to confirm an object out of view of the audience? Is it to create a sense of foreboding? Is it to give us a clue to the internal mind of a character? If the design is meant to establish a soundscape to help the audience understand location—then it must have certain elements contained within. Let us say that we are creating a soundscape that describes an offstage effect giving the sense of proximity to a large body of water such as the ocean. What are the *structural tones*? What are the elements that we need to tell the story, the ones that are absolutely essential for the audience to hear our "soundscape melody"? Perhaps the surf as it hits the rocks, wind, some seagulls. These elements, the "structural sounding objects," will create the soundscape. From here we can begin to establish a rhythm with the *structural tones*—the wave hits once, then again, a gust of wind, another wave and a bird flies overhead while squawking. We add another element—a large boat with a steam whistle—it is a *passing tone* between the waves and the wind, and now it changes the rhythm of the waves as they hit the shore.

8.4. Variation

The ear and brain are very clever when it comes to recognizing repetition. Think about how long it would take for you to identify a one-measure loop, for example. Composers can use repetition to give an audience a sense of familiarity with the melodic line and then use variation techniques to keep the audience engaged because they don't quite know where the piece is going next. In addition, after wandering out in the wilderness, it is always great to return home to the known. Variation techniques can be applied to transform the very short "sounding object motif" into the longer "melodic soundscape." These techniques provide a way to stretch out an idea, to wield the material, shape it and give it variation and interest. Variation techniques can relate to modifications in rhythm, pitch or a combination of both.

Rhythmical *displacement* makes changes to when the pattern begins; this is especially useful if there are implied strong downbeats. Elements that were on a strong beat are now displaced to a weaker beat or reside between beats. *Augmentation* is simply an extension of the length of the sounding object. Conversely, *diminution* reduces the length. Augmentation and diminution, in differing degrees of time and variation along with displacement and repetition, can add rhythmical variation to the soundscape even if no other new elements are added. *Inflection* adds or subtracts an accidental, thereby raising or lowering the pitch by a semitone. *Inversion* corresponds to an analysis of the interval and subsequently inverts it. All intervals add up to the value nine, so inversion is simply a mathematical process. A simple illustration of this is that a perfect fourth will invert to a perfect fifth. (figure 8.5) The technique of *retrograde* puts the elements in reverse order. This could be done rhythmically, tonally or in a combination of the two. (figure 8.6)

When applying these variation techniques to a "sounding object motif," one is not limited to just variations in pitch and rhythm. In 1969, Tenney wrote of developments in music and "a new importance of parameters other than pitch (and time) in determining shape-relations between the *clangs* in a sequence." Taking Tenney's observation as inspiration, we can then apply these compositional techniques to other parameters. For example, one could choose a single fundamental element of the sounding object such as the volume envelope. Is the attack quick and sharp or blurring and long? Does the object have a long release time, or is it abrupt?

Figure 8.5   Intervals and Inversion Examples.

Figure 8.6   Examples of Variation Through Retrograde Technique.

Define these as the structural elements of the sounding object. From here, begin to change one of these elements at a time. Analyze the frequency spectrum of a sound object; through a program like *Spear* you change a single frequency in the object just as you would change the note in a piece of music. Apply the techniques of inversion or diminution to the sonic object.

8.5.   Designing From the Tonic

Next up on the list of core compositional techniques is *establishing the tonic*. Even Arnold Schoenberg, an avid proponent of atonal music, admits to the necessity of understanding the tonic. Before his works on the 12-tone method, earlier in his career, he wrote:

> Tonality is a formal possibility that emerges from the nature of the tonal material, a possibility of attaining a certain completeness or closure by means of a certain uniformity. To realize this possibility it is necessary to use in the course of a piece only those sounds and successions of sounds, and these only in a suitable arrangement, whose relations to the fundamental tone of the key to the tonic of the piece, can be grasped without difficulty.
>
> (Schoenberg, 1978, p. 27)

When developing a melody in a certain mode, it is not solely the key signature that defines the mode but also the tonic in relation to the key signature. For example, if we have a key signature with no flats or sharps and we continue to return to C, then we are likely in the key of C major—the Ionian mode. If we continually return to A, then we are most likely in the natural minor or the Aeolian mode. This returning to the tonic establishes our tonal center—it defines our relationship to the tones that we are hearing. Our emotional relationship to the mode is deeply embedded both within the cultural context and the within the laws of physics. Extrapolating from Pythagoras and Descartes, Rameau in his *Treatise on Harmony* (1772) defines the octave and the fifth as the most harmonious intervals in all of music. They are so because they exist within the sounding of the fundamental pitch. He further points out the importance of returning to the fundamental while using the dissonance of the previous leading tone to finish the piece with finality.

> The note which completes the perfect cadence is called the tonic note, for it is with this note that we begin and end, and it is within its octave that all modulation is determined. The sound which

precedes the octave and forms all the major dissonances is called the leading tone, because we never hear one of these major dissonances without feeling that either the tonic note or its octave should follow immediately. This name is thus eminently suitable for the sound which leads us to that sound which is the center of all modulation.

(Rameau, 1772, ch. 5)

The technique of *establishing the tonic* and the subsequent use of the *leading tone* could relate largely to the overall conceptual design or, more granularly, in how the tonic exists within an individual sounding object. It could be done with reference to a certain pitch that establishes the tonic; however, in respect to sound design, it could be thought of more broadly. The tonic is sometimes referred to as the *root*. It is the most basic element needed to establish both the mode and the harmony within it. The root is essential to the core of the melodic idea.

When beginning the process of developing a conceptual design for storytelling, we must begin with the text; from the text we can extract clues. These clues might be in the form of literal suggestions for diegetic source material like the cock crowing in Shakespeare's *Hamlet* or an abstract suggestion such as Chekov's enigmatic "string break" in *The Cherry Orchard*. More likely they are to be found in references that the characters make to the quality or color of light, the dust on the road, the hunger or longing for connection, a dark hidden secret from the past, the pace of the city in which they live, a loud regime's oppressive clanging, or the breath of a love lost. Clues for building a tonic exist in all of these examples. Our tonic might be a lonely cricket in Willy Loman's Brooklyn backyard circa 1949: we hear it every time he returns home after a long day on the road—everyone else is asleep—he is alone in his kitchen with his visions—the screen door is open. There are not many crickets as you would find in the country or the affluent suburbs but a single lonely cricket striving against the odds within the Brooklyn night-scape. The tonic could be the element of breath—a horse breathing—or instruments enlivened with breath. It is the tonic to which we return—that grounds us conceptually.

The elements within the tonic could be the sonic texture, timbre, perspective, frequency, space. The tonic could be the reverberation of the museum space—the one where our story begins, where the main character imagines her mind palace. How the piece develops in relation to the tonic over time will influence the form of the piece. James Tenney, in his exploration of forms, declares that the large formal rhetorical model "is most clearly expressed in traditional sonata form, with its exposition, development, and recapitulation and its excursion away from and back to a tonic" (Tenney, 1969, p. 161). In this sense, the tonic is key to establishing form.

It is also essential in establishing harmony and by extension consonance and dissonance. In a very basic way, even just by using pitch as our parameter, let's say we lay down two drone-like elements in a room tone—some kind of suspended tone like the hum of the city filtered through glass and the air handling system. We use these to create a harmonious room tone by having one drone be a sounding object with the pitch C as its fundamental and the other be a perfect fourth or fifth above. Then as the scene shifts and conflict emerges, we could create tension by choosing intervals that are further away in the harmonic series, perhaps the seventh, the leading tone that tries to pull us back to the tonic.

8.6. Modes and Methods

The composer uses modes[3] to orient the listener. The simple relationship between one note and the next in a series of notes—a half step, a whole step, sometimes more, or less in the case of microtonal music—establishes the mode. At its fundamental core, the *mode* is defined by how we experience the movement from one sound to the next over a series of tones moving upward and downward and the subsequent repetition of this idea. In essence this is a pattern based on established intervals. The use of a mode allows us to follow the musical journey, to understand the path we are on and to enjoy the occasional turn in the road when the mode is changed. As an improvisational tool, it is at the core of every jam session. As children we are aware of the power of music to enliven us when we want to dance, to console us when we feel sad or to soothe us to sleep at night. We are aware of our own culture's modes and recognize that some modes are unfamiliar to us. As we listen, the music evokes images of far-off lands, and we are transported. Modes have the ability to represent emotion and by their use in context our cultures as well. They are a powerful sonic device to be sure.

Barry Truax in his writings on soundscape composition states that through the repetition of both unprocessed and manipulated versions of "found sound" over the course of the piece, the listener may "explore harmonic character and their symbolic associations" (Truax, 2002, p. 6). In the same way that modes create relationships that evoke emotional and cultural landscapes, so too can the soundscape utilize relationships and patterns to evoke emotion and a sense of time and place. The sound object can be used for its meaning taken from the original recording and context. In addition, it can also be used to create harmonic content. This harmonic content—with an awareness of tonal modality can add a further layer of intricacy to the sonic imagery. Say we are about to participate in a sound-based improvisational session. How would we define the mode to which

Table 8.1 Sonic Relationships.

Sonic Relationships		
Attack	Fast and sharp	Dull and slow
Duration	Short	Long
Frequency	High	Low
Texture	Smooth	Rough
Color	Bright	Dark
Proximity to source	Close	Far
Space	Reverberant	Anechoic
Movement	Static	Ever changing
Tone	Pure tone	Multiharmonic
Direction	Pointed	Scattered
Iteration	Single	Multiple
Position	Foreground	Background
Material	Organic	Synthetic

our collaborators would contribute? Would it be place dependent—an urban scene, a windswept frozen tundra—or would it be something more abstract—perhaps a single original seed of material manipulated through time stretching, reverberation, multiplication or filtering. What gives our piece cohesion? Having an awareness of the power of establishing a mode can be an asset to developing a cohesive conceptual sound design. All the better if our mode relates back in some way to the original text.

Another method for establishing a mode is to define and develop relationships between and among the sound objects. Looking at the characteristics of sound and deciding where they exist on this spectrum is helpful in developing the soundscape. In the same way a composer chooses specific articulations for a phrase on an instrument, so too can sound designers utilize the characteristics of sound to develop their ideas. (table 8.1)

8.7. Timbre and Texture

The compositional process begins with the generation of the sound itself. We must have some material from which to begin. From the moment we start to think about how the soundscape is heard, we start to formulate ideas about how we will capture, manipulate and build our sonic objects. If our "instrument" is derived from recorded material, the way in which it is recorded, the microphones used, the distance from the source, the movement of the source and the microphone itself are all part of the compositional process. If the desired outcome is to recreate a realistic scene, our approach will be markedly different than if we are creating an abstract world. Every sound we chose will itself begin to define how it is expressed. The listener is paramount in this process as well.

> The object is the object of our listening *alone*, and it is relative to it. We can physically act on the tape, cut into it, change its speed. Only a given listener's act of listening can give us an account of the perceptible result of these manipulations. Coming from a world in which we can intervene, the sound object is nonetheless *entirely contained within our perceptual consciousness*.
>
> (Schaeffer, North and Dack, 2017, p. 67)

In these digitally connected times, we may not even be aware of the origins of some of the materials at our disposal. Yet their properties dictate how they function within the context of a design. If a recording of a thunderclap exists embedded in a rainstorm, our ability to extract the thunder on its own is limited. The thunder is an example of a unique event or *gesture*, and the rain is the generalized aggregate or the *texture* (Schafer, 1994). The rain itself is the amalgamation of all of the tiny gestures of raindrops. Working with *sonic textures* can be an excellent way to connect with other design areas of storytelling. The fabric choices from the costume designer and the texture of the scenery could all be inspiration for the sound design.

The fidelity of the material is also part of our pallet. If we are looking for a certain amount of transparency—the sound "itself"—then the recording should be pristine, particularly if the amplitude will be increased for effect. If it is a less than stellar recording with a great deal of background noise, this will be perceived as part of the sound experience whether or not that was our original intention.

8.8. Graphical Scores: Another Way of Looking at Things

Until this point, we have been looking at standard musical notation and how certain compositional techniques could be applied to the process of conceptual sound design. Since the 1950s, composers have explored *graphical notation* as a means to represent musical thought outside the bounds of traditional music notation schemes. This type of notation could be an excellent tool for the sound designer as these forms are not limited to note-based interpretations. Beyond their sonic manifestations, graphical scores can be so visually compelling that some works, particularly those of Karlheinz Stockhausen and George Crumb, have become equally well-known in their visual format. This way of communicating a time-based medium on a single canvas, something that can give the viewer a sense of the whole in a single instant, can be a powerful tool in communicating sonic ideas. In many ways, the sky's the limit for this type of notation, particularly in regard to the sound designer. Often the creative conversation is between the sound designer and their collaborative visual partners, not an ensemble of musicians who are required to read and interpret these unusual and complex scores. In this way, the sound designer has an advantage over the composer. This graphical notation can be a good way to get across ideas in the initial stages of the design process and a way to brainstorm with visual artists.

Another possible tool for notation is in the use of "sonic mind maps." Traditionally these maps have been used as an assessment tool, a way to collect data on naturally occurring soundscapes. They are drawings of the space as listeners have experienced them. The drawing in Figure 8.7 was made in situ on the Chesapeake Bay in Virginia, USA. The air was humid with a slight breeze, the cyclical sound of waves punctuated by birdsong. The listener's perspective is at the bottom of the page with thick lines

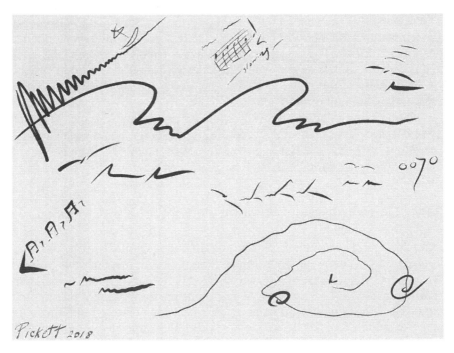

Figure 8.7   Sonic Mind Map of the Chesapeake Bay.

representing greater presence in the sonic field. This drawing is a sort of hybrid between a sonic mind map and a graphical score as it contains both spatial relationships and notation information.

The use of drawings instead of words helps to create a very different representation. The perspective is more clearly defined and the relationship between the listener, reflected surfaces and distance from the source is more evident (Marry, 2011, p. 254). In addition to using a sonic mind map as an assessment tool as part of a *sound walk*, these drawings can be used to create a virtual soundscape in a graphical notation format before it is created sonically. It is a chance to work out ideas on paper that can describe location in the sonic field and perspective of the listener—whether the final format is 7.1, binaural or a designed sound system in an outdoor amphitheater.

8.9.   Conclusion

In 1937 John Cage wrote: "Whereas, in the past, the point of disagreement has been between dissonance and consonance, it will be, in the immediate future, between noise and so-called musical sound" (Cage, 2011, p. 4). Because he was breaking new ground and exploring the idea of the composer as "organizer of sound," he was pushing against the classical context of music, which saw noise outside of its traditional instrumentation. Sound designers use "noise" as one of their primary means of expression. They can play with the rules of harmony, notions of consonance and dissonance in the same way that composers do. We as sound designers are not burdened by the responsibility of labeling our creations with the term "music." And yet, when we work on a project and our collaborators ask for "music," I think we have a common understanding as to what they are asking for and what we can expect to deliver. There is some connection between sound design for storytelling and cultural expectation. At the root of this is our perception of music and sound as it relates to consonance versus dissonance and music versus noise. Because the sound designer must support the story they must be able to understand the theories of the past and the present, the archaic and the modern, the reflecting back and the imagining of future worlds. All things that we create are filtered through these theories; we are as much influenced by Rameau as we are by Cage. Why not embrace all of these possibilities in how we approach and manipulate sound?

Notes

1.   It is my earnest belief that were these theoreticians writing today, they would have used inclusive language. Their writings were meant to share their ideas with all students and

teachers of composition. With that in mind, I have taken the liberty of replacing the pronoun "he" (when it is used to reference the generic student composer/student) with gender-neutral pronouns—"they," "them," etc.

2. Referring to a relative silence, for as Cage (p. 13) has illustrated, there is no true silence; even our bodies make sounds unintentionally that we can only hear when we are in an anechoic chamber.

3. This use of the word "mode" in this context extends beyond the classical church modes and includes everything from the Indian raga, to the Arabic maqam, from the Aeolian mode to the Yo scale.

References

Cage, J. (2011). *Silence: Lectures and writings*. 50th anniversary ed. Foreword by Kyle Gann. Middletown, CT: Wesleyan University Press.

Kennedy, M., Kennedy, J. B. and Rutherford-Johnson, T. (2013). *The Oxford dictionary of music*.6th ed. Oxford: Oxford University Press. Online Edition.

Lowinsky, E. E. (1964). Musical genius—Evolution and origins of a concept. *The Musical Quarterly*, 50(3), pp. 321–340.

Marry, S. (2011). Assessment of urban soundscapes. *Organised Sound*, 16(3), pp. 245–255.

Rameau, J. P. (1772). *A treatise of music*. 1971 ed. [Kindle version]. Translated from the French with an Introduction and Notes by Philip Gossett. New York: Dover Publications.

Schoenberg, A. (1978). *Theory of harmony*. Berkeley: University of California Press.

Schaeffer, P., North, C. and Dack, J. (2017). *Treatise on musical objects: An essay across disciplines*. [Kindle version]. Oakland: University of California Press.

Schafer, R. M. (1994). *Our sonic environment and the soundscape: The tuning of the world*. Rochester, VT: Destiny Book.

Suskin, S. (2009). *The sound of Broadway music: A book of orchestrators and orchestrations*. Cary: Oxford University Press.

Tenney, J. (2015). *From scratch: Writings in music theory*. Urbana: University of Illinois Press.

Truax, B. (2002). Genres and techniques of soundscape composition as developed at Simon Fraser university. *Organized Sound*, 7(1), pp. 5–13. Cambridge: Cambridge University Press.

9

Music Theory for Sound Designers

Adam Melvin and Brian Bridges

9.1. Introduction

One obvious but still important context for sound designers who wish to approach music composition is that musical structures are part of a wider field of (perceptually organized) sound. Frequently, music theory is equated with the particulars of Western music notation and its terminology. However, while such particulars might not be immediately familiar to sound designers, issues around pacing, points of rest, lines, contour and texture can all provide accessible points of departure. Furthermore, the basic structural axes of music (stability/instability and opacity/clarity) find obvious counterparts within sound design's textural dynamics.

Beyond these basic parallels, other aspects of music-theoretical modeling can be considered as details: the particular accents and weightings of rhythms or harmonic progressions as anticipating or prolonging auditory or visual events, and the manner in which the basic archetype of a harmonic timbre is extrapolated into a chord, arpeggio or progression. This chapter will therefore approach music theory from perspectives that may be familiar to (or accessible to) sound designers. It will follow some of the perceptual and organizational principles with which sound designers may be familiar, drawing, in particular, on ideas of musical "motion" that can be found in various interdisciplinary accounts of music (Arnheim, 1974, pp. 372–377; Johnson, 2007, pp. 246–248). The organization of musical events can be thought of as largely concerned with manipulating and reconciling music's horizontal (temporal progression) and vertical (combination of pitches, or harmony) parameters. As will become apparent, the two are often intrinsically linked to the extent that any single-line melody acquires a sense of harmonic structure simply by the order and placement of pitches it employs, while a sense of line is fundamental to the functionality of "block" chord changes.

9.2. Musical Pitch, Scales, Contour and Motion

9.2.1. The Relationship Between Musical Pitch and Sound Design

In his guide to music theory for computer musicians, Hewitt (2008) provides a useful starting point with which to begin our understanding of music theory, identifying three interrelated parameters that make up the most fundamental of building blocks, that of a musical tone: *pitch*, *tone quality* (or *timbre*) and *intensity* (2008, p. 3). What may be immediately apparent to the sound designer is that *sonic objects* might also be considered as comprising these same three components. However, whereas sound design tends to afford, if not equal status, then certainly equal variability to all three parameters, in a musical tone it is largely pitch—or more specifically, the stability of pitch—that is paramount. Stable pitches are clearly discernible rather than those perceptually absorbed into a more complex spectrum of pitches and are largely constant as a result of the uniformity of periodic sound waves. That is not to say, of course, that stable pitch is not present in sound design contexts; simply that it is *pitch stability* that characterizes an aural event as a specifically musical tone rather than a sonic object. This difference is demonstrated in the famous (and much parodied) THX "Deep Note" ident in cinema: a mass of swirling, unstable pitches (sound design) steadily glide and converge toward a stable spread of largely unison pitches (music).

Thus, it is pitch—and its organization—with which the rules of music theory are primarily concerned; in standard Western musical practices, timbre and intensity are governed elsewhere, notably via performance directions.

9.2.2. Musical Pitch, Intervals and Western Scale Structures

In Western music, pitch is organized via a system of evenly distributed pitch divisions, each of which is labeled using the first seven letters of the Latin alphabet (*A–G*). Even for the non–piano player, this can be easily visualized by looking at the white notes of the piano keyboard (Figure 9.1):

Figure 9.1 Pitch Divisions of the Piano Keyboard.

As is evident in Figure 9.1, the cycle of letter names repeats every eight notes. The distance—or *interval*—between two notes of the same letter name is therefore known as an *octave* and corresponds to a ratio between pitch frequencies of 2:1 (or a doubling or halving of pitch frequencies in either direction). For example, a concert *A* (the note to which many will have heard orchestras tune) has a frequency of 440 Hz (440 cycles per second). The *A* one octave above vibrates twice as fast (880 Hz), while the A one octave below vibrates at 220 Hz, and so on. In perceptual terms, this is known as *octave equivalence*: this phenomenon causes us to perceive each note with the same letter name as constituting essentially the same musical note even if its absolute pitch frequency is different. It can be demonstrated by humming the first two notes of Arlen's *Somewhere over the Rainbow*; the two notes of the "some-where" are an octave apart. As Dowling and Harwood (1986, p. 93) note, the vast majority of world cultures treat octaves as basically equivalent in their musical practices. The various intervallic relationships formed between pitch divisions of differing letter names are described in comparable terms: *A* to *G* spans seven pitch divisions (letter names) and is therefore referred to as a seventh, *G* to *E* forms a sixth, etc.

The smallest interval is that between adjacent pitches and is referred to as a second, of which there are two types: the *minor second*, also known as a *half-step* or *semitone*, and the *major second*, also known as a *whole-step* or *whole tone*. Both intervals are used in specific combinations with one another to form linear pitch sequences known as *scales*. The difference between the two is easily discerned by referring to the piano keyboard. Adjacent white keys forming whole tone intervals are separated by one of the piano's five black keys, e.g. *C* and *D*. Each black key divides the interval into two semitones; together with the white keys, they divide the octave evenly into 12 semitones. Where no black key is present between two white keys, e.g. *E* to *F* and *B* to *C*, the interval is also a semitone. Black keys are named using either the term *sharp* (notated #), when referring to the white note it immediately succeeds, or *flat* (notated b) when referring to the note it immediately precedes. Both are correct; the choice of which to use depends on the harmonic context of the music at that particular moment. The black key immediately above *C* is thus known as *C#* (sharp) or alternatively *Db* (flat); the *C#* sounds a semitone higher than *C*, and the *D* sounds a semitone higher than *C#/Db* and so on. By this same rationale, the interval between *C#* and *D#* is a whole tone (two semitones), as is the interval between *E* and *F#* (also two semitones). The prefixes, *minor* and *major* are also used to differentiate larger and smaller versions of some other intervals; e.g. the interval between *G* and *B* (a distance of two whole tones or four semitones) is known as a *major third*, while *G* to *Bb* (spanning a whole tone plus a semitone or three semitones) is called a *minor third*.

The rising two-note motif that characterizes the opening to John Williams's theme from the movie *Jaws* (Spielberg, 1975) is often used to aurally demonstrate the interval of a semitone. The interval's intense, narrow trajectory can be differentiated from the more open, clearly delineated "steps" of the oscillating major second intervals that make up the first line of *Strangers in the Night* (Bert Kaempfert, 1966). The opening of Henry Mancini's *Pink Panther* (1963) theme, on the other hand (Figure 9.2), employs both intervals in the form of two, successive rising semitone gestures ("ba-dum, ba-dum") placed a whole tone apart.

The system of whole tones and semitone materials can also be conceptualized as a grid, sometimes termed a "piano roll" (after early 20th-century player piano technology), which is the dominant format for composing note materials within a standard digital audio workstation (DAW) such as Cubase, Logic, etc. Wishart (1996) referred to this in critical terms as a *lattice* structure, seeing it as limiting musical expressivity. However, the use of such a grid, with vertical axis for pitch divisions and a horizontal axis for rhythmic divisions, can provide an accessible alternative to standard Western musical notation when considering compositional structure.

9.2.3. *Musical Pitch, Intervals and Tonal Attraction*

Like octave equivalence, the functioning of other musical intervals can be ascribed to acoustics in interaction with perceptual and cognitive factors. Each of our equal pitch divisions in Western music is, in fact, an approximation that is taken to be convenient for musicians and instrument designers. The origin of our intervals can be traced to acoustic phenomena, namely the simple (whole-numbered) ratios of lengths of vibrating strings or resonating tubes or the *harmonic* (overtone) *series* of whole-numbered frequency multiples of a common fundamental frequency. However, these acoustically based intervals do not easily reduce to a single, repeated interval, upon which others can be based. So another, smaller "modular interval" (Dowling and Harwood, 1986, p. 4), the aforementioned semitone, is created. This allows us to have scales with regular step sizes, though at the expense of small deviations from acoustically or mathematically "pure" tunings based on the whole-numbered ratios; this is known as *12-tone equal temperament*.

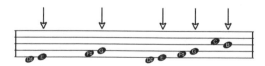

Figure 9.2 *Pink Panther* (Mancini)—Melodic Reduction.

Creating a scale based entirely upon semitones (known as a *chromatic scale*) produces 11 distinct intervals. The lack of difference in size between these intervals will mean that it is not feasible for listeners to "track" their position within the scale based on the presence of neighboring intervals of different sizes, (a scale consisting entirely of whole tones, e.g. *C–D– E–F#–G#–A#–C* suffers from a similar problem). If all note divisions are used to construct a melody or scale, the overall impression will be of a lack of a clear center; no note will be significantly more "attractive" than another in terms or where a melody or scale might lead.

Diatonic scales, on the other hand, employ different patterns of whole tones and semitones, with these patterns allowing for senses of central pitches based on the difference in step size. For example, the major scale rises/descends as tone, tone, semitone, tone, tone, tone, semitone (the white notes between *C* and *C* on the piano), a pattern that can be seen to split the octave in half (Arnheim, 1969, pp. 218–219). The rising semitone between the major third (that note that essentially characterizes a major scale) to the fourth note provides a more incisive intervallic movement than the whole tones that precede it, accelerating the "pull" of the pitch progression away from the "home" note (called the *tonic*). This is mirrored by the same pattern in the second half of the ascent that, from the fifth degree of the scale, essentially pulls the progression back home. This illustrates a basic principle of musical scales and musical melodies: that all pitch categories are generally not created equal. In most musical compositions, a particular pitch, often repeated, will tend to be perceived as more important or significant than others; based on how it is approached, there exists a sense of dynamic imperative that particular notes might tend to require *resolution* to a "central" pitch.

The acoustic origin of most larger musical intervals (i.e. beyond single-scale steps) can be found within the early parts of the harmonic series. The *overtone* frequencies (those above the lowest component, or *fundamental*) can be considered to be related to a series of musical intervals, proceeding with each interval becoming smaller as the series progresses: *C–G–E–B(b)–D*, etc. The earlier intervals of the series are often the most frequently used notes within tonally based compositions.

Leaving aside the octave as an interval that can be treated as equivalent (see Shepard, 1964,) the musical intervals within Western scales may be thought of in distinct groups based on their relative stability/dynamism in a musical context. A psychological study by Krumhansl (1979) presented various intervals to listeners so that judgments about relative similarity/ proximity could be investigated (see Figure 9.3). The structure seemed to give rise to a number of distinct groupings (which can be seen as relatable to the early part of the harmonic series), with the fifth and major third seen as being particularly close, with other stepwise diatonic and chromatic

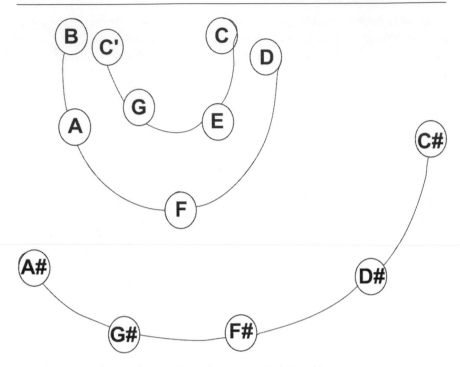

Figure 9.3   Pitch Category and Tonal Proximity Relationships.

Source: After Krumhansl (1979)

divisions being assigned to two other, more distant groupings. This pattern and grouping of materials from the harmonic (overtone) series may therefore contribute an explanation as to how they relate to one another within the context of *functional harmony* (the syntax of note combinations); we use "harmonic" for both cases.

A three-dimensional representation of this structure (Figure 9.4) makes the utility of intervals based on these pitches even more clearly apparent: the conical representation sees *C–E–G* (the notes tracing out a perfect fifth and a major third) as particularly close to a perceptual (tonal) center based on the root note (*C*), with other notes from within a diatonic scale occupying the next distinct grouping or "level," and chromatic intervals (the black notes of *C–C* on a piano keyboard) occupying the last, most distant level. This hierarchical structure is based on a tonal center of *C*; structures based on other tonal centers would have the same types of patterns but be shifted up or down by the interval difference between them and *C* (e.g. a tonal center of *D* would feature *A* (the fifth) and *F#* (the major third) as the closest intervals).

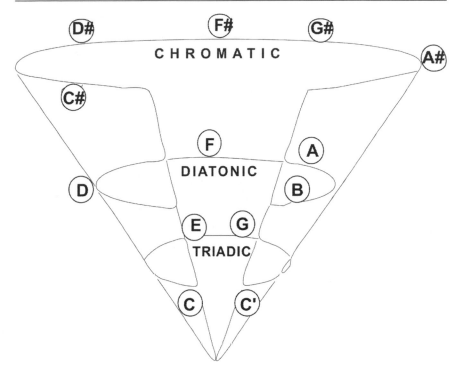

Figure 9.4 Pitch Category and Tonal Hierarchy Relationships in Three-Dimensional Representation.

Source: After Krumhansl (1979) with added functional divisions (triadic, diatonic, chromatic)

The opening at the center of this cone indicates other pitches that are close to the *tonal center* of C in the context of requiring resolution; *B*, for example, is unstable/tonally distant but is adjacent to *C* in stepwise movement and resolves easily. *C#*, as a *chromatic alteration* of *C*, while the same distance in sequential movement, is more distant in terms of the tonal hierarchy. This highlights an important central factor in much Western music composition: that the dynamics of simultaneous combinations of pitch (i.e. the vertical dynamic) may compete with sequential dynamics of scalar (horizontal) melodic motion. Lerdahl (2001, p. 50) examines how contextual factors such as direction of approach may alter the basic spatial structure of a tonal hierarchy, identifying a musical "force" described as being akin to gravitational attraction.

Finally, those with some previous musical experience will note that this structure does not fully account for tonal structures based on minor intervals (such as a minor third, in this case an *Eb* rather than an *E*). For the present, such cases are best conceptualized as a parallel system of sonorities with basically similar structural dynamics, with the shifts in interval sizes resulting in changes in sonorous texture but not in its overall tonal-positional function.

9.3.   Interval, Melody and Motion: Rhythm and Pulse

9.3.1.   Melodic Contour: Pitch in Succession

As we have previously noted, music is dependent on the presence of sound materials with broadly stable, periodic vibrations. However, in contrast to the broadly periodic waveforms that produce them, the linear organization of pitch materials, whether it be a short motivic gesture or a more extended melodic passage, rarely employs such stability. As the composer Charles Ives once remarked, "[W]e like our melody not straight, but blended" (1969, p. 110). Most treatises on linear composition tend to acknowledge the need to employ both conjunct, stepwise pitch progressions along with larger intervallic leaps in order to achieve a sense of contrast and ultimately compositional interest within a musical line. Melodies rarely move in one direction but in waves (Schoenberg, 1967, p. 103) that vary in size and shape, creating contrasting points of melodic/harmonic motion and rest, tension and release. This "pattern of ups and downs" can be referred to as *melodic contour* (Dowling, 1978, p. 341).

In a sense, melodic contour is established the moment two stable pitches are brought together in succession. A single, isolated pitch is a somewhat benign audio event. Once a second pitch is added, a dynamic is immediately established, creating a sense of melodic motion that destabilizes the state of repose of the initial pitch (Schoenberg, 1967, p. 102); we hear the second pitch within the context of the first and vice versa. The exact nature of how we perceive this hierarchy depends on the intervallic distance between the two pitches and the direction of travel, i.e. whether the interval rises or falls.

As more notes are added, the melodic contour becomes more defined, as do the relationships between pitches, forming a clearer sense of structure as to where particular points of instability and stability occur. Within diatonic melodies, the aesthetic quality of melodic contour is informed by the harmonic conventions built upon the overtone series previously mentioned. Music that is less bound by diatonic convention (e.g. music that is heavily based upon chromatic divisions) tends to operate somewhat more organically; "intervals can follow each other in any order and may be arranged to form any pattern of tension interplay" (Persichetti, 1962, p. 15). The absence of a clear diatonic tonal center means that attractional dynamics may be based more upon repetition and localized melodic patterns rather than more prescriptive hierarchical models of tonal relationships.

9.3.2. Melodic Contour and Motion

A fundamental aspect of how contour functions is the aforementioned construction of intervallic relationships, more specifically, the effects that conjunct intervals and larger intervallic leaps have on the perception of musical contours. Several theorists have argued that melodic pitches heard in succession leave a residual trace in the memory (Nikolsky, 2015; Larson, 1997). An intervallic leap (i.e. equal or greater than a third) between two melodic pitches provides a gap between notes that facilitates this phenomenon, allowing the former pitch to perceptually resonate as we experience the second. Although part of the same linear construct, they are perceived as somewhat distinct. In contrast, where there is a stepwise motion between pitches (the interval of a second), the new pitch supplants and effectively erases the tonal trace of the previous one; pitches are bound into one continuous linear stream (Nikolsky, 2015, p. 4).

Linear pitch progressions can therefore be seen to function in a similar way to Klee's free line, i.e. as a single active point/dot taken "on a walk" (1953, p. 16). A pattern of conjunct intervals facilitates a cohesive sense of linear fluency, shifting the proverbial starting point forward in one continuous, self-contained movement (the metaphorical pencil remains in constant contact with the paper)! Although the linear gradient is relatively gentle—the intervallic distances are short—the sense of melodic motion is more horizontally progressive than a pattern employing wider intervals. The common trope found in both cartoons and early (8-bit) gaming of ascending and descending *scalic* runs (i.e. like a scale, involving only 2nds) to isomorphically represent a character's movements, e.g. the use of an arc of conjunct intervals to accompany a character jumping in early platform games such as *Chuckie Egg* (1983) is demonstrative of this linear fluency. By contrast, a fanfare-like melody employing only leaps (of various sizes), such as *The Last Post*, operates more like Klee's linear-active model, i.e. a series of fixed points (Klee, 1953, p. 18) that are navigated between: "rather a series of appointments than a walk" (Klee, 1961, p. 109). Although encompassing a steeper intervallic movement between notes, the sense of horizontal motion is more limited. Contour is discernible yet more disjunct, rather like a join-the-dots rendering of an image.

9.3.3. Musical Motion, Stability and Metaphors of Motion and Force

An intrinsic part of how musical motion is perceived is the aforementioned tonal pitch hierarchy and resultant harmonic reasoning that emerges from each melodic line as it progresses or, in other words, where we perceive points of tonal instability and stability to reside. Both of the preceding examples depend upon identifiable points of (harmonic) rest as a functional aspect of their respective pitch contours. The melodic arc of the jumping trope is anchored by the same "home" pitch at both ends confirming the departure and return of the character on-screen to terra firma; this same single pitch—often rapidly repeated—is typically used to signify the character's movement across the level plane of the platform. The fanfare example, on the other hand, employs only the lower partials of the harmonic series, i.e. those close to the tonal center in Figure 9.4. Because all of the pitches in a fanfare are, comparatively speaking, harmonically stable, the sense of forward (harmonic) motion is more static, capturing a necessarily steadfast quality.

Various authors have sought to describe compositional structures and aspects of music theory using concepts derived from physical, embodied experience. Possibly the most succinct expression of this conceptual framing can be found in Johnson's (2007) *music-as-moving-force* metaphor: that musical progressions can be thought of as based on the apparent movement of (virtual) objects across dimensions of pitch and time. Larson (1997, p. 102) offers further considerations with respect to melodic stability in the form of his own *musical forces* model, which, while specific to diatonic music, can be applied to some extent to all pitch progressions. He identifies three underlying principles of gestural force inherent within a musical line: (1) the tendency for pitches perceived as unstable to descend ("gravity"), (2) the tendency for pitches perceived as unstable pitches to move to the nearest (harmonically) stable pitch ("magnetism") and (3) the tendency for patterns of musical motion, once established, to continue in the same vein ("inertia").

The "Frère Jacques" example cited in the previous chapter by Sarah Pickett provides a simple demonstration of these forces in practice. Note how the unstable notes, i.e. those further away from the tune's tonal center of *G* (notated with an "x"), resolve by step ("magnetism") to more stable pitches initially by ascent, once the rising three-note pattern of the melody is repeated and thus established ("inertia") or, later, more generally downward as the melody begins to gravitate toward the home pitch of *G* ("gravity"). In measures 1 and 2, the ascending stream of conjunct intervals advances the melody away from "home," before an intervallic leap down from the relatively stable third returns it to the root (*G*); Klee's metaphorical pencil is removed from the page, and the stable pitches of the tune's tonal center resonate. Measures 3 and 4 rearticulate the ascending contour, this time between the third and fifth; the tune is still stable but slightly further away from home as the *G* is not present. Measures 5 and 6 revisit the line-and-fall pattern of 1 and 2 but with a greater sense of motion and linear interest; the undulating line exemplifies Schoenberg's wave, propelling the melody back to *G*. The final "Ding, Dang, Dong" gestures simply serve to reinforce the home key; the largest leap of the tune allows its two most stable pitches of *G* and *D* to resonate.

More volatile musical examples can demonstrate the application of these properties within a less tonally stable framework. Rimsky-Korsakov's famous *Flight of the Bumblebee* (Figure 9.5) harnesses the chromatic scale's aforementioned lack of tonal center to create a linear contour that sounds somewhat frantic. An initial leap and subsequent changes of direction to the otherwise semitonal melodic line disrupt the natural "gravity" and "inertia" of the pitch contour, providing structural "corners" rather than moments of rest to the overall trajectory of the music, thus replicating the erratic movements of the piece's insect subject. Meanwhile, Brad Friedel's "Tunnel Chase" cue in *The Terminator* (Cameron, 1984) (see Figure 9.6) achieves an even more unstable effect via a repeated bass synth pattern employing ascending and descending semitonal movement to decorate what is essentially an ascending major-scale arpeggio to such an extent that the stability of the diatonic scaffolding is erased, leaving only sudden leaps and brief linear fragments.

9.3.4. Rhythm and Pulse, Rhythm and Pitch

Thus far, the content of this chapter has been primarily concerned with pitch. However, a significant factor within many of the theories just discussed and one that holds equal importance for the sound designer is that of rhythm.

At its most fundamental, rhythm is simply regulated time (Lefebvre, 2004, p. 9), a phenomenon that, as many writers have argued, is found in abundance in the cyclical patterns of our environment and bodily experience (Hewitt, 2008, p. 40; Lefebvre, 2004, p. 9). Just as we experience the various contrasting rhythms of the external world in the context of our own body's natural, often orderly regulated, internal rhythms—as "superimposed" onto them (Lefebvre, 2004, p. 9)—music, too, incorporates two interrelated yet distinct temporal strata: beat, or pulse, and (in its broader definition) rhythm (Gracyk, 1996, p. 131).

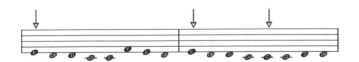

Figure 9.5 *Flight of the Bumblebee* (Rimsky-Korsakov) Reduction.

Figure 9.6 "Tunnel Chase" (Fiedel) Reduction.

In keeping with its shared anatomical meaning, pulse is quite simply the uniform periodic placement of individual beats that provides the basic marking of musical time, the internal "clock" of a particular passage of music. The frequency of these beats is known as *tempo*. In popular and jazz music, pulse is usually articulated by the rhythm section, notably, the drumkit, and therefore forms an intrinsic part of the music itself (Gracyk, 1996, p. 132). In classical music, pulse tends to be implied rather than made audible. Information regarding a piece of music's pulse and tempo are contained within the mensural aspects of musical notation; pulse is then felt/counted internally by the players and/or indicated by a conductor. The term "rhythm," on the other hand, is often used more broadly and may refer to any given discernible grouping of (equal or contrasting) durations, including beat. Thus, while two passages of music may have the same *pulse/beat*, they will usually employ different *rhythms* according to the various musical events that are structured around the former (Gracyk, 1996, p. 132).

In Western music, beats are organized into groups called measures (or, in notation, bars) that form repeated cycles. This is known as a metric cycle, or simply *meter*. Fundamental to the experience of meter is the perceived presence of a pattern of stronger/louder and weaker/softer stresses on the beats contained within each metric cycle. As outlined in the previous chapter, all metric cycles can be broken down into groups of two and three. The simplest meter is a single, repeated group of two (one-two), known as duple time or a march (left-right). As in all metric groupings, the first beat of the cycle—the *downbeat*—is perceived as the strongest, while the second—the *upbeat*—is weaker (ONE-two). A group of three is known as triple time, or a waltz. Here, beat one is still the strongest, however, of the two remaining beats, we tend to perceive the upbeat at the end of the measure as slightly stronger than beat two as it provides the *anacrusis*, or pickup (the musical equivalent of a backswing in, say, golf or tennis that precedes the "downbeat" of the forward swing and subsequent ball strike) to the downbeat of the next cycle: ONE-two-three. The stresses in a four-beat measure—known as quadruple or common time—follow a similar pattern: beats one and three are the strong beats, with beat three perceived as somewhat weaker. Beats two and four are the weak beats, with beat four (the anacrusis) perceived as somewhat stronger; like triple time, the beat immediately succeeding the downbeat is the weakest: ONE-two-THREE-four.

The inherent pattern of *accents* contained within standard metric cycles forms the basis of a dynamic of rhythmic stability and tension that is, in many ways, comparable to that of pitch. Rhythmic gestures whose accents coincide with those of the meter against which they are set will tend to have a rhythmically stable effect as they affirm the inherent accentual properties of the music. Patterns that place stresses on weaker beats within a given meter will have the effect of disrupting the music's natural pulse, creating rhythmic tension. This is known as *syncopation.*

Besides these accentual properties, the other important trait of rhythm is that of duration. A single beat within a metric cycle maybe be subdivided or extended to create shorter and longer rhythmic durations. A note that lasts for four beats, i.e. the full duration of a common time measure, is known as a whole note. A two-beat note is known as a half-note, and a single beat, a quarter-note. Similarly, a single beat may also be split into two, equal-length 8th notes (the "1 and, 2 and" pattern mentioned in the previous chapter) and in half again to form four 16th notes (and so on). These various values can be fused—or tied—together to form longer durations, e.g. three, 8-notes. Beats may also be subdivided into three. This is known as an 8th-note triplet and can be understood by counting "1–&–a, 2–&–a."

A further examination of some of the musical examples mentioned previously reveals several ways in which rhythm (and accent) can be applied to pitch to enhance the musical forces underpinning melodic contour. In "Frère Jacques," the rhythm of the melodic line articulates and reinforces the tune's underlying four-beat (common time) pulse; the syllables of the song's lyrics align with its accentual properties throughout. Accent is equally important in defining the melodic character of some of the other examples mentioned earlier. Although both the *Jaws* and *Pink Panther* motifs employ the same rising semitone pattern, distinctions in rhythmic placement between the two gestures cause us to perceive their tonal hierarchy differently. In the latter, the second of the two 8th-notes that form the rising semitone motif (the "dum" of "ba-DUM") coincides with the rhythmic downbeat of the music's pulse, while the preceding note is placed in the "&" of the previous the offbeat. As a result, we hear the motif as a move from anacrusis to resolution (reinforced by the fact that each motif resolves to a consonant pitch of the underlying harmony rather than a tension note). In contrast, John Williams's theme places the stress on the first pitch, which is, initially slightly longer and articulated without a clear sense of underlying pulse. The effect is more unsettling. The trajectory of the rising motif contravenes Larson's conventional forces, presenting a sense of departure, of something pulling away from a position of stasis (a shark emerging from the shadows?). The rhythmic jerkiness of *Flight of the Bumblebee* is caused by the fact that the initial leap after the first melodic descent of 16th-notes occurs on the weakest beat—beat two—of the measure's second four-note pattern. The more even, four-down/four-up pattern of the second measure and accompanying downbeats help steady the ship. Meanwhile, the bassline of the Brad Fiedel cue contains no discernible accents: it gives way to syncopated stabs rather than being accompanied by them. The lack of a clear rhythmic anchor for its inherently volatile pitch contour gives the riff a headlong, chaotic feel.

9.4. Texture/Timbre/Pitch and Harmony

9.4.1. Sensory Consonance and Dissonance, Notes and Noise

From the foregoing discussions, it will be apparent that pitched materials in Western music exert influences upon one another when heard in both sequential and simultaneous fashion; "harmony" is not simply about the sonorous effect of simultaneous groups of notes (chords, triads) but also about degrees of tonal attraction and types of tonal function based on sequential and scalar logics. However, the sonorous, textural aspect is perhaps its most obvious manifestation. The phenomena of *sensory dissonance* and the influence of *frequency masking* (Fletcher, 1940) on various combinations of sounds is likely to be familiar to the sound designer. Our ear physiology influences the perception of combinations via the patterns of vibrations that occur on the *basilar membrane*, a somewhat resonant structure inside the inner ear, which resonates at different points along its length in response to different frequencies, with associated neural responses. As such, it functions in the manner of a frequency analyzer.

The basilar membrane response entails that very small differences in frequency produce a generally *fused* percept. Above 15 Hz, the periodic tones will result in a somewhat segregated effect, with a pronounced degree of *roughness* associated with the lack of discernible periodicity (this is known as *sensory dissonance*). For intervals such as seconds, sensory dissonance will be particularly pronounced. Beyond the interval of approximately a minor third, the tones will perceptually separate to significant degree, with significantly reduced sensory dissonance effects. These effects—the relationship between musical interval size and degrees of sensory dissonance (or lack thereof)—were studied by Plomp and Levelt (1965), who tested subjects through presenting synthesized sine wave combinations of various musical intervals. The result of their experiments was that, within a given frequency region of the basilar membrane, the *critical band* (Fletcher, 1940), sensory or *tonal dissonance* effects were strong. Outside this region, dissonance effects were minimized.

Although musical (i.e. periodic) tones of different timbres will vary with respect to the position and level of the components of their frequency spectrum, an idealized spectrum for a periodic tone can be computed based on a number of harmonics. As a result, Plomp and Levelt (1965) were thus able to extrapolate their plot of relative sensory dissonance for various musical intervals within the octave. The results, approximated in

Figure 9.7, plot various Western intervals against judgments of relative dissonance, ranging from the smaller intervals within an octave, with low values around the midpoint of the octave (at the fourth and fifth), before increasing again. These results align with interval consonance/dissonance treatment within Western music and were later corroborated in broad terms in experiments with complex tones by Kameoka and Kuriyagawa (1969), though the exact details of the sensory consonance/dissonance values were shown to vary due to differing patterns of harmonics.

It will be apparent that increased complexity within a sound source's frequency content will give rise to increased sensory dissonance effects. As more and more frequency components are added, certain sources will become "noisier," both in and of themselves and in their combination. Indeed, the combination of such "inherently noisy" timbres can also be compared with the combination of a number of more coherent, periodic tones within a small frequency region. Within a *tone cluster* (an intervallic combination in which multiple adjacent intervals are present; generally,

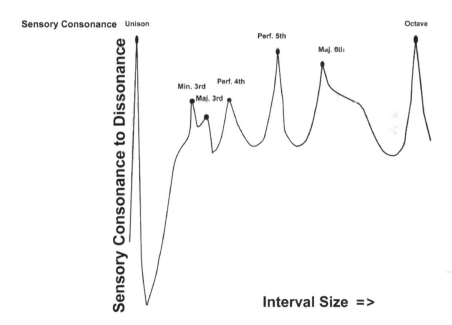

Figure 9.7 Relative Sensory Consonance/Dissonance Values Extrapolated from Critical Bandwidth Data.

Source: Diagram after Plomp and Levelt (1965)

these are at least three in number and are separated by semitones), the density of materials makes it difficult for the hearing mechanisms to clearly segregate frequency/pitch information, with the additional sensory dissonance effects previously noted e. Whichever the source type, whether a smaller number of inherently noisy timbres, the result, for the basilar membrane and for our perception, is a confusing mass of vibrations without discernible individual pitch identities. This textural effect can be used to great effect in compositional and soundtrack contexts. "Lux Aeterna" (1966) and "Atmosphères" (1961), the compositions by György Ligeti featured in the soundtrack of *2001: A Space Odyssey* (Kubrick 1968), employ gradually accumulating tone-clusters in the creation of unusual sonorities and different degrees of sensory dissonance, which are used to evoke a sense of otherworldly environments within the film.

9.4.2. Tonal Consonance and Harmonic Function

A key feature of Western harmony is a tendency to use a functional, three-note unit of simultaneous pitch combinations, known as *triads*. Triadic groups of notes (often, more colloquially, known as chords) can be built on different degrees of a scale (Figure 9.8) and, in combination, can provide pronounced differences in their consonant/dissonant sensory effects but also directional imperatives. Many of the simplest triads consist of intervals of the (perfect) fifth and major/minor third on various scale steps. In some cases, however, triads built on a given scale step may deviate from this; a triad built on *C* in *C* major will consist of *C–E–G* (major third/perfect fifth), but a triad built on *B* in *C* major will consist of *B–D–F* (a minor third and a fifth reduced by a semitone, a *diminished fifth*).

This building block–based approach to harmony acts as a competing dynamic with the melodic function of individual musical lines and became more dominant in the West from the 18th century onward. The influential theorist of the Classical period, Rameau, author of a 1722 *Treatise on Harmony*, began to consider how musical motion and structure related to the notes contained within these triads but not their specific ordering. For example, if *E* and *G* were present in a chord, but the chord is *inverted* by placing the *C* at the top (for reasons of melodic contour), the *E*, *G* and *C* chords would be taken to have a *fundamental bass* of a lower *C*, and the

Figure 9.8   Examples of Triads on Different Scale Positions in C Major.

chord would have a similar function to a standard triad built on *C*: *C-E-G*. Although, in the *E-G-C* case, the top interval (*G-C*) is a fourth as opposed to a *C-G* (fifth) in the standard *C* triad, they are treated as essentially the same for functional purposes. This particular functional approach is borne out within standard usage in Western music; such altered-order triads (inversions) are taken as functionally equivalent.

In order to understand how chords function, it is preferable to consider them in terms of their component notes/proportions (as previously) and how these transition rather than as fixed (bottom-up) vertical entities. All chordal movement essentially involves combinations of stepwise motion, comparable to the forces models mentioned earlier, along with some element of stasis or inertia. Chord progressions will often seek to maintain continuity or stepwise motion of individual voices (lines) where possible. Please note that in most musical styles, the lowest note of the chord tends to convey a sense of harmonic progression while the more distinguishable highest pitch is often perceived within the context of melodic movement. The simple II-V-I progression in Figure 9.9 illustrates this. In each case, two notes of the chord move by step while one remains static. Note, too, how the home note of *C* is approached more acutely, i.e. by a half step rather than whole step.

In common with this dynamic, the harmonic content of a triad on the first step of a scale will include clear harmonic content around the fifth scale step (the *dominant*); as such, a move to the dominant position is a relatively easy transition. As this interval is also close to the tonal center within our tonal hierarchy, transitioning from the chord built on the fifth to one on the first degree of a scale can provide a strong sense of resolution, reinforcing the scale and its associated key (i.e. tonal center and whether it is major or minor via the type of third used). This type of movement (fifth degree, or V to first degree, or I), is called a *perfect cadence*, with the I chord treated as a point of rest. Many Western Classical–period harmonies are built upon extensions and ornamentations of I–V–I, often including the chord built on the fourth step (IV), as it provides mostly stepwise movement in certain inversions (along with a common note, reinforcing the tonal center) between I and V. Chord progressions during this period also made more occasional use of "weaker" triads built on other scale divisions.

Figure 9.9   C Major II-V-I Progression with Step-Wise Voicing.

9.4.3. Counterpoint and Contrast

Thus, harmony is about more than the momentary sensations of sensory consonance and dissonance, just as melodic relationships are about more than the momentary concern for whether to proceed by step or leap or even the more localized aesthetic issues of the form of a contour. As previously stated, an understanding of music theory is largely based on developing an awareness of how these two parameters are mutually evolved over a prolonged temporal frame.

The term "counterpoint" holds particular significance in this regard. Derived from the Latin for "point against point" (and prompting comparisons with the earlier Klee analogy), *counterpoint* may be used in a variety of musical contexts to describe the method and resultant dynamic of combining independent musical layers to achieve harmonic and melodic coherence within a given passage of music—the reconciliation of horizontal and vertical referred to earlier. The traditional singing of "Frère Jacques," i.e. as a single melody layered by staggered entry points (known as a *canon*, or *round*), represents a simple form of counterpoint.

The term is also often appropriated to refer more broadly to a comparable sense of interplay between contrasting forces at work outside of purely musical models, e.g. between music and sound or within mixed-media formats. The opening sequence to *Star Trek: The Motion Picture* (Wise, 1979), for example, features a number of contrasting yet interdependent musical and sonic ideas—including several of the examples discussed earlier in this chapter. The resultant "counterpoint," this time between opposing musical/sonic events rather than individual pitches serves to establish the film's fundamental narrative themes.

As three Klingon starships fly steadily in formation across the screen, we hear the sequence's principal musical theme: a syncopated fanfare melody based on the first- and fifth-degrees of the scale—the tonic and dominant notes—sounded over a steady duple time, march-like pulse. The harmonic and melodic stability of the music immediately establishes the Klingon ships as heroic, while the pulse underpins the military, processional quality of what we are observing (the score even features a clip-clop-like percussion echo of the fanfare melody).

In contrast, the alien cloud with which the Klingons battle is announced with sustained klaxon-like electronic pitches of three whole tones (or a diminished fifth) apart. This is answered with ominous low-bass chords based around semitonal movement that later develops into a *Jaws*-like, chugging semitone oscillation and displaces the undulating dominant tonic patterns of the Klingon march once the mysterious cloud repels the Klingon bombardment. The electronic timbre of the former gesture provides a marked contrast with the "purer" (in the musical sense) acoustic, orchestral sounds of the Klingon's music, to the extent that the audience is even left wondering if the sound may be emanating from the cloud itself (a diegetic sound). Similarly, the choice of harmonically dissonant intervals draws on the established musical technique of employing the more distant pitch relationships outlined in section 9.2.3 to provide a sense of musical and consequently dramatic unease. This reaches its climax once the audience learns that the alien cloud is earthbound. The electronic pitches merge with the orchestral texture sounding a menacingly overblown two-chord statement based around a rising minor 6th (eight semitones).

Meanwhile, the third party in the sequence—the *Epsilon Nine* space station that observes the initial confrontation—is assigned music that seems to share elements of the other two. Softer orchestral textures sustain the oscillations of the music first in the form of alternating major 3rds and falling 4ths. The latter restates the tonic-dominant pattern of the opening cue while together the distance between 3rd and 4th retains the semitone movement. The music then narrows to whole tone steps and finally semitones. A more "blended" melodic brass motif, featuring a rising leap followed by falling steps, outlines the initial shape of the oscillating accompaniment. We also hear a dipping scalic run à la *Chuckie Egg* (the character is present on-screen in the form of an astronaut, although terra firma is absent!); the smooth contour of both reinforces the on-screen calm "after the storm." Together, the blending of musical ideas enables the soundtrack to articulate oppositional contrasts while simultaneously achieving sonic—and, in turn, narrative—coherence.

9.5. Conclusion

In common Western practice, harmony, then, involves the perception of sensory consonance and dissonance alongside the treatment of this consonance/dissonance axis as the basis of a syntax of progressions that enhance or undermine a sense of forward momentum and temporal development. Similarly, melodic relationships are dominated by the contours of melodic lines as much as their component leaps and steps. Counterpoint integrates these dynamics; its semiautonomous voices are nonetheless grounded by recourse to harmonic syntax, the roles of given intervals within the tonal hierarchies of a given musical key.

More broadly, all of these factors in music composition and its associated theoretical ideas are competing dynamics. However, conveniently for anyone who is starting to engage with music composition, many of the key concepts can be thought of based on relationships with either the structures inherent within the notes themselves (i.e. overtone/harmonic structures) or through metaphors based on familiar physical experience (ideas of tonal and melodic force and inertia and center and periphery). Music theory, for all of its specific vocabulary and occasional specialist concepts, can be interpreted through the lens of other aesthetic practices, affording sound designers an entry point into a wider field of organized sound.

References

Alderman, N. (1983). *Chuckie Egg*. UK: A&F Software.

Arlen, H. and Harburg, E. Y. (1939). *Somewhere over the rainbow*. New York: Leo Feist.

Arnheim, R. (1974). *Art and visual perception: A psychology for the creative eye – the new version*. Berkeley and Los Angeles: California University Press.

Arnheim, R. (1969). *Visual thinking*. Berkeley and Los Angeles: California University Press.

Cameron, J. (dir.) (1984). *The terminator* [film]. United States: Orion Pictures.

Dowling, W. J. (1978). Scale and contour: Two components of a theory of memory for melodies. *Psychological Review*, 85(4), pp. 341–354.

Dowling, W. J. and Harwood, D. L. (1986). *Music cognition*. London: Academic Press.

Fletcher, H. F. (1940). Auditory patterns. *Reviews of Modern Phys*ics, 12, pp. 47–65.

Gracyk, T. (1996). *Rhythm and noise: An aesthetics of rock*. London and New York: I. B. Tauris & Co.

Hewitt, M. (2008). *Music theory for computer musicians*. Boston: Course Technology, Cengage Learning.

Ives, C. (1969). *Essays before a sonata and other writings*. London: Calder and Boyars.

Johnson, M. (2007). *The meaning of the body: Aesthetics of human understanding*. Chicago: University of Chicago Press.

Kameoka, A. and Kuriyagawa, M. (1969). Consonance theory part II: Consonance of complex tones and its calculation method. *Journal of the Acoustical Society of America*, 45(6), pp. 1460–1469.

Keampfert, B. (1966). Strangers in the night. *Polydor*.

Klee, P. (1953). *Pedagogical sketchbook*. Translated by S. Moholy-Nagy. London: Faber & Faber.

Klee, P. (1961). *The thinking eye*. Translated by R. Manheim. London: Lund Humphries.

Krumhansl, C. L. (1979). The psychological representation of musical pitch in a tonal context. *Cognitive Psychology*, 11(3), pp. 346–374.

Kubrick, S. (dir.) (1968). *2001: A space odyssey* [film]. United States: Metro-Goldwyn-Mayer Corp.

Larson, S. (1997). The problem of prolongation in "Tonal" music: Terminology, perception and expressive meaning. *Journal of Music Theory*, 41(1), pp. 101–136.

Lefebvre, H. (2004). *Rhythmanalysis: Space, time and everyday life*. London and New York: Continuum.

Lerdahl, F. (2001). *Tonal pitch space*. New York: Oxford University Press.

Mancini, H. (1963). *The pink panther theme*. RCA Victor.

Nikolsky, A. (2015). Evolution of tonal organization in music mirrors symbolic representation of perceptual reality. Part-1: Prehistoric. *Frontiers in Psychology*, 6(1405). Available at: https://doi.org/10.3389/fpsyg.2015.01405 [Accessed January 2018].

Persichetti, V. (1962). *Twentieth century harmony*. London: Faber & Faber.

Plomp, R. and Levelt, W. J. M. (1965). Tonal consonance and critical bandwidth. *Journal of the Acoustical Society of America*, 38(4), pp. 548–560.

Rimsky-Korsakov, N. and Emerson, G. (1994). *The Flight of the Bumblebee*. Ampleforth: Emerson Edition.

Schoenberg, A. (1967). *Fundamentals of musical composition*. London: Faber & Faber.

Shepard, R. N. (1964). Circularity in judgments of relative pitch. *The Journal of the Acoustical Society of America*, 36(12), pp. 2346–2353.

Wise, R (dir.) (1979). *Star trek: The motion picture* [film]. United States: Paramount Pictures

Wishart, T. (1996). *On sonic art*. Amsterdam: Harwood Academic Publishers.

Leveraging Online Audio Commons Content for Media Production

Anna Xambó, Frederic Font, György Fazekas,
and Mathieu Barthet

10.1. Introduction

Since the popularization of the Internet in the late 1990s, within the World Wide Web (WWW) ecosystem, there has been an exponential growth in storage capacities, in *semantic technologies* (i.e. data structured to be understood by both human and machine agents) and in social activities (Gruber, 2008; Shadbolt et al., 2006). There exists a range of online services that offer both free or paid access to a varied range of multimedia content (e.g. SoundCloud[1] for music, Freesound[2] for sounds, YouTube[3] for videos, Flickr[4] for photos, and so on). New ways of managing this content have emerged (e.g. sharing, reusing, remixing and repurposing), leading to a new community of *prosumers* who both produce and consume online digital content (Ritzer and Jurgenson, 2010). In the field of audio recording, prosumers are those who work in *project studios*, which are professional studios built at home using affordable digital technologies, where prosumers both consume the equipment when purchasing it and produce content from using this equipment (Cole, 2011). Another change brought by the Internet has been Creative Commons (CC). CC is a mechanism founded in 2001 to establish a legal and technical infrastructure for sharing content. CC offers a range of licenses and has helped to foster the WWW as we know it nowadays (Merkley, 2015). The development of CC licenses has offered a finer-grained level of licensing possibilities, compared to the classical copyright model, which was too strict for the new practices around the generation and reuse of digital content (Lessig, 2004).

In this chapter, we present general concepts and technologies aimed at consuming or producing crowdsourced online audio content in the context of linear media production. We will cover CC sound and music content, which we refer to as Audio Commons content (Font et al., 2016). This chapter targets sound designers, music composers, researchers, developers and anyone passionate about sounds and the Internet, with a will to learn how Audio Commons content can be leveraged for media production. In particular, this chapter focuses on the challenges and opportunities of using Audio Commons content and what it can bring to the traditional digital audio workstation (DAW). It is worth noting that Audio Commons content can also be repurposed for interactive media. For those interested in sound interaction, this chapter can be complemented by the second volume from this series, *Foundations in Sound Design for Interactive Media*.

The remainder of the chapter is organized as follows. First, the concepts of uploading, retrieving, licensing and attributing Audio Commons content are presented, before discussing new workflows enabled by online audio content and services. The second part of the chapter presents five use cases showing how Audio Commons can serve the generation of sound textures, music production, soundscape composition, live coding and collaborative music making. Lastly, future directions and challenges for sound design and music production are discussed. By the end of this chapter, the reader should be able to (1) identify the key concepts related to Audio Commons and how media production can be changed by using online audio content, (2) get inspired by existing tools and practices repurposing Audio Commons content, and (3) create, share and reuse Audio Commons content. Although equally interesting, it is out of the scope of this chapter to present the web technologies behind online audio databases (e.g. search engines, client-server architecture, semantic web and so on); more information on these can be found in online tutorials and textbooks.

10.2. Uploading, Retrieving and Consuming Online Audio Content

In this section, we present the main concepts related to sound design and music production using online audio content. We will discuss terms linked to the description of audio content (*metadata* and *folksonomies*), their retrieval (semantic audio, text-based queries versus content-based queries), Creative Commons licenses, as well as online digital audio workstations (DAWs) and web application programming interfaces (APIs).

10.2.1. Uploading Audio Content

Storing sound and music online and sharing it instantly with people all over the world might have sounded like science fiction a few years ago. Nevertheless, after the *social media revolution* (Smith, 2009), it has become part of our daily routine. Examples of this practice are demonstrated by the success of online sound and music sharing websites like SoundCloud,

Bandcamp, Soundsnap, Looperman, Jamendo, Freesound and many others. These sites host hundreds of thousands of audio files that need to be indexed to become accessible and reusable.

Audio files stored online can be of different audio quality and may be encoded using different file formats. We can distinguish between lossless formats (the audio is preserved in its full quality, typically formats with extensions like. wav,. aiff and. flac) and lossy formats (the least important information in the audio is removed so that the file size can be reduced, typically formats with extensions like. ogg or. mp3). To avoid online sound and music collections becoming long lists of audio files that are only identifiable only by their file names, extra *metadata* must be provided. In other words, they need to be described so that sound and music sharing platforms can allow users to find them. Even though some metadata can be automatically derived by computers analyzing audio files (e.g. file format properties, duration, number of channels), richer descriptions still need to be manually provided by humans (e.g. music genre for a song, the microphone used for recording a sound effect or the location where a sound was recorded). This is also true for other kinds of multimedia items without an intrinsic textual representation, such as video and images (Bischoff et al., 2008).

As one can imagine, there is no single and definitive way to describe all kinds of sounds and music content. For example, describing the recordings of a musical instrument, a sound effect or a speech will likely require different sets of information related to production, context and perception.

Sound and music sharing platforms will generally allow users to provide metadata for their sounds in the form of, at least, some keywords (or *tags*) and a textual description. Some platforms let users provide specific metadata fields such as music genre, tempo expressed in beats per minute (BPM) or the artist's name. Nevertheless, it remains a real challenge to come up with useful keywords and textual descriptions that summarize well the contents of an audio file, especially in the case of nonmusical audio content where the most important perceptual aspects are less well established compared to the case of music, which is more easily associated with notation and music theory systems.

A good practice when describing a sound is to look at the following information levels:[5]

- *Semantic [S]*—information about the sound sources corresponding to the different events appearing in the sound, the actions that generate the sounds, and the meaning of the sounds for the listener
- *Perceptual [P]*—description of the sounds' perceptual qualities (e.g. timbre, timing), not necessarily tied to the source(s) of the sound
- *Technical [T]*—information about the gear and techniques that were used to create or record the sound

- *Contextual [C]*—other aspects like the location where a sound was recorded (if relevant at all) and the purpose of the sound

All levels are complementary and bear relevant information for indexing and retrieval purposes (Marcell et al., 2001). Figure 10.1 shows an example of a textual description for a sound where the preceding information levels are indicated next to each sentence:

A description such as the one in Figure 10.1 can be summarized with a number of tags falling into the semantic, perceptual, technical and contextual categories. An example of a list of tags related to the sound previously described is shown in Figure 10.2. One of the benefits of describing a sound with tags is that it can become easier to find similar sounds, e.g. through interactive tag clouds. It can also facilitate visualizing tag patterns among a user's own sounds, as discussed in section 10.2.2.

10.2.2. Retrieving Audio Content

Existing services for sound and music distribution, including those mentioned in the previous section, use a variety of search and retrieval methods. In current online audio search engines, the most common model for

An interesting field recording of birds, led by Blue Jays, doing their alarm calls in a peaceful forest ambience [S,C]. High-pitched sounds and chirps reverberating in the woods [P,C]. Recorded with a Sound Devices recorder and stereo microphone (Rode NT4) [T]. I was in the middle of the woods when I noticed that the birds seemed to be alarmed like there was a predator [C,S]. I finally spotted a big Barred Owl sitting high in a tree about 50 yards from my recorder. Pretty amazing how the bird community sounds the alarm [S,C].

Figure 10.1 Example of Rich Sound Description Including Semantic (S), Perceptual (P), Technical (T) and Contextual (C) Information.

Source: Adapted from https://freesound.org/s/151599/

field-recording   birds   bird-calls   blue-jays
ambience peaceful forest woods sound-devices
rode-nt4

Figure 10.2   Example of Tags Describing a Sound.

Image credit: Freesound

retrieval is the use of a single search box that provides free-text entry. In this approach, keywords, search terms or more general text entered by end users are matched with descriptive file names and metadata, such as title, artist or contributor where appropriate (see section 10.2.1). Interactive *tag clouds* with weighted keywords related to occurrences are also employed to help the user finding sounds by more visual means (see Figure 10.3). It is also common to provide a mechanism for filtering results, i.e. provide a way to show only items in the search results that match designated categories. For instance, this can restrict result sets to the kind of sound library or package the user is interested in or focus on other metadata elements such as genre, theme or license (see section 10.2.3). The result of queries is most commonly presented as a flat list of media items (e.g. sounds, music files) that are typically ranked by a set of relatively simple criteria. Figure 10.4 illustrates an example of a sequential list of sounds retrieved with Jamendo's Music and Sound Search Tool (MuSST).[6] These items often include popularity, for instance, the number of downloads, some quality rating provided by the users, the duration of the sound files, or the date at which they were uploaded or added to the catalogue of the provider.

While these search mechanisms appear sufficiently rich, navigating large sound or music libraries can become daunting, especially when we consider the large amount of content in existing platforms. In addition, audio is time based and linear, as opposed to many other forms of media. Consequently, the quality and/or fit for purposes of a retrieved item cannot be judged at a glance, similarly to looking at an image, gleaning over a short piece of text or reviewing a couple of key frames in video. Many services provide an embedded media player that allows users to audition items displayed in the search results prior to downloading or listening to the entire recording. Showing information related to the audio content is also relatively common. This may include a reduced representation of the audio waveform or its frequency content, such as a time-frequency representation or spectrogram. These displays help to judge the complexity of the recording for those who know how to interpret them, e.g., show how rich or sparse the content is in time and frequency, and to allow for guessing whether it is a short sound effect, speech or music, for example. However these visual displays tell us surprisingly little about many important

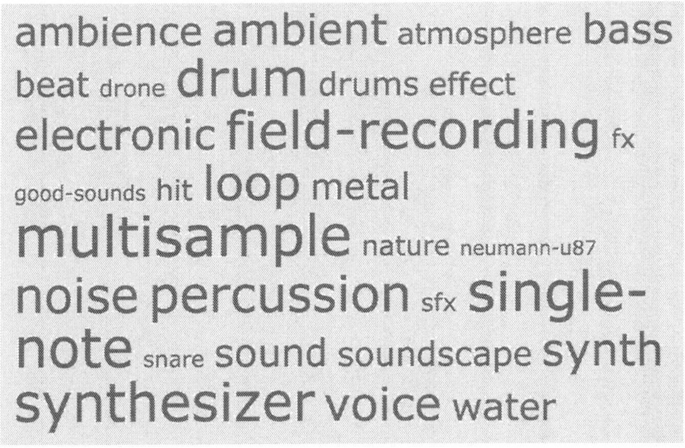

ambience **ambient** atmosphere **bass**
beat drone **drum** drums effect
electronic field-recording fx
good-sounds hit **loop** metal
multisample nature neumann-u87
noise percussion sfx **single-**
note snare **sound** soundscape **synth**
synthesizer voice water

Figure 10.3 Example of an Interactive (Weighted) Tag Cloud from Freesound's Front Page Website: The size of the tags are proportional to their popularity in Freesound. Selecting a tag yields a list of all the sounds described with this tag.

Source: Freesound.org

criteria of interest, such as the audio quality, the sound sources present in a recording, the meaning of a spoken word content or the mood expressed in a recording.

These problems are commonly addressed using free-form tags aiming to describe audio content thoroughly, a technique discussed in section 10.2.1. Audio may be tagged by experts or collected from contributors or end users of sound sharing services. Tags collected this way are often referred to as *folksonomies* (Lamere, 2008). Expert annotation is a slow and expensive process that does not scale with rapidly growing collections. Crowdsourcing allows for solving this issue, but it has its own problems, most prominently, folksonomies are plagued by lack of consistency, biases and other sources of "noise" in the tagging process (Choi, 2018).

Automatic analyses of the audio content and its meaning, a procedure often identified as *semantic audio*, allows for the extraction of metadata such as tempo, instruments or perceptual qualities (e.g. rough or soft) directly from the audio recordings. This is becoming increasingly common in sound sharing services, despite audio feature extraction; the representation of the resulting metadata is still an active field of research (Fazekas et al., 2011). Although either many high-level or complex semantic audio descriptors cannot reliably be extracted yet, several services employ automatic analysis tools. These tools provide search or filtering functionality based on tempo or key estimation and even higher-level analyses aiming to determine musical genre or mood.

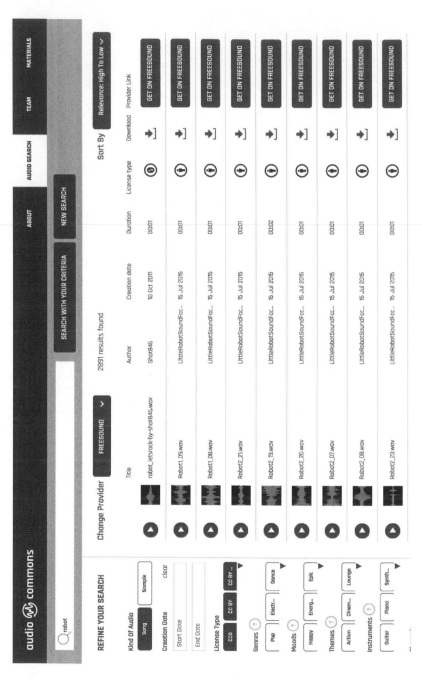

Figure 10.4 Screenshot Example of the Query Results from Jamendo's MuSST.

A possible way to overcome the limitations of complex semantic feature extraction previously explained is to navigate by similarity to a selected seed item from a collection. There are machine learning and visualization techniques that enable placing similar items close to one another in low-dimensional spaces. When these items are visualized, they can facilitate retrieval by navigating the collection by similarity among its items. For example, Freesound Explorer[7] is a visual interface for making queries to Freesound and exploring the results in a two-dimensional space where sounds are organized according to timbral similarity (see Figure 10.5). The map is computed using a dimensionality reduction technique over automatically extracted spectral audio features (Font and Bandiera, 2017). In this way, closer sounds in the timbre space tend to have similar timbre, and clusters of search results naturally emerge in different parts of the visual space. This allows users to navigate Freesound content by combining a standard text-based search mechanism and a visual method for exploring the results.

Studies on the needs of modern music consumers have revealed a strong interest in being able to search and browse music by mood (Lee and Downie, 2004). Researchers have investigated semantic mood models aimed at music recommender systems (e.g. Barthet et al., 2013). A common psychological model devised to characterize human emotions is Russell's arousal/valence (A/V) model (1980), which represents levels

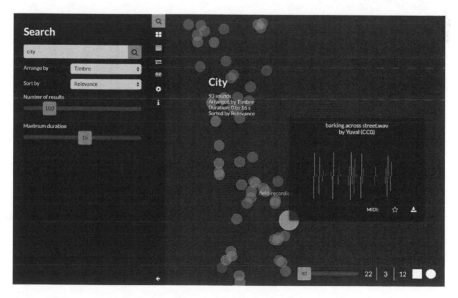

Figure 10.5   Screenshot Example of Freesound Explorer.

of activation or excitation (*arousal*) and positiveness (*valence*). The A/V model has been applied to music retrieval in a number of works. We provide here examples related to the Moodplay web-based social music player described in Barthet et al. (2016). Figure 10.6 shows an example of a search interface letting users search for songs according to emotional indications expressed using the two-dimensional A/V model (in the model used, "uplifting" music is represented as light positive music). Figure 10.7 displays a map of the songs available in the Moodplay music player's library, resulting from predictions of the emotions expressed by the songs in the A/V space using semantic audio.

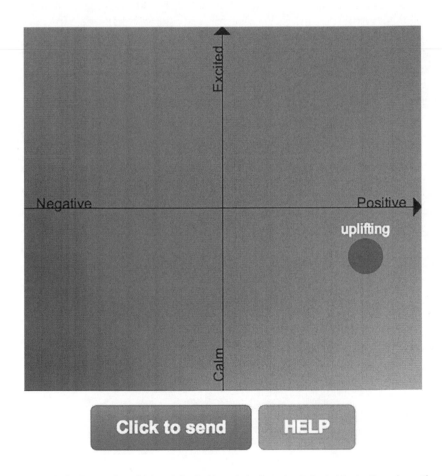

Figure 10.6 Example of Moodplay's User Interface to Select Music Based on the Arousal/Valence Emotion Model.

Source: Barthet et al. (2016)

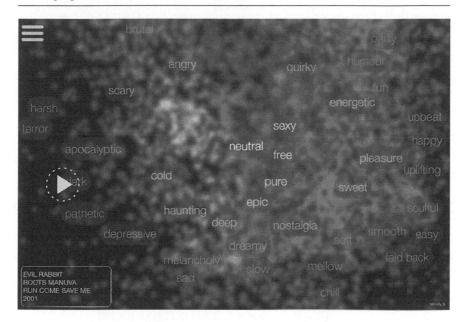

Figure 10.7 Example of Moodplay's User Interface Displaying Music Tracks in the Two Dimensional Arousal/Valence Space.

Source: Barthet et al. (2016)

10.2.3. Licensing and Attribution

The "right to copy," or *copyright*, is a legal right that exists in many countries in relation to an original piece of work and its creator(s). The legal framework covers aspects such as reproduction, derivative works, distribution and public exhibit of an original work. The copyright has a limited amount of time (typically the *copyright length* is 50–100 years after the passing of the creator, a number of years that varies depending on the country). For example, since 1998, the copyright in the United States lasts a creator's life span plus 70 years, while this extends to 95 years for copyrights owned by corporations. This means that royalties need to be paid for the use of intellectual works during this period (Lessig, 2000). Once copyright expires, the original work becomes part of the public domain, and the public right of common use takes effect enabling its free and unrestricted use. Copyright has a number of benefits, namely giving the creator(s) credit to their original work, legal protection against plagiarism and potential revenues from distribution and derivative works. However, copyright law can limit the growth of creative arts and culture due to the long wait required until having free access to these works. The use of derivative works (e.g. unpaid, unnegotiated) can be penalized unless it is released to the public domain. The risk that Lessig refers to as *copyright perpetuity* (Lessig, 2000), which prevents an original piece to become part of the public domain and the free circulation of creative ideas, was a relevant topic of debate during the foundations of the WWW. This debate, motivated by the advent of digital content and network capabilities, fostered the creation of Creative Commons (CC).

CC licensing relates to the concept of *free culture*. Here, "free" is understood not as in "free beer" but as in e.g. "free speech," "free markets" and "free will" (Lessig, 2004, p. xiv). Inspired by Stallman's (2002) ideas on free software and free society, Lessig distinguishes a *free culture* (the desired CC model) from a *permissions culture* (the existing copyright model). In a free culture, creators and innovators are supported and protected by granting intellectual property rights. In a permissions culture, creators and innovators can only create with the permission of creators from the past. CC licensing is thus connected to the notion of providing as much freedom as possible to the creators within a legal framework. It is worth mentioning that Lessig emphasizes that free culture is not a culture without property or where creators are not paid. It is instead "a balance between anarchy and control" (Lessig, 2004 p. xvi).

As noted in Merkley (2015), CC-licensed work has nearly tripled between 2010 (400 million CC-licensed works) and 2015 (over 1 billion). CC content type includes images (photos, artworks), videos, research (journal articles), open educational resources, texts (articles, stories, documents), audio tracks (4 million reported in 2015 from 16 platforms) and other (multimedia, 3D) (Merkley, 2015). The various licenses are described in the CC organization website.[8] Table 10.1 outlines the four main components in CC licenses that are typically combined. Figure 10.8 shows the combination of these four components resulting in six different license options that span, in a continuum, from less to more permissive licenses from the point of view of free culture.[9] These licenses are providing various degrees of freedom

Table 10.1 Description of Available CC Elements.

CC Elements	Description
ⓘ *BY*	Attribution or the need to credit the original creation
Ⓢ *NC*	Non Commercial or building upon the original work noncommercially
⊜ *ND*	Non Derivatives or keeping unchanged the original creation
ⓢ *SA*	ShareAlike or license the new creations under identical terms as with the original creation

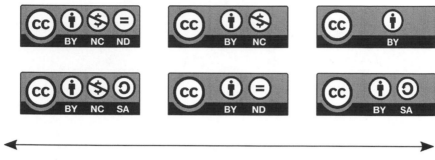

Less Permissive   More Permissive

Figure 10.8 Continuum of the Available CC Licenses from Less to the More Permissive Levels.

when releasing a creative work. A more detailed explanation of the licenses illustrated with examples can be found on the CC webpage[10] with a service that helps to choose the most suitable license.[11]

When using CC-licensed audio content, it is important to make sure that the planned application matches the license requirements. For instance, reusing a CC-licensed musical track for commercial purposes may require licensing fees (e.g. see the Jamendo Licensing stock music for commercial use).[12] Placing a work in the public domain or *CC0* (no rights reserved) is also possible. As explained in the Freesound website,[13] apart from CC0 licenses, CC-licensed audio content should be credited by citing the title of the sound/music, the author and a link to the resource. Figure 10.9 exemplifies how to attribute correctly a CC-licensed audio item. If the list of sounds is too long to be displayed in the credits section, alternatively it is suitable to provide a link to a separate document with the list (e.g. "for the full list see here: www.mysite.com/work-credits.html").

A *remix*, or the creation of an adaptation from existing CC-licensed content, needs to be done carefully to comply with CC licensing. Typically, it is possible to remix CC content when it is licensed without the Non Derivative element. As a rule of thumb, it is possible to create an adaptation and release it under a similar, compatible license or a more restrictive license (unless it has the ShareAlike element), but never the other way around. Table 10.2 highlights the license restrictions when publishing new sounds or music that include, modify or remix others' sounds or music.[14],[15]

10.2.4. *Transforming Linear Audio Production Using Online and Web Applications and Resources*

Before venturing into how the Internet can transform linear audio production, some of the differences among online, web and cloud applications

> This [video/performance/...] uses these sounds
> from Freesound:
>
> • sound1 by user1
>   (http://freesound.org/people/user1)
> • sound2, sound3 by user2
>   (http://freesound.org/people/user2)
> • (...)

Figure 10.9   Example of How to Attribute Sounds from Freesound.

Source: Freesound.org

Table 10.2   License Restrictions When Creating a Remix.

Type of License for User A's Work	Type of License for User B's Remix	Can User B Remix A's Work?
CC0	CC0	Yes
CC0	BY	Yes (*)
CC0	BY-NC	Yes (*)
BY	CC0	No
BY	BY	Yes (**)
BY	BY-NC	Yes (**)
BY-NC	CC0	No
BY-NC	BY	No
BY-NC	BY-NC	Yes (**)

Source: Freesound.org

(*) If a third user C adapts the creation from user B, they must attribute it to user B.
(**) User B must attribute the creation to user A. If a third user C uses the creation from user B, they must attribute both users A and B.

are worth mentioning. An *online application* or tool is a software installed locally on a computer, which relies on an Internet connection to access information (e.g. the Skype communication tool). A *web* (or *web-based*) *application* or tool is a software based on a client-server architecture for which the client side runs in a web browser (e.g. the Google Docs word processor). For *cloud-based applications*, the majority of the processing and data storage takes place in remote servers and data centers. *Cloud applications* are designed so as to be operational most of the time for potentially large numbers of concurrent users. They can be operated from the web browser and/or be installed on desktops (e.g. the Dropbox file hosting software). These different architectures can lead to a varied collection of interaction models to leverage online audio content and services for media production. Next we review how web technologies can be applied to tools used for sound design, music production and live performance, such as DAWs, plugins and live coding environments.

10.2.4.1.   Internet-Connected DAWs

Historically, digital audio workstations (DAWs) and digital musical interfaces have been conceived to operate with local resources and content (e.g. local databases of sounds or music) by being for the major part disconnected from the Internet (apart for updates and license authentication). DAWs have traditionally operated in isolation from the Internet both by design, for example for security and data protection reasons, and due to technological limitations, for example, due to the lack of mature web standards. However, several connected audio production tools have emerged recently, paving the way for the Internet to transform audio production similarly to other domains such as communication.

Turning a DAW into an online tool changes its status of a "closed box" to one where the amount of information and audio content available to end users can be made more open-ended. Here, connected DAWs refers to desktop DAWs connected to the Internet, in contrast to browser-based DAWs, which are discussed in the following section "Web-based music production applications and resources." When looking for specific books, searching a personal local library collection may be limited compared to having access to the content from a well curated public library. Likewise, when searching for specific sounds, personal audio collections or libraries of music samples and loops featured in some modern DAWs may not provide content that is topical enough for a given creative task. Online audio collections provide an opportunity to diversify the audio content that one can access to support creative needs. Storing large collections of audio files locally also takes space compared to downloading online content only when it is needed.

In addition to gaining access to a wider range of audio content, a connected DAW may also leverage web services providing information facilitating music analysis and production. These services can expand the intrinsic capabilities of native DAWs programmed for specific platforms. For example, artificial intelligence services could be deployed to provide DAWs with the capability to analyze multitrack audio to infer knowledge about musical attributes, e.g. instruments, chords, structure (Fazekas et al.,

2011) or to make synthetic renderings more expressive by modifying timbral patterns (Barthet et al., 2007). Cloud services can also be tailored to making recommendations to find similar sounding tracks (Fazekas et al., 2013), which could be used to find reference tracks in a mastering session or to find songs to learn for pedagogical purposes (Barthet et al., 2011). Web services may also be designed to share creative information generated within the DAW with online communities, for example metadata characterizing associations between equalization (EQ) parameters and timbre, in order to investigate various mixing techniques (Stables et al., 2016). Internet-connected DAWs also enable switching the workflow model from individual to collaborative music production. Tools such as Avid Cloud Collaboration for Pro Tools[16] or Ohm Studio[17] let multiple users access and work together on the same production project.

In summary, from the perspective of linear media production, Internet-connected DAWs have the following (nonexhaustive) interesting features:

- To support creativity by enabling access to audio content coming from a wide diversity of online audio content providers
- To benefit from web services based on artificial intelligence algorithms that expand the capabilities of native DAWs by providing users with additional metadata
- To share creative content with online user communities
- To collaborate on a production with remote users

Figure 10.10 illustrates how web technologies may benefit desktop DAWs by enriching features related to sound content, music analysis, synthesis and distribution.

Desktop DAWs can benefit from features brought by the Internet in different ways. One approach is to have web functionalities integrated directly within the DAW by means of appropriate extensions that expose features through the graphical user interface (GUI). For example, some manufacturers enable this through their Software Development Kit, see e.g. Reaper's Extensions SDK.[18] Another approach consists in installing web-enabled plugins developed using interfaces compatible with the DAW (e.g. Virtual Studio Technology, Audio Units). We present in sections 10.3.1 and 10.3.2 two examples of web-enabled audio plugins allowing users to search for, retrieve and process samples from online audio content providers (AudioGaming's AudioTexture and Waves Audio's SampleSurfer). A third, less direct approach involving apps external to the DAW, is to use online applications bridging the gap between online audio content and standard DAWs. This is the case for the Splice Studio[19] application, which enables users to search for sounds from an online database and import them to the DAW using drag-and-drop actions.

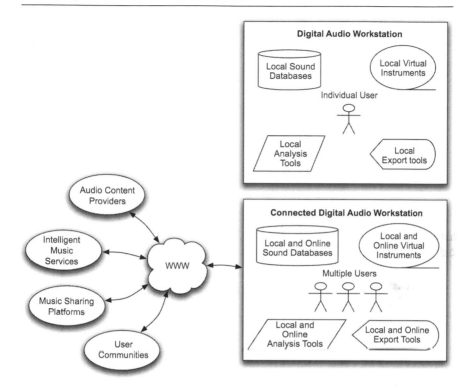

Figure 10.10 Shift from standard (top) to Internet-connected (bottom) digital audio workstations.

10.2.4.2. Web-Based Music Production Applications and Resources

Contrary to online desktop applications, web-based applications run in the browser and generally do not require users to install additional software components locally. The evolution of audio frameworks and APIs for the web, such as Web Audio, a high-level JavaScript API for processing and synthesizing audio,[20] has made possible the development of web applications supporting major features from desktop DAWs such as audio signal routing, sample-accurate sound playback with low latency, high dynamic range and mixing processing techniques. Chapter 10 in this series' second volume discusses how Web Audio can be used to create interactive music applications on the web.

Soundtrap[21] is an example of browser-based collaborative music production platform exploiting Web Audio (Lind and MacPherson, 2017). CCMixter[22] acts as a social music platform connecting instrumentalists, vocalists and producers to create music collaboratively using audio content licensed under CC (see section 10.2.3). This platform facilitates the sharing of *stems*, which are digital audio files containing a group of instruments or vocal tracks serving a specific function in a musical arrangement (e.g. bass, rhythmic, singing, accompaniment parts) and that can be used in combined ways to create novel music compositions (often called *remixes*). Playsound.space,[23] which we describe further in section 10.3.5, is a web application designed to let users mix CC audio content retrieved from the Freesound online provider by using semantic terms (Stolfi et al., 2018). The application can be used to compose soundscapes (e.g. Pigrem and Barthet, 2017) or to play free music improvisations.

10.2.4.3.   Cloud-Based Live Coding

Live coding is a musical practice involving the use of computer programming to generate sounds by writing and executing code on the fly (Collins et al., 2003). If the practice has traditionally relied on sound synthesis and samples stored on local sound databases or, more rarely, online databases, recent approaches have investigated how multiple online audio content could be repurposed during live coding (Xambó et al., 2018). Software modules can interface with a dedicated API, like Audio Commons', so that live coders can pull and further process online sounds during a live performance. This approach is examined further in section 10.3.4.

10.2.5.   Audio Commons Ecosystem

In the previous section, we reviewed several models with which online audio content and services could benefit linear media production. In order to bridge the existing gap between online audio content providers, music production software and end users, there is a need to establish a networked architecture and tools enabling the exchange of audio and licensing information at different points of the sound design and music production chain. The Audio Commons Initiative (Font et al., 2016) has started as a European-funded project investigating these issues with a particular focus on crowdsourced online audio content licensed under Creative Commons.

The Audio Commons Ecosystem (ACE), described in Figure 10.11, refers to the complex network made up of interconnected audio content, users (e.g. creators, consumers) and software systems, designed to support the aims of the Audio Commons Initiative. The ACE is designed so that content providers can expose CC audio content to amateurs and professionals from the creative industries alike and to provide an infrastructure for users to seamlessly integrate such CC content in creative workflows. The use cases presented in section 10.3 introduce technologies implementing this ecosystem.

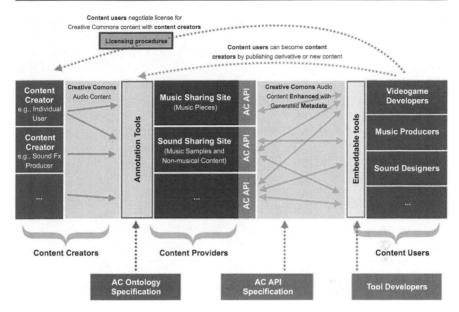

Figure 10.11 Conceptual Diagram of the Different Components That Are Interconnected in the Audio Commons Ecosystem.

To date, the Audio Commons Ecosystem includes the following content providers:

- *Freesound*[24]—providing a crowdsourced audio collection of several hundreds of thousands of nonmusical and musical sounds released under CC licenses
- *Jamendo*[25]—providing curated collections of several hundreds of thousands of songs from independent artists for free streaming and downloads (Jamendo Music) and commercial use through licensing (Jamendo Licensing)
- *Europeana*—hosting a collection of music and sounds related to European cultural heritage under CC licenses[26]

A portal to the ACE content can be found online (Audio Commons Music and Sound Search Tool, 2018).

10.2.6. Integrating Audio Commons Content Using Web APIs

The Audio Commons Ecosystem provides a web API that can be used to integrate Audio Commons content from the aforementioned sound and music providers in third-party applications. The Audio Commons API is offered as a REST API[27] that acts as an intermediary for individual content

provider APIs (e.g. Freesound, Jamendo and Europeana) and offers a unified interface to access all services. In the rest of this section, we show
simple "hello world" examples for using the Audio Commons API. A basic
knowledge of web technologies is assumed. It is out of the scope of this
chapter to describe how to build a platform for sound sharing and retrieval,
which the reader can learn from other sources (e.g. see Font et al., 2017).

In order to use the Audio Commons API, an account needs to be created at
the website of the Audio Commons Mediator.[28] Then, API credentials must
be generated through the Developers page. The reader can find details about
how to generate API credentials (and other Audio Commons API usage topics not covered in this section) in the Audio Commons API documentation.[29]
Assuming API credentials have been obtained and are properly included in
the requests sent to the API, we can use the Search URL to query all content
providers at once and get all responses in a single JSON file:

```
REQUEST:

https://m.audiocommons.org/api/v1/search/text/
?q=cars

RESPONSE:

{
  "meta": {
    "response_id": "8f64cbcd-47d2-4ac8-bebf-
    2c2d9f416cef",
    "status": "PR",
    "n_expected_responses": 3,
    "n_received_responses": 0,
    "sent_timestamp": "2018-07-03 13:19:20.091232",
    "collect_url": "https://m.audiocommons.org/
    api/v1/collect/?rid=8f64cbcd",
    "current_timestamp": "2018-07-03 13:19:20.
    101594"
  },
  "contents": {}
}
```

This will return a response containing the URL of where the search results can be retrieved as soon as individual content providers return a search result response. Accessing this URL will show the list of results retrieved so far:

REQUEST:

```
https://m.audiocommons.org/api/v1/collect/?rid=
8f64cbcd
```

RESPONSE:

```
{
  "meta": { . . . },
  "contents": {
    "Jamendo": {
      "num_results": 48,
      "results": [{
        "ac:id": "Jamendo:1317252",
        "ac:url": "http://www.jamendo.com/track/
        1317252",
        "ac:name": "1000miglia",
        "ac:author": "Naturalbodyartist",
        "ac:license": "BY-NC-SA",
        "ac:preview_url": "https://mp3d.
        jamendo.com/download/track/13172..."
      }, . . . ]},
    "Freesound": {
      "num_results": 7153,
      "results": [{
        "ac:id": "Freesound:326146",
        "ac:url": "https://freesound.org/s/326146/",
        "ac:name": "Inside Car Ambience Next to
        School . . .",
        "ac:author": "15050_Francois",
```

```
        "ac:license": "BY-NC",

        "ac:preview_url": "https://freesound.
        org/data/previews/326/32614..."

    }, . . . ]},

  "Europeana": {

    "num_results": 25,

    "results": [{

        "ac:id": "Europeana:/916107/wws_object_
        2164",

        "ac:url": "http://www.europeana.eu/
        portal/record/91610...",

        "ac:name": "Brokindsleden - The sounds
        of traffic",

        "ac:author": null,

        "ac:license": "BY",

        "ac:preview_url": "http://www.
        workwithsounds.eu/soundfiles/5a7/5..."

    }, ... ]}

  }

}
```

Note that in each of these audio results, there is an ac:preview_url metadata field, which points to a preview version of the actual sound file. This preview version can be played and downloaded. Queries can be narrowed down using filters. For example, the following query will return the same results as the preceding one but will only include sounds with Creative Commons Attribution license.

REQUEST:

```
https://m.audiocommons.org/api/v1/search/text/
?q=cars&f=ac:license:BY
```

The Audio Commons API provides a unified access to all of the services of the Audio Commons Ecosystem, but sometimes these services offer specific functionalities that are not supported by the Audio Commons API and can only be used by directly accessing the service's own API. As an example, Freesound supports similarity-based queries that, at the time of this writing, are not supported by the Audio Commons API. To use such a service, requests will need to be addressed to Freesound instead of Audio Commons. For example, assuming API credentials for Freesound have been obtained,[30] the following request will return a list of sounds that sound similar to the target Freesound sound with ID 291164:

REQUEST:

`https://freesound.org/apiv2/sounds/291164/similar/`

10.3.   Repurposing Audio Commons Content: Use Cases

In this section, we present five use cases from sound design to music production and live performance applications that exploit crowdsourced sounds differently: (1) sound texture generation using Le Sound's AudioTexture plugin, (2) music production using Waves Audio's SampleSurfer plugin, (3) soundscape composition leveraging online audio content, (4) live coding with the MIRLC library for SuperCollider, and (5) compositions through semantic ideation using the web-based app Playsound.space. This section should provide the reader with the width and breadth of potential novel media production applications relying on cloud-based audio databases.

10.3.1.   Sound Texture Generation With AudioTexture

Examples of sound textures include the sound of rain, crowd, wind and applauses (Saint-Arnaud and Popat, 1995; Strobl et al., 2006). Although there is no consensus on the definition of *sound texture*, an agreed working definition includes two main features: constant long-term characteristics and short attention span (Saint-Arnaud and Popat, 1995; Strobl et al., 2006). *Constant long-term characteristics* refer to a sound that emits similar characteristics over time (e.g. sustained pitch and rhythm) irrespective of the presence of local randomness and variation. This means that if two snippets are randomly picked from the same sound texture, they should sound similar. Therefore, in a sound texture, the sound is constantly sustained. *Attention span* refers to the time needed to characterize the texture, typically a few seconds. It is also worth mentioning that in conjunction with the high-level characteristics, looking more closely at a lower level, we find that sound textures are formed of *atoms* (basic sound snippets) that are repeated periodically, randomly or both, a behavior that is defined by the high-level characteristics. Sound textures can have multiple applications, ranging from background music and game music to audio synthesis and audio signal restoration (Lu et al., 2004).

Developed by Le Sound, AudioTexture is a plugin prototype for sound texture synthesis that leverages Audio Commons by bringing CC-licensed audio content into the DAW. The AudioTexture plugin lets users generate sound textures from audio recordings from either online or local databases within a DAW environment, such as Logic Pro X, Ableton Live or Reaper. In particular, the plugin integrates Audio Commons content for creative sonic/musical explorations in the form of sample-based synthesis or *concatenative synthesis* (see Schwarz, 2007) for a discussion on concatenative synthesis), which refers to the computational generation of sounds using existing sound samples. The plugin is particularly suited for environmental sounds with short-term repetitive units or atoms (e.g. water drops, rock falls, construction work and so on). It is, however, possible to use the plugin with musical sounds, which can also lead to interesting textures.

To operate AudioTexture, first users need to select a sound from either an online content provider from the Audio Commons Ecosystem such as Freesound (see Figure 10.12) or from a local database. The sound is

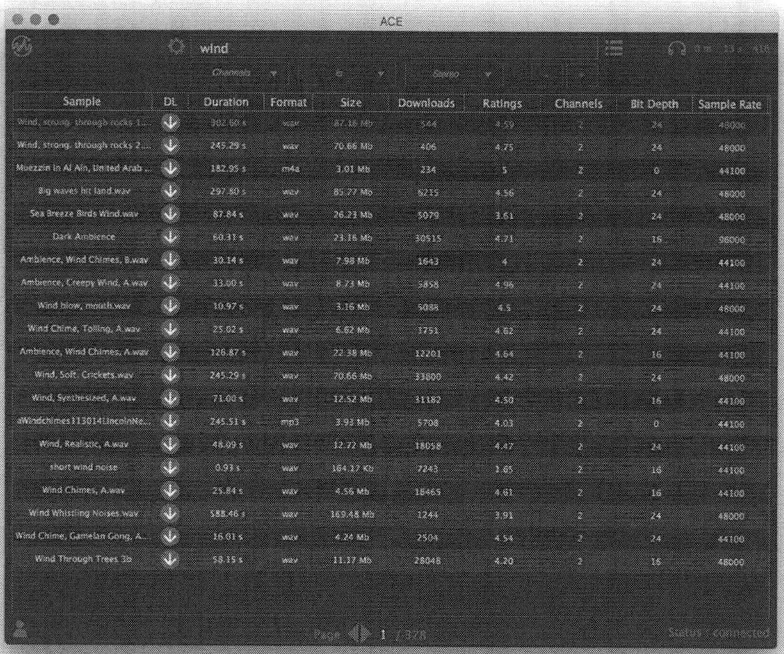

Figure 10.12 Screenshot of Audio Commons Retrieval Interface in Le Sound's AudioTexture Plugin.

Image credit: AudioGaming.

represented by a waveform providing overall temporal and envelope cues and a larger spectrogram for frequency and energy cues over time (see Figure 10.13). With automatic segmentation, the plugin decomposes the audio signal into adaptively defined atoms that are not equal in size. Markers can be visually activated to display the positions of the atoms and understand the behavior of the sound synthesis algorithm. The unit size of the atoms can be changed ranging from multiple small atoms to fewer, large bigger atoms. It is possible to select a range of the full sound using the horizontal slider X, which can be moved to start at any position of the sound. The control of the sound synthesis (the logic of how the atoms are played) is mainly determined by the values of three semantic (low-level audio) descriptors grouped in the vertical slider Y: energy, noisiness and brightness. AudioTexture includes factory presets that can guide users with suitable values (e.g. rain, fire, footsteps, birds, mechanics, waterfall, wave, applause, music).

The resulting sound texture can be recorded in the DAW by routing the output of a track with the plugin to another track armed for recording. A set of examples are provided in the companion website, which showcases the

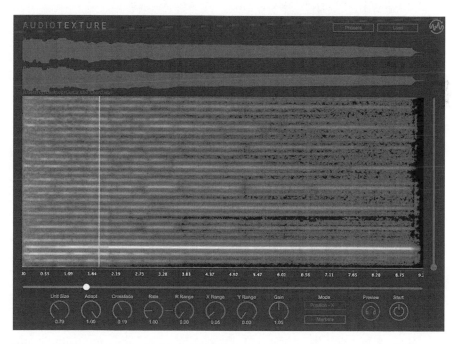

Figure 10.13 Screenshot of Sound Editing Interface in "Le Sound's AudioTexture Plugin.

Image credit: AudioGaming.

sonic and musical possibilities of the plugin using crowdsourced sounds (Audio Commons Routledge Website, 2018).

10.3.2.   Music Production With SampleSurfer

Music production is a whole field within music technology, which includes sound recording (Huber and Runstein, 2013), mixing (Owsinski, 2013) and sometimes audio mastering (Katz, 2003). Music production software is typically referred to as a digital audio workstation (DAW), which we discussed in section 10.2.4.

SampleSurfer has been developed by Waves Audio LTD and is another plugin for the ACE that serves as an audio content search engine based on semantic metadata and musical features. The plugin is designed to integrate Audio Commons sound and music samples in a DAW-based environment by providing basic editing capabilities (e.g. fades, trims) to optimize the music production workflow. It supports well established DAW applications, such as Logic Pro X and Ableton Live.

As shown in Figure 10.14, the plugin lets the user choose from a set of CC-licensed audio content providers (including Freesound, Jamendo), search by a set of filters including type of CC license, harmony (key, scale), beat (BPM, duration) and sample characteristics (sample format, sample rate, bit depth, number of channels). From a query, the user gets back a filtered list, from which sounds can be selected and downloaded. The user will be able to access the download history. The plugin offers edit tools, such as fade in and out, change of the amplitude level or trim of the sound to a shorter range from the original audio clip. Another feature offered by the plugin is the possibility of connecting to a folder in the local drive, scan and analyze the content and then be able to search the local material. The companion website presents a set of examples that illustrate the capabilities of the plugin (Audio Commons Routledge Website, 2018).

10.3.3.   Soundscape Composition Using Online Audio Content

Soundscape composition has grown from acoustic ecology and soundscape studies, which are fields seeking to document, archive and analyze the evolving sounds of our world (Schafer, 1993). Barry Truax defines *soundscape composition* as a form of electroacoustic music "characterised by the presence of recognizable environmental sounds and contexts, the purpose being to invoke the listener's associations, memories, and imagination related to the soundscape."[31] Reviews on soundscape composition approaches and their applications can be found in Truax (2002) and Pigrem and Barthet (2017), respectively. A large amount of crowdsourced online audio content is based on recordings of human- and nature-related environmental sounds. About two-thirds of Freesound's content falls

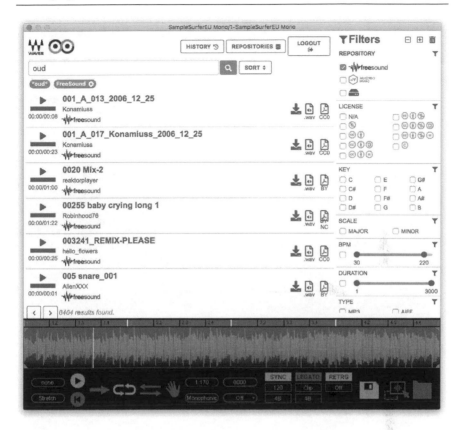

Figure 10.14 Screenshot of Search Interface and Edit Tools from Waves Audio's SampleSurfer Plugin.

Image credit: Waves Audio Ltd.

within these categories, and cultural heritage resources such as Europeana also offer environmental recordings.[32] Crowdsourced CC audio content can hence be a rich resource for soundscape composers looking to convey meanings about place and time through audio.

In the fall of 2017, students from the Sound Recording and Production Techniques module led by Mathieu Barthet at Queen Mary University of London were invited to produce short soundscapes leveraging Audio Commons online audio content and tools. Soundscape themes were ideated in class inspired by the *bootlegging* participatory design technique introduced in (Holmquist, 2008). Students had to write down on Post-its two ideas in each of the four following categories: character, place/environment, situation/action and mood. The Post-its were shuffled, and students had to pick up randomly one idea per category (see Figure 10.15).

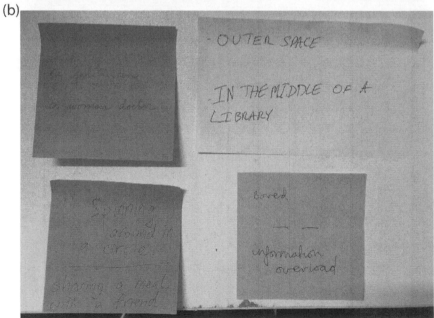

Figure 10.15 Participatory Ideation of Soundscape Themes: Students pick Post-its in character/place/situation/mood categories (left) and examples of categories (right).

Source: Photo by Mathieu Barthet.

After a brainstorming session, original soundscape themes emerged resulting from the combination of ideas from each category. Students were given three weeks to produce their short soundscapes. They could use an opened range of approaches, from figurative to abstract, and were given as a creative constraint to use only found sounds or loops retrieved/ processed using Freesound and AudioTexture (previously introduced) or Apple Loops (a royalty-free collection of prerecorded musical patterns and sound effects). Some examples of soundscapes produced by the students can be found on SoundCloud.[33] Although it mainly served as an academic purpose, this exercise demonstrated the ability of the ACE to respond to a wide range of audio production creative needs. More information can be found on our companion website (Audio Commons Routledge Website, 2018).

10.3.4. Live Coding With MIRLC

As pointed out in section 10.2.4.3, the musical practice of *live coding* is based on improvization and generation of code in real time (Collins et al., 2003). This can be done with several live coding environments, such as SuperCollider (McCartney, 2002), a platform for audio synthesis and algorithmic composition. In live coding, the integration of music information retrieval (MIR) techniques for sound retrieval and the use of Audio Commons content have been little explored. MIRLC (see Figure 10.16) is a

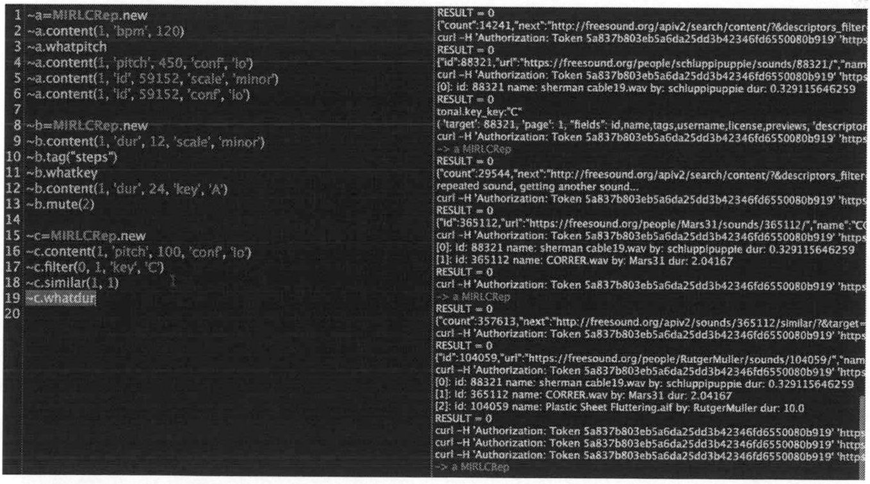

Figure 10.16 Screenshot Illustrating How the MIRLC Library Can Be Used During a Live Coding Session.

library designed to repurpose audio samples from Freesound, which can also be applied to local databases, by providing human-like queries and real-time performance capabilities (Xambó et al., 2018). The system is built within the SuperCollider environment by leveraging the Freesound API.

The novelty of this approach lies in exploiting high-level MIR methods (e.g. query by pitch or rhythmic cues) and ACE content using live coding techniques. Both content-based (e.g. similarity) and text-based (e.g. tags) queries are possible. Sounds are loaded in user-defined groups and played in a loop. Sounds of each group can be triggered either simultaneously or sequentially. This approach allows the user to load groups of related sounds and operate them with a higher level of control, which contrasts with operating single sounds. It is also possible to operate playback controls on a group of sounds (e.g. solo, mute, pause). Textual feedback is given of the different processes (e.g. queries, results).

The capacity of accessing a large amount of MIR parameters and sounds invites the live coder to explore and create subspaces of sounds through code. Feedback from four expert users reported that the tool has an interesting level of unpredictability and experimentation in the musical process and that querying was perceived as a nonlinear process, where sounds are retrieved organically following their own downloading times (Xambó et al., 2018). A credit list of the downloaded sounds is automatically created for each live coding session. A number of examples of live coding using crowdsourced online sounds are provided in the companion website (Audio Commons Routledge Website, 2018).

10.3.5.   Live Collaborative Music Making With Playsound

As discussed in section 10.2.1, sounds can be described with high-level semantic attributes representing how they were generated or what they express, in lieu of or in addition to musical characteristics such as pitch or chords. The Playsound.space[34] platform (Stolfi et al., 2018) was designed using Web Audio to let users mix Audio Commons content using semantic searches without requiring specific musical knowledge. The ACE provides ways to query sounds using descriptive metadata through its API. Playsound provides a fast access to the Freesound audio content and allows users to play and loop multiple audio files with basic editing capabilities, including segment selection, panning, playback rate and volume controls. Figure 10.17 shows the GUI of Playsound, with the list of selected sounds to the left and the search interface to the right, displaying the search text box, and the list of retrieved sound items with metadata and visual *spectrogram*[35] representations. Credits to authors of selected CC sounds are displayed at the bottom of the interface. Live interactions with Playsound can be recorded and exported.

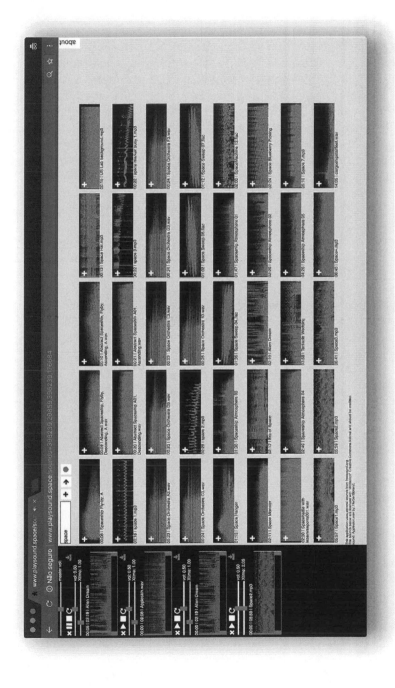

Figure 10.17 Graphical User Interface from the Playsound.space Web Application.

In Stolfi et al. (2018), the authors present a study where Playsound was used by small ensembles of laptop musicians to play free live music improvizations. Free music improvizations are not necessarily bound to predefined musical attributes and structure as in score-based compositions (e.g. key, chords, meter), and the activity emphasizes the performing process and the interaction between musicians (Bergstroem-Nielsen, 2016). Results indicated that it was easy for both musicians and nonmusicians to play live, collaboratively using the tool. The semantic sound search functionality facilitated verbal and nonverbal interaction between musicians and led to interesting musical situations through the use of similar or contrasting materials at different times and rich variation of timbres and rhythms. Users were able to express "sound ideas" even without technical expertise and musical technique. The companion website includes several examples of productions made with Playsound (Audio Commons Routledge Website, 2018).

10.4.   Conclusion

In this chapter, we reviewed some of the salient opportunities and challenges related to the use of Audio Commons online content in linear media production. We first outlined key concepts related to the uploading, retrieval or consumption of online audio content. These key concepts include principles to describe the technical and creative characteristics of online audio content (metadata, folksonomies), technologies that enable the search for sounds and navigate collections (semantic audio, text-based queries, content-based queries, graphic-based interfaces), and licenses. We then described how Audio Commons content can be reused and repurposed either through Internet-connected DAWs or browser-based applications. We introduced the Audio Commons application programming interface (API), which enables developers to access resources from the Audio Commons Ecosystem. Finally, we illustrated through five use cases how online Audio Commons content can be leveraged by sound designers and musicians. We presented a set of audio plugins and web-based interfaces to generate novel sound textures, integrate CC sounds into compositions, create soundscapes collaboratively, and augment live music performance with audio content from the web.

As shown in the preceding sections, combining crowdsourced and cloud-based online audio resources with the traditional DAW is a promising approach that can enrich the creation of media content, both original and remixed. Sound designers inclined to develop creative coding skills may also learn how to design their own web-based tools leveraging CC audio ecosystems (e.g. web development, practice-based research). These new skills can be combined with the traditional skill set of the professional sound designer in innovative ways leading to novel technologies and workflows yet to be discovered. The aim of this chapter was to highlight existing methods, tools and techniques that can be seen as a starting point in this promising new domain.

Notes

1. https://soundcloud.com
2. https://freesound.org
3. www.youtube.com
4. www.flickr.com
5. The sound description guidelines presented here are based on those provided by Freesound: https://freesound.org/help/faq/#hey-i-got-this-bad-description-moderated-file-can-you-help-me-create-a-better-description
6. http://audiocommons.jamendo.com
7. https://labs.freesound.org/fse/
8. https://creativecommons.org/licenses/
9. This flowchart by CC Australia is helpful for choosing the most suitable CC license: http://creativecommons.org.au/content/licensing-flowchart.pdf
10. https://creativecommons.org/share-your-work/licensing-types-examples/licensing-examples/
11. https://creativecommons.org/choose/
12. https://licensing.jamendo.com
13. A detailed explanation on how to credit properly can be found in the Freesound's FAQ section: https://freesound.org/help/faq/#how-do-i-creditattribute
14. A detailed explanation on how to remix or repurpose new sounds or music can be found in the Freesound's FAQ section: https://freesound.org/help/faq/#license-restrictions-when-publishing-new-sounds-that-includemodifyremix-other-sounds.
15. A complete chart and further explanations on how to legally remix CC-licensed material can be found in the CC website: https://creativecommons.org/faq/
16. www.avid.com/pro-tools/cloud-collaboration
17. www.ohmforce.com/OhmStudio.do
18. www.reaper.fm/sdk/plugin/plugin.php
19. https://splice.com/features/studio
20. www.w3.org/TR/webaudio/
21. www.soundtrap.com
22. http://ccmixter.org
23. www.playsound.space
24. https://freesound.org
25. www.jamendo.com
26. www.europeana.eu/portal/en/collections/music
27. https://en.wikipedia.org/wiki/Representational_state_transfer
28. https://m.audiocommons.org
29. https://m.audiocommons.org/docs/api.html

30. Documentation for the Freesound API can be found in http://freesound.org/docs/api/
31. www.sfu.ca/~truax/scomp.html
32. www.europeana.eu/portal/en/collections/music
33. Examples of soundscapes produced using CC sounds: https://soundcloud.com/qmulsrpt/ sets/qmul-short-soundscapes-2017-18
34. www.playsound.space
35. A spectrogram is a time-frequency representation of a sound indicating how the energy of frequency components evolve over time.

References

Audio Commons Routledge Website. (2018). Available at: www.audiocommons.org/rout ledge/ [Accessed July 2018].

Audio Commons Music and Sound Search Tool. (2018). Available at: http://audiocom mons.jamendo.com/ [Accessed July 2018].

Barthet, M., Kronland-Martinet, R. and Ystad, S. (2007). Improving musical expressive-ness by time-varying brightness shaping. In: *International symposium on computer music modeling and retrieval*. Berlin and Heidelberg: Springer, pp. 313–336.

Barthet, M., Anglade, A., Fazekas, G., Kolozali, S. and Macrae, R. (2 011). Music recom-mendation for music learning: Hotttabs, a multimedia guitar tutor. In: *Workshop on music recommendation and discovery*, ACM RecSys, pp. 7–13.

Barthet, M., Marston, D., Baume, C., Fazekas, G. and Sandler, M. (2013). Design and evaluation of semantic mood models for music recommendation. In: *Proceedings of the international society for music information retrieval conference*, International Society for Music Information Retrieval, pp. 421–426.

Barthet, M., Fazekas, G., Allik, A., Thalmann, F. and Sandler, M. B. (2016). From interac-tive to adaptive mood-based music listening experiences in social or personal con-texts. *Journal of Audio Engineering Society*, 64(9), pp. 673–682.

Bergstroem-Nielsen, C. (2016). Keywords in musical free improvisation. *Music and Arts in Action*, 5(1), pp. 11–18.

Bischoff, K., Firan, C. S., Nejdl, W. and Paiu, R. (2 008). Can all tags be used for search? In: *ACM conference on information and knowledge management*. ACM, pp. 193–202.

Choi, K., Fazekas, G., Sandler, M. and Cho, K. (2018). The effects of noisy labels on deep convolutional neural networks for music tagging. *IEEE Transactions on Emerging Topics in Computational Intelligence*, 2(2), pp. 139–149.

Cole, S. J. (2011). The prosumer and the project studio: The battle for distinction in the field of music recording. *Sociology*, 45(3), pp. 447–463.

Collins, N., McLean, A., Rohrhuber, J. and Ward, A. (2003). Live coding in laptop perfor-mance. *Organised Sound*, 8(3), pp. 321–330.

Fazekas, G., Barthet, M. and Sandler, M. (2013). The BBC desktop jukebox music recom-mendation system: A large scale trial with professional users. In: *IEEE international conference on multimedia and expo workshops*, pp. 1–2. Available at: https://iee explore.ieee.org/document/6618235 [Accessed March 2019].

Fazekas, G., Raimond, Y., Jakobson, K. and Sandler, M. (2011). An overview of semantic web activities in the OMRAS2 project. *Journal of New Music Research Special Issue on Music Informatics and the OMRAS2 Project*, 39(4), pp. 295–311.

Font, F. and Bandiera G. (2017). Freesound explorer: Make music while discovering Free-sound! In: *Proceedings of the web audio conference*. London: Queen Mary University of London.

Font, F., Brookes, T., Fazekas, G., Guerber, M., Burthe, A. L., Plans, D., Plumbley, M. D., Shaashua, M., Wang, W. and Serra, X. (2016). Audio commons: Bringing creative commons audio content to the creative industries. In: *Proceedings of the 61st AES international conference: Audio for games*, Audio Engineering Society.

Font, F., Roma, G. and Serra, X. (2017). Sound sharing and retrieval. In: T. Virtanen, M. D. Plumbley and D. Ellis, eds., *Computational analysis of sound scenes and events*. Cham, Switzerland: Springer International Publishing, pp. 279–301.

Gruber, T. (2008). Collective knowledge systems: Where the social web meets the semantic web. *Web Semantics: Science, Services and Agents on the World Wide Web*, 6(1), pp. 4–13.

Holmquist, L. E. (2008). Bootlegging: Multidisciplinary brainstorming with cut-ups. In: *Proceedings of the tenth anniversary conference on participatory design 2008*. Indianapolis, IN: Indiana University, p p. 158–161.

Huber, D. M. and Runstein, R. E. (2013). *Modern recording techniques*. New York: Focal Press.

Katz, B. (2003). *Mastering audio: The art and the science*. Oxford: Butterworth-Heinemann.

Lamere, P. (2008). Social tagging and music information retrieval. *Journal of New Music Research*, 37(2), pp. 101–114.

Lee, J. A. and Downie, J. A. (2004). Survey of music information needs, uses, and seeking behaviors: Preliminary findings. In: *Proceedings of the 5th international society for music information retrieval conference*, Universitat Pompeu Fabra.

Lessig, L. (2000). Copyright's first amendment. *UCLA Law Review*, 48, pp. 1057–1073.

Lessig, L. (2004). *Free culture: How big media uses technology and the law to lock down culture and control creativity*. New York: The Penguin Press.

Lind, F. and MacPherson, A. (2017). Soundtrap: A collaborative music studio with web audio. In: *Proceedings of 3rd web audio conference*. London: Queen Mary University of London.

Lu, L., Wenyin, L. and Zhang, H. J. (2004). Audio textures: Theory and applications. *IEEE Transactions on Speech and Audio Processing*, 12(2), pp. 156–167.

Marcell, M. M., Borella, D., Greene, M., Kerr, E. and Rogers, S. (2001). Confrontation naming of environmental sounds. *Journal of Clinical and Experimental Neuropsychology*, 22(6), pp. 830–864.

McCartney, J. (2002). Rethinking the computer music language: SuperCollider. *Computer Music Journal*, 26(4), pp. 61–68.

Merkley, R. (2015). *State of the commons*. Available at: https://apo.org.au/node/60681/ [Accessed July 2018].

Owsinski, B. (2013). *The mixing engineer's handbook*. Boston: Nelson Education.

Pigrem, J. and Barthet, M. (2017). Datascaping: Data sonification as a narrative device in soundscape composition. In: *Proceedings of the 12th international audio mostly conference on augmented and participatory sound and music experiences*. New York: ACM, pp. 431–438.

Ritzer, G. and Jurgenson, N. (2010). Production, consumption, prosumption: The nature of capitalism in the age of the digital 'prosumer.' *Journal of Consumer Culture*, 10(1), pp. 13–36.

Russell, J. A. (1980). A circumplex model of affect. *Personality and Social Psychology*, 39(6), pp. 1161–1178.

Saint-Arnaud, N. and Popat, K. (19 98). Analysis and synthesis of sound textures. In: D. F. Rosenthal and H. G. Okuno, eds., *Computational auditory scene analysis: Proceedings of the IJCAI-95 workshop,* CRC Press, pp. 293–308.

Schafer, R. M. (1993). *The soundscape: Our sonic environment and the tuning of the world*. Rochester, VT: Inner Traditions, Bear & Co.

Schwarz, D. (2007). Corpus-based concatenative synthesis. *IEEE Signal Processing Magazine*, 24(2), pp. 92–104.

Shadbolt, N., Berners-Lee, T. and Hall, W. (2006). The semantic web revisited. *IEEE Intelligent Systems* 21(3), pp. 96–101.

Smith, T. (2009). The social media revolution. *International Journal of Market Research*, 51(4), pp. 559–561.

Stables, R., De Man, B., Enderby, S., Reiss, J. D., Fazekas, G. and Wilmering, T. (2016). Semantic description of timbral transformations in music production. In: *Proceedings of the 2016 ACM on multimedia conference*. ACM, pp. 337–341.

Stallman, R. (2002). *Free software, free society: Selected essays of Richard M. Stallman*. Boston: GNU Press.

Stolfi, A. de Souza, Ceriani, M., Turchet, L. and Barthet, M. (2 018). Playsound space: Inclusive free music improvisations using audio commons. In: *Proceedings of the international conference on new interfaces for musical expression*, pp. 228–233.

Strobl, G., Eckel, G., Rocchesso, D. and Le Grazie, S. (2006). Sound texture modeling: A survey. In: *Proceedings of the 2006 sound and music computing international conference*.

Truax, B. (2002). Genres and techniques of soundscape composition as developed at Simon Fraser University. *Organised Sound*, 7(1), pp. 5–14.

Xambó, A., Roma, G., Lerch, A., Barthet, M. and Fakekas, G. (2 018). Live repurposing of sounds: MIR explorations with personal and crowdsourced databases. In: *Proceedings of the international conference on new interfaces for musical expression*, pp. 364–369.

11

Sound Ontologies

Methods and Approaches for the Description of Sound

Davide Andrea Mauro and Andrea Valle

11.1. Introduction

Categorizing and classifying sounds is a useful tool to organize a palette for sound designers and composers. If doing this for acoustic and traditional instruments is a relatively well established task, e.g. Von Hornbostel and Sachs (1961), the same cannot be said for any sound in general. With sounds that are not originated by traditional instruments (or even produced by using traditional instruments in nontraditional ways), we might lack the same rigorous set of criteria, e.g. the originating mechanism: vibrating strings or resonating membranes. For this reason, employing the same classification rules will not necessarily lead to the desired outcomes.

Categories and ontologies can be used to relax the strict requirements of a formal classification and allow users to organize their personal sonic space. As the color palette prepared by a painter contains only a subset of all the possible colors, the tools that we use to synthesize sounds present us with a limited set of options, so rather than being neutral or agnostic tools, they are contributing in shaping our own creativity. Furthermore, ontological representations of sounds are required in order to support a semantic retrieval of sound resources, as in accessing sounds from a library.

Our goal is to present notable attempts in these directions highlighting both the technical/technological substrate and their philosophical approach.

11.2. Phenomenological Approaches

Phenomenological approaches to sound description (e.g. Erickson, 1975) are intended to elicit categories that are perceptually relevant to the listener without an explicit reference to their acoustic properties. While such categories may be culturally biased, still a careful definition can provide useful ways to identify and describe a variety of sounds. In relation to such an approach, the most relevant proposal is the one by Schaeffer (2017), dating back to 1966. Schaeffer has proposed a double-sided analytic device—a typo-morphology—intended as a multifaceted tool for the description of all the objects of the audible domain (*sound objects*). In particular, the typology is meant as the description of a sound object in relation to other objects, while the morphology is intended as a description of the sound object per se. Starting from the latter, morphological criteria are defined as a set of seven analytical properties (i.e. parameters having different values) characterizing a sound object. These criteria are (Chion, 2009):

1. *Mass*—mode of occupation of the pitch-field by the sound. Differently from pitch, mass takes into account two notions: site as a position on the continuum (i.e. as the actual register of the sound object) and caliber, indicating properly a range of occupation. Pitched sounds thus have a limited caliber, while noisy sounds have a greater caliber.
2. *Harmonic timbre*—diffuse halos of the sound and associated qualities that seem to be linked with mass
3. *Dynamic*—development of sound in the intensity-field
4. *Grain*—micro-structure of the matter of the sound, suggesting the texture of a cloth or mineral
5. *Allure*—oscillation, characteristic vibrato of the sustainment of sound
6. *Melodic profile*—general profile of a sound developing in tessitura
7. *Mass profile*—general profile of a sound where the mass is sculpted by internal variations

Taken together, these criteria are able to describe in detail many qualitative aspects of a sound. The morphological point of view has been widely reconsidered by Smalley (1986 and 1997), who has proposed a *spectro-morphology*. The term clearly refers to spectral content of sound, but the proposal does not take into account physical notions or measures. In a spectral typology, the continuum *note-noise* includes *node* as its middle term. Spectral content can be basically categorized along the axis *gesture-texture*. In relation to sound, gesture indicates a figurative bonding toward an acoustic model and a clear direction in development. Texture is related to internal behavior patterning. Sound gestures are described according to three *morphological archetypes* that can be combined into *morphological models*: *attack, attack-decay, graduated continuant*. Spectral motion, i.e. the way in which sound evolves, can be organized into a typology based on five basic types: *unidirectional, bidirectional, reciprocal, centric/cyclic, eccentric/multidirectional*. At a higher level (i.e. in relation to complex, evolving sounds), Smalley (1986) proposes a classification of 15 structural functions, modeled following the morphological archetypes. An example of description (related to spectral density) is shown in Figure 11.1.

Differently from morphology's analytical criteria, Schaeffer (2017)'s typology is meant as a way to define each sound object in relation to other sound objects. Six typological categories for sound description are identified (*mass, variation, duration, sustain, facture, balance*), then they are tentatively combined in a two-dimensional space for sake of simplicity. This space is a sort of cartography of potential sounds (Risset, 1999), and each object can be described by assigning it to a position. In its final arrangement, the typological space is divided in 28 areas, representing typological labeled *classes*, and the areas are grouped into three *regions* (balanced, slightly original, too original). Every sound (object) thus belongs to a certain class and consequently to a certain region.

Valle (2015 and 2016) has suggested a simplified revision of Schaeffer's space by isolating four (rather than six) categories: *sustain, profile, mass, variation*. Sustain describes a sound object's internal temporality. Thus, in relation to sustain, it is possible to individuate three cases:

1. *Sustained*—constant activity over time
2. *Impulsive*—activity as a singular moment
3. *Iterative*—activity as a series of repeated contributions

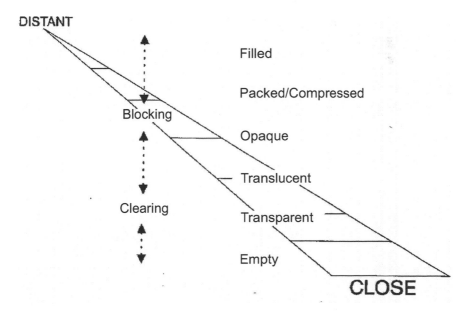

Figure 11.1 Description of Spectral Density.

Source: From Smalley (1997)

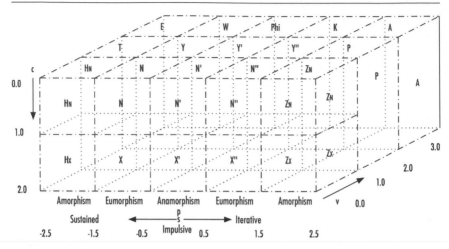

Figure 11.2 Typological Space for Sound Objects.

Source: From Valle (2015)

While *sustain* defines the way in which a sound object is maintained into duration, *profile* describes its external temporal form, in relation to beginning, duration, ending:

1. *Eumorphism*—relevance of all the three categories. The sound object has a well-defined temporal shape.
2. *Amorphism*—Duration is relevant, while beginning and end are not (amorphous sounds lasts indefinitely or do not depend on beginning/end).
3. *Anamorphism*—Profile is compressed, and duration is not relevant (sound objects as events).

Sustain and *profile* are orthogonal categories that collapse in the case of impulsive sustain and amorphous profile. Finally, temporality in specific relation to mass is articulated by Schaeffer by introducing *variation* as a criterion, which allows us to describe how much the mass (site/caliber) changes in time (from stable to varying objects). The previous dimensions can be combined into a three-dimensional space (Figure 11.2), where letters represent classes referring to Schaeffer's usage, and the axes receive arbitrary numerical ranges that have the only means of providing a reference for an explicit annotation.

11.3. Voice-Based Approaches

The qualitative features of sound may find a reference in acoustic instruments and everyday objects. In this sense, the human voice provides a basic, embodied tool against which to describe and categorize sounds. *Articulatory phonetics* deals with the descriptions of the mechanics of sound production in speech. While indeed not all sounds are available for human vocal production, a wide variety is in use in languages and can act as a reference for sound identification. The International Phonetic Association (IPA) has proposed over the years an International Phonetic Alphabet (also IPA) to annotate speech sound.[1] The main distinction is between consonants (unpitched, noisy) and vowels (pitched). In the IPA, consonants are organized in a two-dimensional chart (Figure 11.3) by their *place* (which part of the vocal tract is obstructed) and *manner* (how it is obstructed).

The resulting chart allows for the description and annotation of many sounds and for the possible identification of their similarity. Vowels are described by IPA by means of a space that couples *height* and *backness* (Figure 11.4).

The first is related to the aperture of the jaw (close-open); the second indicates the position of the tongue relative to the back of the mouth. Such a space is continuous and may provide the sound designer hints to describe harmonic, pitched sounds in terms of vowel qualities. As an example, Takada et al. (2010) use IPA transcriptions to annotate environmental sounds. The study of speech has prompted other general investigations on the description of sound qualities. Since Jakobson et al. (1952), acoustic phonetics has been instrumental in exploiting sonograms as compact time/frequency representations for sounds and in proposing descriptive categories to differentiate spectral mixtures. Moving from such studies, Cogan (1984) has proposed 13 categories for the general interpretation of spectral phenomena (Figure 11.5). In the annotation of sound spectra, these categories can receive four different values: negative, positive (if respectively the first or the second one is dominant), mixed (if both are present) or neutral (if not relevant). It can be observed that both frequency content and time are taken into account.

CONSONANTS (PULMONIC) © 2015 IPA

	Bilabial	Labiodental	Dental	Alveolar	Postalveolar	Retroflex	Palatal	Velar	Uvular	Pharyngeal	Glottal
Plosive	p b			t d		ʈ ɖ	c ɟ	k ɡ	q ɢ		ʔ
Nasal	m	ɱ		n		ɳ	ɲ	ŋ	N		
Trill	ʙ			r					ʀ		
Tap or Flap		ⱱ		ɾ		ɽ					
Fricative	ɸ β	f v	θ ð	s z	ʃ ʒ	ʂ ʐ	ç ʝ	x ɣ	χ ʁ	ħ ʕ	h ɦ
Lateral fricative				ɬ ɮ							
Approximant		ʋ		ɹ		ɻ	j	ɰ			
Lateral approximant				l		ɭ	ʎ	L			

Symbols to the right in a cell are voiced, to the left are voiceless. Shaded areas denote articulations judged impossible.

Figure 11.3 IPA alphabet: Consonants.

VOWELS

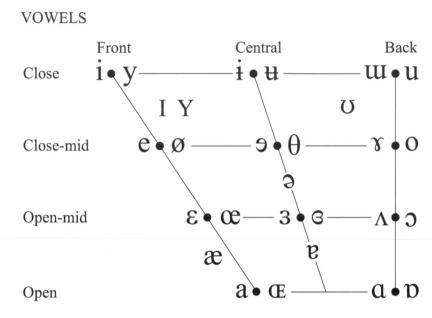

Where symbols appera in pairs, the one
to the right represents a rounded vowel.

Figure 11.4   IPA Alphabet, Vowels.

On the same path, Slawson (1981, 1985) explicitly distinguishes timbre
from sound color. While in Cogan the temporal dimension is still relevant,
in Slawson sound color is what remains of sound qualities once time features are eliminated. Starting from phonological categories by Jakobson
et al. (1952) and Jakobson and Halle (1956), Slawson (1985) has proposed
a two-dimensional space for a specific subset of timbre, named *sound
color*, and inspired by vowel formant space. A formant space is constructed
by coupling on its two axes the frequencies of the first two formants, i.e.
spectral peaks of a vowel (typically indicated as F1 and F2). Such an organization is able to provide a clear definition of vowels in terms of positions
into specific regions of the formant space (Fant, 1960). Starting from such
a space, Slawson hypothesizes three dimensions for sound color: *openness, acuteness, laxness*. They are correlated to the physical characteristics of the speech filter through the two formant frequencies. Openness
indicates the opening of the oral cavity as a filter deformation ([i]–[æ]);
acuteness grows according to the second resonance ([u]–[i]); laxness
indicates the state of relaxation of muscle tension ([u]–[ə], a vowel produced by no constriction at all). A fourth dimension, *smallness*, models the
length of the vowel tube and can be thought as the height involved in the
vowel (resulting from the difference between the two formants, [u]–[a]).

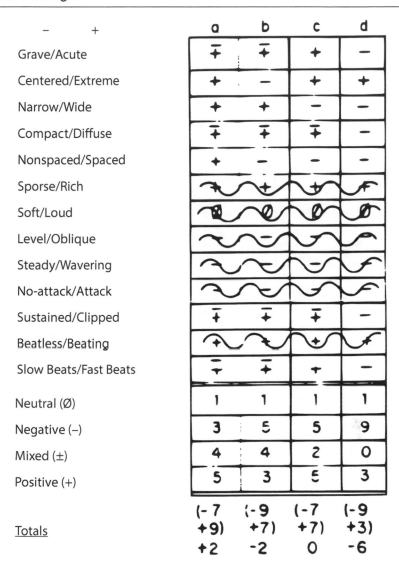

Figure 11.5   An Example of Analysis Chart by Annotation of Spectral Categories.

Source: From Cogan (1984)

Precisely as a consequence of vocal abstraction, the formant space F1/ F2 is a strictly continuous one. For each of the four dimensions, Slawson defines a family of *loci* in which the value of the dimension is invariant, i.e. isometric contours of equal openness, equal acuteness, equal laxness and equal smallness (Figure 11.6, in which the author uses "ne" for [ə]; equal smallness is omitted). A relevant feature of this space is that, even

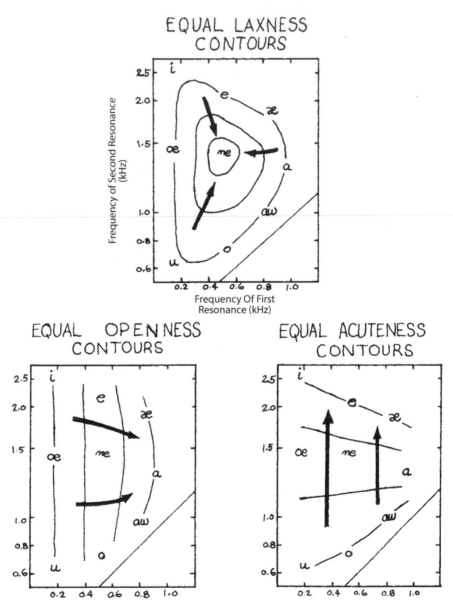

Figure 11.6 Three Dimensions of Sound Color in a F1/F2 Space.

Source: From Slawson (1981)

if inspired from acoustic measurements of formant frequencies, in the end it refers to the speech articulation. In short, a sound color is defined as a vowel color. In this sense, Slawson's space may be thought of also as a phenomenological one, as timbre can be described with reference to the human voice. An operational aspect of the space is that it allows timbre transpositions as geometrical translations. Along the same path, McAdams and Saariaho (1991) have proposed a voice-based organization of timbre.

11.4. Psychoacoustic Approaches

Classification of sounds has been pursued in psychoacoustic studies mainly in relation to *timbre*. The latter is intended as a qualitative feature of sound, i.e. the subjective counterpart of the spectral composition of tones, even if it has been proved that the temporal behavior crucially contributes to such a qualitative assessment (Rasch and Plomp, 1982).

Timbre cannot be ordered on a single scale, as it is a multidimensional attribute of the perception of sound. Hence the need to identify various attributes and to arrange them via multidimensional scaling. A common technique is to collect similarity judgments (Plomp, 1976; Rasch and Plomp, 1982; Grey, 1977; Wessel, 1979) and to arrange them into a space (a *timbral space*), in which geometry respects their similarities. The arrangement per se does not provide semantic categories. Yet, as many times the sounds are collected from acoustic instruments (e.g. organ stops in Plomp, 1976 and Rasch and Plomp, 1982 or orchestral instruments in Grey, 1977 and Wessel, 1979), the reference to commonly used musical instruments may act as reference for sound identity. The visual representation of timbral spaces provides per se a sort of similarity map that can be explored and exploited in the production context. This is explicitly the aim of Wessel (1979) in defining a parallelogram model that is able to predict timbre analogies by means of geometrical patterns in the space, to be used in sound design (analogously to what happens in Slawson, as previously explained). In timbre studies, many semantic categories (i.e. couples of verbal terms intended to be opposite) have been proposed to categorize sounds, to be possibly matched with (hence causally motivated by) acoustic features. As an example, Bismarck (1974) provides to listeners 30 verbal categories (like hard-soft, sharp-dull, coarse-fine). While such categories are proposed in input to the listeners, other verbal categories result from the interpretation of timbral spaces, mostly by individuating common acoustic features (Handel, 1989). Bismarck (1974) sums up his research proposing *sharpness* and *compactness*. The first is related to the distribution of spectral energy around the higher-frequency region, the second is a factor distinguishing between tonal (compact) and noisy (noncompact) aspects of sound (Rasch and Plomp, 1982). Plomp (1976) suggests few versus many strong higher harmonics for his space. Poli and Prandoni (1997) indicate *brightness* (boosting of the fundamental frequency) and *presence* (spectrum midband enhancement). The three axes in Grey (1977)'s space (Figure 11.7) can be interpreted as spectral energy distribution (narrow-wide), synchronicity (in the collective attacks and decay of upper harmonics, i.e. spectrally stable versus fluctuating), attack dispersion (high-frequency, scattered energy versus energy concentrated on the fundamental frequency, i.e. buzz-like versus soft attack).

It is apparent how, together with spectral features, the temporal dimension (i.e. the attack) is a crucial factor in identifying sound. Accordingly, Wessel (1979) describes the two axes of his two-dimensional arrangement (Figure 11.8) in relation to the spectral energy distribution of the tones and to the nature of the onset transient. The resulting categories can be

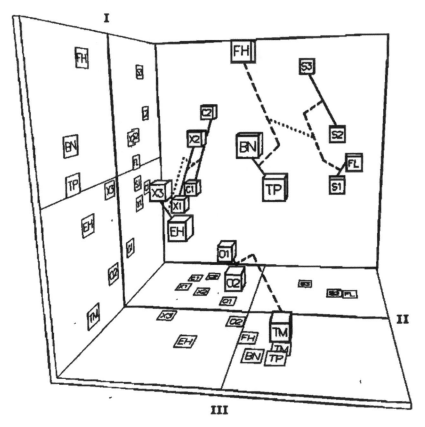

Figure 11.7 Three-Dimensional Spatial Solution for 35 Sounds.

Source: From Grey (1977)

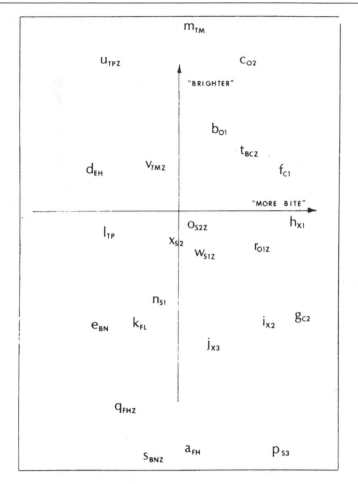

Figure 11.8 Two-Dimensional Timbre Space Representation of 24 Instrument-Like Sounds.

Source: From Wessel (1979)

labeled, respectively, as *bright/mellow* and *soft/biting*. McAdams (1999) individuates three dimensions. Spectral centroid is the center of gravity of the spectrum, spectral flux is intended as the degree of variation of spectrum in time, and log attack time is indeed related to onset time. To these, spectral smoothness can be added, as the degree of amplitude variation between adjacent partials.

To sum up, various categories have been proposed to describe timbre as an overall quality of sounds.

While temporal aspects have proven to be relevant, still it seems a general feature can be found that can be extracted from most of the previous discussion and seems to be a semantic correlative of an energetic acoustic quality. *Brightness* (versus *dullness*) is thus the one dimension that tends to be found across a variety of studies (Bregman, 1990).

11.5. Ecological Approaches

Ecological approaches to sound have introduced specific ways to describe sounds in relation to their environmental context. The *soundscape* is thus the sonic counterpart of the landscape.

In his foundational work, Murray Schafer (1977) introduced the lo-/ hi-fi categorization and a tripartite classification of sound material into *keynote sounds*, *sound signals*, and *soundmarks*. Starting from the former, keynote sounds are the sounds heard by a particular society continuously or frequently enough to form a background against which other sounds are perceived (e.g. the sound of the sea for a maritime community). Signals stand to keynote sounds as a figure stands to a background: they emerge as isolated sounds against a keynote background (e.g. a fire alarm). Sound-marks are socially or historically relevant signals (e.g. the ringing of the historical bell tower of a city).

In relation to soundscape, Böhme (2000) has proposed an aesthetics of *atmospheres*. Every soundscape has indeed a specific scenic atmosphere, which includes explicitly an emotional and cultural dimension. An atmo-sphere is an overall layer of sound that cannot be analytically decomposed into single sound objects, as no particular sound object emerges from it. While keynote sounds are intended as background sounds (i.e. they are a layer of the soundscape), atmospheres identify the whole sound complex.

The *lo-/hi-fi* categorization differentiates soundscape in relation to the presence of large masking sounds. It is both a theoretical and a historical classification. Premodern soundscape, where low-intensity sonic details were audible, is typically hi-fi, while modern soundscape, after the elec-tromechanic revolution, is typically lo-fi. Apart from this historical con-text, the distinction is useful as a general classification of soundscapes.

Krause (2015) has extensively used the tripartition *geophony/biopho-ny/anthrophony* to characterize environmental sounds in relation to three layers of production, respectively natural nonliving phenomena, living beings and humans.

11.6. Analytical Approaches

It is important to note that, rather than a single unique classification approach to sound, many possible perspectives can emerge, each one highlighting a specific mechanism behind it (e.g. what types of similarities are used) and giving birth to a peculiar set of categories.

Houix et al. (2012) present experiments aimed at classifying environmental sounds and strategies for their categorization. Their analysis starts with the first attempt of Vanderveer (1979) that shows participants grouping sounds together by analyzing the cause that produced such interaction or according to some acoustic properties.

Marcell et al. (2000) report a classification of 120 environmental sounds within 27 very heterogeneous categories corresponding to sound sources (e.g. four-legged animal, air transportation, human, tool, water/liquid), locations or contexts (kitchen, bathroom) or more abstract concepts (hygiene, sickness).

Gygi et al. (2007) report similar results, finding 13 major categories based on 50 sounds. The most frequently used categories referred to the type of sources (e.g. animals/people, vehicles/mechanical, musical and water). In a lesser proportion, sounds were grouped by context (e.g. outdoor sports) or location (e.g. household, office, bar).

Guyot et al. (1997) propose a framework for the classification of environmental sounds based on two strategies: the first is based on psychoacoustic criteria (e.g. pitch, temporal evolution), while the second is based on the identification of the source. They use the three levels of abstraction formalized by Rosch and Lloyd (1978): superordinate, base and subordinate levels. At the superordinate level, listeners identify the abstract mechanism of sound production. At the base level, they identify actions. And at the subordinate level, they identify the source. The different types of categories are not mutually exclusive and can be mixed during a classification task (across participants and/or for a single participant) because a sound can belong to multiple categories corresponding to different conceptual organizations. This cognitive process has been called "cross-classifications" in Ross and Murphy (1999).

A notable attempt at classifying everyday sounds has been proposed by Gaver (1993a, 1993b) representing different classes of physical interactions (solids, liquids, gases). The system has a hierarchical structure (similar to a taxonomy) and is based on the physics of sound-producing events. Gaver himself observes that the framework is not exhaustive and that entirely different ways of organizing the materials are indeed possible.

To read more on this topic, see also Stefano Delle Monache and Davide Rocchesso's chapter, "Sketching Sonic Interactions," in the third volume of this series, *Foundations in Sound Design for Embedded Media*, where the authors present a way to organize sonic material to build a personalized sonic sketchbook.

A common problem with classification is that different users, even adopting the same taxonomy, can classify the same object into different classes, and they might want to do that assigning a different "degree of membership." In Ferrara et al. (2006), the authors investigate an ontology aimed at describing music pieces and address this specific problem in terms of genre classification. Assigning a degree of membership is typical of fuzzy logics, and different strategies can be employed to modify standard tools to support the aforementioned concept. The problem can be addressed either by extending the tools to fully support those concepts at the cost of losing compliance and support of the main standard tools used for working with ontologies or by finding ways of expressing the same concepts from within the standard. For the latter solution, extra work is required in order to tweak the tools at the risk of losing a bit of simplicity but retaining compliance with the standards.

An important contribution to the definition of sound ontologies comes from the field of computational auditory scene analysis (CASA) where the definition of suitable ontologies is a requirement in order to provide a meaningful classification of the events. In Nakatani and Okuno (1998), the sound ontology is composed of three elements: *sound classes*, definitions of *individual sound attributes*, and their *relationships*. The ontology is defined hierarchically by using:

1. *Part-of*—a hierarchy based on the inclusion relation between sounds.
2. *Is-a*—a hierarchy based on the abstraction level of sound.

The Part-of hierarchy of basic sound classes is composed of four layers of sound classes. A *sound source* is a temporal sequence of sounds generated by a single sound source. A *sound source group* is a set of sound sources that share some common characteristics as music. A *single tone* is a sound that continues without any durations of silence, and it has some low-level attributes such as harmonic structure. In each layer, an upper class is composed of lower classes that are components sharing some common characteristics. For example, a harmonic stream is composed of frequency components that have harmonic relationships. The Is-a hierarchy can be constructed using any abstraction level. For example, voice, female voice, the voice of a particular woman and the woman's nasal voice form make up an Is-a hierarchy. With sound ontology, each class has some attributes, such as fundamental frequency, rhythm and timbre. A lower class in the Is-a hierarchy inherits the attributes of its upper classes by default. In other words, an abstract sound class has attributes that are common to more concrete sound classes.

Burger et al. (2012), without explicitly referring to ontologies, define 42 "noisemes" as fundamental atomic units of sound capturing objective properties of the acoustic signal. The labels are used in a classification task for environmental noise sounds.

Fields (2007) reflects on the difference between a top-down approach (ontology), and a bottom-up approach (folksonomies) and applies this conceptualization framework to building an audio ontology describing the *Computer Music Tutorial* by Curtis Roads using the Protégé tool (Musen, 2015).

In Lobanova et al. (2007), the authors deal with sounds that cannot be categorized by linking them to a source concept. Their implementation is based on WordNet (Miller, 1995), a widely used lexical resource in computational linguistics.

In Bones et al. (2018), the authors emphasize how categorization of sounds is based upon different strategies depending on context and the availability of cues. The study is focused on the categorization of three different types of environmental sound: dog, engine and water sounds, for which subjects were able to describe sounds with three types of attributes: the *source-event* (referring to the inferred source of the sound), the *acoustic signal* (explicitly referring to the sound itself) or a *subjective-state* (describing an emotional response caused by the sound or the sound source).

11.7. Technical Tools and Applications

In order to semantically enhance the retrieval of sound files (or profiles), many have attempted to define an annotation and classification schema. The first technical problem is the language of such scheme. With the advent of semantic technologies for the web and RDF-based vocabularies, semantic interoperability has become one of the desiderata of data models. In this regard, a tendency that showed up is the confluence of different domain-specific vocabularies in more general ontologies used for data integration. In 2012, the World Wide Web Consortium (W3C) mapped some of the most used schemas for media objects in the Ontology for Media Resources: OWL Web Ontology Language (VV.AA., 2012), recommending it for the annotation of digital media on the web. The W3C specification is not bound to a serialization in a particular language, so it can be used as a general schema.

In Hatala et al. (2004), the authors use ontologies to retrieve sound objects in an augmented reality (AR) application for museums. This implementation uses DAML+OIL (a standard now superseded by OWL) but highlights how the designers valued the possibility of performing reasoning with the system to automatically retrieve digital objects. The auditory interface follows an ecological approach to sound composition. Three areas are taken into account: psychoacoustic, cognitive and compositional. Psychoacoustic features of the ecological balance include spectral balancing of audible layers. Cognitive aspects of listening are represented by content-based criteria. Compositional aspects are addressed in the form of the orchestration of an ambient informational soundscape of immersion and flow that allows for the interactive involvement of the visitor.

In the SoDA project, Valle et al. (2014) organize a collection of sounds and implement a semantic search engine based on classical techniques borrowed from information retrieval (IR), whose main task is finding relevant "documents" on the basis of the user's information needs, expressed to the system by a query. The annotation is based on an OWL schema inspired by annotation in state-of-the-art sound libraries. Among the libraries used by sound designers taken into consideration are Sound Ideas Series (6000, 7000 and 10000), World Series of Sound, Renaissance SFX. The search tools for audio documents taken into account were SoundMiner,[2] Library Monkey,[3] Basehead,[4] Audiofinder.[5]

Audio Set, presented in Gemmeke et al. (2017), is a large-scale data set of manually annotated audio events using a structured hierarchical ontology of 632 audio classes. One of the aims of the authors is to bridge the relatively large gap that still exists between image recognition and sound recognition, providing a comprehensive coverage of real-world sound. The ontology is released as a JSON file.[6] In the creation of the ontology, in order to avoid biasing the categories, the authors started from a neutral, large-scale analysis of web text.

Even if not formally presented as ontologies, a number of tools have attempted to organize a sonic space in order to enable the composer/performer to search and act on such a potentially vast space.

In Rocchesso et al. (2016), the authors propose to represent the sonic space of a sound model as a plane where a number of prototype synthetic sounds are positioned. The spatial organization is based on a dimensionality reduction on the set of available sound, each represented by a high-dimensional feature vector. Two-dimensional spaces are particularly relevant for sound designers (see the preceding discussion) because they can be used as sonic maps, possibly accompanied by few landmarks that are highlighted and serve the role of prototypical sounds for a certain "class."

For similar tasks, the first problem is how to represent and describe our sounds: whereas digital signals are described by sequences of many values, we want to obtain compact descriptions that can be better manipulated. In the area of music information retrieval, a lot of research has been devoted to automatically extract descriptors (or features; the MIRToolbox presented in Lartillot and Toiviainen 2007 is widely adopted for these tasks) that could concisely represent sounds. Once the sounds are associated with a compact representation, it is possible to try to organize them in a low-dimensional space (typically two or three). A classic way to do that is by means of principal component analysis (PCA), which is based on singular value decomposition (SVD). An in-depth description of all these methodologies is beyond the scope of this chapter, but Drioli et al. (2009), Scavone et al. (2001), and Fernström and Brazil (2001) can be used as starting point to explore how to apply those techniques to this task.

The main goal of the previously discussed tools is to allow the users to use the power of IR techniques to be able to search and navigate through a collection of sounds. It is important to point out that a main limitation is that some of the processes presented here are manual or at least partially supervised by human input. The reason behind this characteristic resides in the already mentioned polymorphic nature of classification strategies.

11.8. Conclusions

In this chapter, we presented an overview on the topic of ontologies and, more in general, the classification and organization of sounds. We immediately realized how it is not possible to define a single unifying approach to sound organization because many different criteria may be useful in the classification tasks depending on various factors. As it is not either possible or useful to define a single classification strategy for sounds, in sound design a multifaceted approach is the most apt to cope with the variety of design contexts. Following these assumptions, we thus presented the reader a number of approaches grouped into five different categories (phenomenological, voice-based, psychoacoustic, ecological and analytical) highlighting the various possible mechanisms that can be employed by the sound designer when she or he wants to attempt a personal classification/organization of a sound materials. In short, the tools that the sound designer seeks to use and develop need to be flexible enough to reflect this variety of contexts and possibilities. The application examples that we introduced in the final section are thus intended to show how to exploit, at various levels, the aforementioned strategies.

Notes

1. www.internationalphoneticassociation.org
2. http://store.soundminer.com
3. www.monkey-tools.com/products/library-monkey/
4. www.baseheadinc.com
5. www.icedaudio.com
6. http://g.co/audioset

References

Böhme, G. (2000). Acoustic atmospheres. A contribution to the study of ecological aesthetics. *Soundscape: The Journal of Acoustic Ecology*, I(1), pp. 14–18.

Bones, O., Cox, T. J. and Davies, W. J. (2018). Distinct categorization strategies for different types of environmental sounds. In: *Euronoise 2018*, Crete, EEA-HELINA.

Bregman, A. (1990). *Auditory scene analysis: The perceptual organization of sound*. Cambridge, MA and London: MIT Press.

Burger, S., Jin, Q., Schulam, P. F. and Metze, F. (2012). *Noisemes: Manual annotation of environmental noise in audio streams*. Technical Report CMU-LTI-12–07, Carnegie Mellon University.

Chion, M. (2009). *Guide to sound objects*. Unpublished.

Cogan, R. (1984). *New images of musical sound*. Cambridge, MA: Harvard University Press.

Drioli, C., Polotti, P., Rocchesso, D., Delle Monache, S., Adiloglu, K., Annies, R. and Obermayer, K. (2009). Auditory representations as landmarks in the sound design space. In: *Proceedings of the sixth sound and music computing conference*. Porto, Portugal, pp. 315–320.

Erickson, R. (1975). *Sound structure in music*. Berkeley, Los Angeles and London: University of California Press.

Fant, G. (1960). *Acoustic theory of speech production*. The Hague, Netherlands: Mouton & Co N.V, Publishers.

Fernström, M. and Brazil, E. (2001, July 29–August 1). Sonic browsing: An auditory tool for multimedia asset management. In: J. Hiipakka, N. Zacharov and T. Takala, eds., *Proceedings of the seventh international conference on auditory display (ICAD)*, Espoo, Finland.

Ferrara, A., Ludovico, L. A., Montanelli, S., Castano, S. and Haus, G. (2006). A semantic web ontology for context-based classification and retrieval of music resources. *ACM Transactions on Multimedia Computing, Communications, and Applications (TOMM)*, 2(3), pp. 177–198.

Fields, K. (2007). Ontologies, categories, folksonomies: An organised language of sound. *Organised Sound*, 12(2), pp. 101–111.

Gaver, W. W. (1993a). How do we hear in the world? Explorations in ecological acoustics. *Ecological Psychology*, 5(4), pp. 285–313.

Gaver, W. W. (1993b). What in the world do we hear? An ecological approach to auditory event perception. *Ecological Psychology*, 5(1), pp. 1–29.

Gemmeke, J. F., Ellis, D. P. W., Freedman, D., Jansen, A., Lawrence, W., Moore, R. C., Plakal, M. and Ritter, M. (2017). Audio set: An ontology and human-labeled dataset for audio events. In: 2017 *IEEE International conference on acoustics, speech and signal processing (ICASSP)*, New Orleans, LA, pp. 776–780.

Grey, J. M. (1977). Multidimensional perceptual scaling of musical timbres. *J. Acoust. Soc. Am*. 61(5), pp. 1270–1277. Reproduced with the permission of the Acoustical Society of America.

Guyot, F., Castellengo, M. and Fabre, B. (1997). Chapitre 2: étude de la catégorisation d'un corpus de bruits domestiques. In: *Catégorisation et cognition: de la perception au discours*. Paris: Editions Kimé, pp. 41–58.

Gygi, B., Kidd, G. R. and Watson, C. S. (2007). Similarity and categorization of environmental sounds. *Perception & Psychophysics*, 69(6), pp. 839–855.

Handel, S. (1989). *Listening. An introduction to the perception of auditory events.* Cambridge, MA and London: MIT Press.

Hatala, M., Kalantari, L., Wakkary, R. and Newby, K. (2004). Ontology and rule based retrieval of sound objects in augmented audio reality system for museum visitors. In: *Proceedings of the 2004 ACM symposium on applied computing.* Nicosia, Cyprus: ACM, pp. 1045–1050.

Houix, O., Lemaitre, G., Misdariis, N., Susini, P. and Urdapilleta, I. (2012). A lexical analysis of environmental sound categories. *Journal of Experimental Psychology: Applied*, 18(1), pp. 52–80.

Jakobson, R., Fant, C. M. and Halle, M. (1952). *Preliminaries to speech analysis. The distinctive features and their correlates.* Cambridge, MA: MIT Press.

Jakobson, R. and Halle, M. (1956). *Fundamentals of language.* Berlin and New York: Mouton de Gruyter.

Krause, B. (2015). *Voices of the wild: Animal songs, human din, and the call to save natural soundscapes.* New Haven, CT: Yale University Press.

Lartillot, O. and Toiviainen, P. (2007). A MATLAB toolbox for musical feature extraction from audio. In: *Proceedings of the 10th international conference on digital audio effects (DAFx-07).* Bordeaux, France, pp. 237–244.

Lobanova, A., Spenader, J. and Valkenier, B. (2007). Lexical and perceptual grounding of a sound ontology. In: J. G. Carbonell and J. Siekmann, eds., *International conference on text, speech and dialogue, lecture notes in artificial intelligence*, Vol. 4629. Berlin and Heidelberg: Springer, pp. 180–187.

Marcell, M. M., Borella, D., Greene, M., Kerr, E. and Rogers, S. (2000). Confrontation naming of environmental sounds. *Journal of Clinical and Experimental Neuropsychology*, 22(6), pp. 830–864.

Miller, G. A. (1995). WordNet: A lexical database for English. *Communications of the ACM*, 38(11), pp. 39–41.

McAdams, S. (1999). Perspectives on the contribution of timbre to musical structure. *Computer Music Journal*, 23(3), pp. 85–102.

McAdams, S. and Saariaho, K. (1991). *Le timbre. Métaphore pour la composition.* Paris: Christian Bourgois-IRCAM. Chapter Qualités et fonctions du timbre musical, pp. 164–180.

Murray Schafer, R. (1977). *The tuning of the world.* New York: Knopf.

Musen, M. A. (2015). The Protégé project: A look back and a look forward. *AI Matters*, 1(4), pp. 4–12.

Nakatani, T. and Okuno, H. G. (1998). Sound ontology for computational auditory scene analysis. In: *Proceeding for the 1998 conference of the American association for artificial intelligence*, AAAI, pp. 1004–1010.

Plomp, R. (1976). *Aspects of tone sensation: A psychophysical study.* London: Academic Press.

Poli, G. D. and Prandoni, P. (1997). Sonological models for timbre characterization. *Journal of New Music Research*, 26(2), pp. 170–197.

Rasch, R. and Plomp, R. (1982). The perception of musical tones. In: Deutsch, D., ed., *The psychology of music.* Orlando, FL: Academic Press, Chapter 1, pp. 1–24.

Risset, J-C. (1999). *Ouïr, entendre, écouter, comprendre après Schaeffer*. Bryn-sur-Marne and Paris: INA-Buchet, Chastel. In: Schaeffer, P.: *Recherche et création musicalese et radiophoniques*, pp. 153–159.

Rocchesso, D., Mauro, D. A. and Drioli, C. (2016). Organizing a sonic space through vocal imitations. *Journal of the Audio Engineering Society*, 64(7/8), pp. 474–483.

Rosch, E. and Lloyd, B. B. (1978). *Cognition and categorization*. Hillsdale, NJ: Lawrence Erlbaum.

Ross, B. H. and Murphy, G. L. (1999). Food for thought: Cross-classification and category organization in a complex real-world domain. *Cognitive Psychology*, 38(4), pp. 495–553.

Scavone, G., Lakatos, S., Cook, P. and Harbke, C. (2001). Perceptual spaces for sound effects obtained with an interactive similarity rating program. In: *Proceedings of international symposium on musical acoustics*. Perugia, Italy.

Schaeffer, P. (2017). *Treatise on musical objects. An essay across disciplines*. Oakland: University of California Press.

Slawson, W. (1981). The color of sound: A theoretical study in musical timbre. *Music Spectrum*, 3, pp. 132–141.

Slawson, W. (1985). *Sound color*. Berkeley and Los Angeles and London: California University Press.

Smalley, D. (1986). *The language of electroacoustic music*. London: Macmillan, Chapter Spectromorphology and Structuring Processes, pp. 61–93.

Smalley, D. (1997). Spectromorphology: Explaining sound-shapes. *Organised Sound*, 2(2), pp. 107–126.

Takada, M., Fujisawa, N., Obata, F. and Iwamiya, S-I. (2010). Comparisons of auditory impressions and auditory imagery associated with onomatopoeic representation for environmental sounds. *EURASIP Journal on Audio, Speech, and Music Processing*, (1), p. 674248. https://doi.org/10.1155/2010/674248.

Valle, A. (2015). Towards a semiotics of the audible. *Signata*, 6.

Valle, A. (2016). Schaeffer reconsidered: A typological space and its analytic applications. *Analitica—Rivista online di studi musicali* 8. Available at: www.gatm.it/analiticaojs/index.php/analitica/article/view/158 [Accessed September 2018].

Valle, A., Armao, P., Casu, M. and Koutsomichalis, M. (2014). SoDA: A sound design accelerator for the automatic generation of soundscapes from an ontologically annotated sound library. In: *Proceedings ICMC-SMC-2014,* Athens, Greece, ICMA, pp. 1610–1617.

Vanderveer, N. J. (1979). *Ecological acoustics: Human perception of environmental sounds*. PhD thesis, Cornell University.

von Bismarck, G. (1974). Timbre of steady sounds: A factorial investigation of its verbal attributes. *Acustica*, 30, pp. 146–159.

Von Hornbostel, E. M. and Sachs, C. (1961). Classification of musical instruments. Translated from the original German by Baines, A. and Wachsmann, K. P. *The Galpin Society Journal*, pp. 3–29.

VV.AA. (2012). *OWL 2 web ontology language document overview*. Available at: https://www.w3.org/2012/pdf/REC-owl2-overview-20121211.pdf [Accessed March 2019].

Wessel, D. (1979). Timbre space as a musical control structure. *Computer Music Journal* 3, no. 2, pp. 45–52.

W3C. Available at: www.w3.org/TR/2012/REC-owl2-overview-20121211/ [Accessed September 2018].

Electroacoustic Music

An Art of Sound

Andrew Knight-Hill

12.1. Introduction

Electroacoustic music is an art of sound. It treats all sound as musically valid material, radically extending the democratization in sound that began with Schönberg's 12-tone technique. The term refers to a wide range of musical practices, each of which emanates from specific aesthetic and technical histories. The term "electroacoustic" is derived from the processes of electromechanical capture and reproduction of sound, and so the term refers less to a specific aesthetic style and more to the methods and practices that make it possible.

Compositional ideas and techniques that have developed within the oeuvre of electroacoustic music are highly relevant to sound design practice, yet there is often a disconnect between these two worlds. This chapter introduces an array of practices and concepts from electroacoustic music, providing an introduction for the reader and a point of reference for further listening, reading and discovery. Such a chapter can never promise to encapsulate all approaches, and indeed our understanding of electroacoustic music continues to develop and evolve. The history of electroacoustic music tends to be dominated by large-scale institutions, and the key figures are often white middle-class men. This is due to both the sociopolitical context of the 20th century and because the early technologies of electronic instruments and computing were very expensive and thus could only be afforded by large institutions. In turn, these large institutions possessed bureaucracies that required the artists and musicians to document and justify their work, thus ensuring a rich historical record and archive. Independent artists and experimenters outside of these systems often, even those with diligent practices of documentation, were marginalized by the legacies of these larger institutes. However, the democratization of music technologies, brought about by the shift to the digital, has opened up electroacoustic music practices to an ever widening sphere, and this increased diversity of practitioners has aroused considerable interest into the works of previously marginalized or forgotten practitioners. It is hoped that this chapter will encourage an ever widening diversity of sound practitioners to engage with the ideas and concepts of electroacoustic music practice, both to inform their own art and to contribute to the practice or electroacoustic music itself.

12.2. Recording and New Musical Possibilities

It is easy to forget how revolutionary the advent of sound recording technology was. Today we are surrounded by tools to capture, edit and replay sound; almost everyone carries around a recording and playback device every day in the form of a mobile phone.

But more important than the technology itself is the way in which recording changed the way people think about and relate to sound. Before sound recording, you had to be present at a sonic event in order to hear it. Sounds were ephemeral, mysterious and fluid. With the advent of recording, sounds could be captured and transported both geographically and temporally. Sound became a "thing" that could be owned, transferred and manipulated. Sounds themselves became directly accessible. This change was to have a revolutionary impact on the composers and artists who grew up in this rapidly changing world.

12.2.1. The Futurists

The futurist movement spanned a wide range of different art forms—from theater to poetry, dance and music—and what united these artists was a desire to create new forms of art, relevant to the age of mechanization. In 1913, Luigi Russolo published a letter *Art of Noises* within which he argued:

> Musical sound is too limited in its variety of timbres. The most complicated orchestras can be reduced to four or five classes of instruments different in timbres of sound: bowed instruments, metal winds, wood winds, and percussion. [. . .] We must break out of this limited circle of sounds and conquer the infinite variety of noise sounds.
>
> (Russolo in Cox and Warner, 2004, p. 11)

To realize his concept, Russolo created an orchestra of noise machines. These "intonarumori" created a series of clanks, whistles and drones that extended the sonic palette by echoing the noises of the city and the machinery of industry.

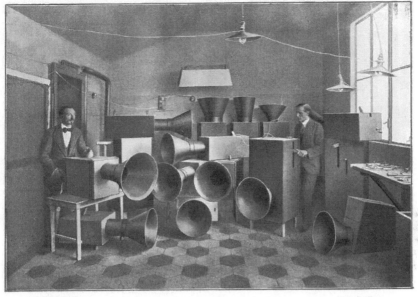

LUIGI RUSSOLO   UGO PIATTI

Nel Laboratorio degli Intonarumori a Milano.

Figure 12.1   Luigi Russolo and His Orchestra of Intonarumori.

Source: *L'Arte dei rumori* (Russolo, 1913)

12.2.2.   Edgard Varèse

At the same time, Edgard Varèse was making exactly the same call for greater sonic diversity, "Our musical alphabet must be enriched. We also need new instruments very badly. [. . .] In my own works I have always felt the need of new mediums of expression [. . .] which can lend themselves to every expression of thought and can keep up with thought" (Varèse, 1916).

Varèse often integrated nontraditional sounds and instrumentation, such as sirens and rattles, into his compositions, but his real desire was to engage with sound as a material itself. He was keen to move away from traditional musical paradigms, choosing to call his work Organized Sound.

"[T]o stubbornly conditioned ears, anything new in music has always been called noise. But after all, what is music but organized noises? And a composer, like all artists, is an organizer of disparate elements. Subjectively, noise is any sound that one doesn't like" (Varèse in Cox & Warner, 2004, p. 20).

This term helped to avoid the traditional connotations of music, opening up new possibilities for creative sonic practice. All sounds were considered musical; rather than thinking of music as constructed of just beats, notes, melodies and keys, Varèse described it in terms of rhythms, frequencies and intensities. He envisaged sound as consisting of sound-masses and shifting planes, within which timbre was a key musical parameter. Prefiguring the development of synthesizers and modern tools for sound production, Varèse imagined new electronic tools for sound creation: "I need an entirely new medium of expression: a sound producing machine" (Varèse in Cox and Warner, 2004, p. 19).

Varèse was a strong proponent of early electronic instruments such as the Theremin and Ondes Martenot, and he was eventually able to compose using recorded sound within the Paris studio of Pierre Schaeffer and, in collaboration with the electronics firm Phillips, to create a multichannel sound installation that was presented at the 1958 World Fair in Brussels. *Poème électronique* was hosted within the Philips Pavilion and visited by over 2 million people.

12.2.3. Experiments in Film: Vertov/Ruttmann

The emerging world of film also provided a site for creatives to experiment and explore the possibilities of working with sound. Russian filmmaker

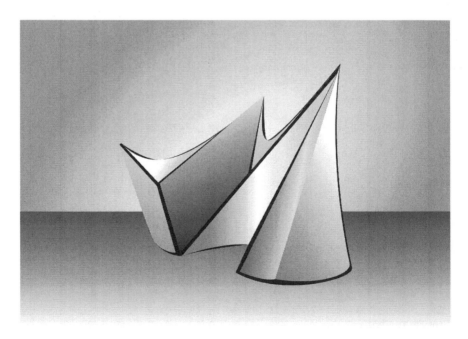

Figure 12.2 An Artist's Rendering of the Philips Pavilion at the 1958 Brussels World Fair.

Source: Graphic by Deborah Iaria and Luca Portik

Dziga Vertov (famous for the film *Man with a Movie Camera*, Vertov, 1929) actually wanted to be a sound artist, but the tools for sound recording and editing were not available to him, so he turned his ideas first to poetry and only later to work with film.

> I had an idea about the need to enlarge our ability for organized hearing. Not limiting this ability to the boundaries of usual music. I decided to include the entire audible world into the concept of "Hearing."
>
> (Vertov, 2008)

With the development of sound film in the late 1920s, Vertov was able to realize some of his musical ideas, for example creating montages and collages from sound and image samples recorded in factories and industrial sites, which formed the basis for the 1931 film *Enthusiasm* (Vertov, 1931). This film makes use of sounds as direct recorded material, exploring their textures, rhythms and intensities.

In Germany, musician and painter Walther Ruttman was part of a group involved in developing experimental abstract films that he described as visual music. These were often silent but aspired to evoke the possibilities of music and sounds in a visual form. Later Ruttman began to wonder whether it would be possible to develop a sonic equivalent; he wrote: "Everything audible in the world becomes material," and to realize this idea, he developed the composition *Wochenende* (1930), a sonic painting of a weekend in (and outside) Berlin. This sonic work was actually composed using optical sound film without any images. This appropriation of film technology allowed an approach to the montage and editing of sound that would not become accessible until magnetic tape became more widely available in the 1950s.

Julio d'Escrivan has argued that certain techniques of sound manipulation and development were utilized in film sound prior to their application in electroacoustic music. He cites examples, such as Rouben Mamoulian's *Jekyll and Hyde* (1932) and Max Steiner's score for the film King Kong (RKO, 1933), as demonstrating the creative use of reverberation, close microphone recording, reversing and layering of sound to develop and morph timbre (d'Escrivan, 2007). These and many other early experiments in film and radio began to explore the potentials of molding recorded sound that were later expanded and developed by musicians and composers working with sound alone.

12.2.4. Pierre Schaeffer and the GRM

Violinist and studio engineer Pierre Schaeffer began his experiments with recorded sound in 1948. At the French National Radio, he founded the experimental music studios, which eventually went on to become the Groupe de Recherches Musicales (GRM).

Focusing upon the manipulation of recorded sound, Schaeffer developed the ideas and practices for a new music that he called *musique concrète*. Schaeffer's early experiments explored the practical potentials of manipulating recorded sound—looping, reversing, transposing—and in tandem, Schaeffer began the process of formalizing the new musical possibilities.

Schaeffer wanted to create an art form of pure sound, in which sounds are freed from their original source associations. Within this process, Schaeffer recognized the inherent challenges: "every sound phenomenon [. . .] can be taken for its relative meaning or for its own substance. As long as meaning predominates, and is the main focus, we have literature and not music" (Schaeffer, 2012, p. 13). Therefore, the goal of *musique concrète* was to transcend the sound source, to access sounds as sonic objects. Sounds as themselves, not anything else.

Experimenting with the turntables that he had at his disposal, Schaeffer identified that it was possible to radically transform the resulting sound by isolating different parts of a recorded sound. Removing the attack, for example, often renders sounds unrecognizable. He also identified that looping a sound erodes its identity and source association. As sounds repeat over and over and over, they become changed and begin to emerge as abstract sound objects.

Through these processes of experimentation and reflection, Schaeffer identified that there is a key distinction between sound as physical signal and sound as a perceived object. The waveform that we observe in the DAW is a representation of sound wave vibrations, but we cannot experience hearing the sound by looking at its waveform. The hearing of the sound may become correlated to the waveform as we recognize some consistencies in shape over time, but the scientific and the perceptual are two very different categories of experience. This concept underpins the notion of acoustics and psychoacoustics, which we explored in the opening chapter on the "Nature of Sound and Recording."

Recognizing that the sound object was perceptual, flexible and not an absolute scientific certainty that could be codified, Schaeffer went on to identify four modes of listening:

1. *Listening [Écouter]*—listening to identify the source, the event, the cause; treating the sound as a sign of this source (e.g. I hear an ambulance)
2. *Perceiving [Ouïr]*—being aware of sounds around you but without focusing on or seeking to understand them (e.g. being vaguely aware of sirens and traffic noise as part of a general city soundscape)

Figure 12.3 Pierre Schaeffer (center) Playing in the Studio with François Bayle (left) and Bernard Parmegiani (right).

Source: Ina GRM. Photo Laszlo Ruska

3. *Hearing [Entendre]*—directed listening toward specific parameters of a sound (e.g. focusing on the pitch/timbre/duration or intensity of the siren)
4. *Comprehending [Comprendre]*—grasping a meaning, treating the sound as a sign, referring this meaning through a language or a code (e.g. there has been an accident or emergency) (adapted from Chion, 2009, pp. 19–20)

These modes are not mutually exclusive, and Schaeffer identifies that one may pass through all of them in the process of listening, but they set out the different ways in which we can interact with sounds and how they can be understood by us as listeners. They remind us that, despite our intentions, there is no absolute pathway between physical signal and perceived object, and that two individuals may hear the same sound in different ways. Thus, context becomes highly significant.

> [The] sound object is the meeting point of an acoustic action and a listening intention.
>
> (Chion, 2009, p. 27)

Schaeffer's ideas have inspired and underpinned many important works on sound and have been reapplied in many different places (for example, Michel Chion, a pupil of Schaeffer, applies his ideas in his writings on film, adopting the listening modes where he collapses Entendre into Ouïr leaving—Causal Listening, Reduced Listening and Semantic Listening (Chion, 1994, pp. 25–31)).

François Bayle, another pupil of Schaeffer who would go on to lead the GRM between 1975 and 1997, describes how the practice of Musique Concrète, working directly with sound to shape and form it, might be considered to be more akin to a plastic art than that of traditional notation-based composition. "When you make music with sounds [. . .] you organise it with your hands—a bit like a painter or sculptor of immaterial things" (Bayle in Obrist, 2013, p. 99). The immediacy of molding sound directly positions this musical practice apart from the distant and dislocated practice of notating with dots and lines on paper and places it directly akin to sound design. However, the contemporary expansion of digital tools for note-based composition can provide contemporary composers with immediate feedback on their sonic ideas, and thus the distance between these worlds is becoming increasingly small.

The GRM has a long and illustrious history of composition and research; currently a part of INA (Institut national de l'audiovisuel), it presents regular concerts of electroacoustic music and develops the GRM Tools, a suite of software plugins for sound transformation and design.

12.2.5. Stockhausen and the WDR

In contrast to the French practice of working with recorded sound, the advocates of experimental music in Germany took synthesis as their starting point. By experimenting with electronic equipment in the studios of what became the West German Radio (WDR) in Cologne, they began an electronic approach to composition, building up sounds from simple tones or noises into compositions of *elektronische Musik*.

With control over the tones generated, they applied serial music techniques to all of the parameters of sound. One of the most famous proponents of this musical approach was Karlheinz Stockhausen. Stockhausen worked with three primary characteristics of sound, originally identified by the mathematician Joseph Fourier, the varying proportions of which are able to define almost every sound and that become accessible through synthesis approaches:

> To synthesise a sound you have to start with something more basic, more simple, than the sounds you encounter in daily life. I started looking in acoustic laboratories for sources of the simplest forms of sound wave, for example sine wave generators, which are used for measurement. And I started very primitively to synthesise individual sounds by superimposing sine waves in harmonic spectra, in order to make sounds like vowels, aaah, oooh, eeew etc., then gradually I found how to use white noise generators and electric filters to produce coloured noise, like consonants: ssss, sssh, fffh, etc. And when I pulsed them it sounded like water dripping.
>
> (Stockhausen in Maconie, 1989, p. 90)

As Stockhausen's experiments demonstrate, each sound is a complex defined by its relative balance of pitch, noise and duration (pulses are very short bursts of sound). Every sound will be an amalgam of these characteristics, evolving over time (or not). These characteristics are a continuum, so that one may move smoothly between one to the other in the creation of new sounds, and it is this fluidity between states that underpins many of Stockhausen's compositional ideas.

His idea of Unified Time Structuring expresses the exploration of this continuum. Through recording sounds on magnetic tape and changing their playback speed, Stockhausen identified that shapes and patterns in time (forms and rhythms) could become timbres and that timbres could become patterns in time. A rhythm, sped up, becomes a tone. While a timbre, slowed down, begins to reveal its inner structure and the pulsing of its oscillation.

Stockhausen identified how this interconnected nature of timbre, rhythm and form "completely challenges the traditional concept of how

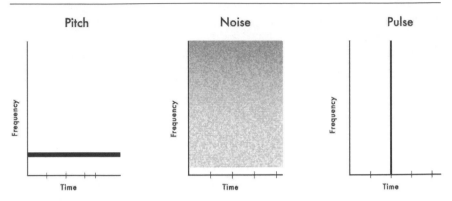

Figure 12.4   Pitch, Noise and Pulse.

Source: Graphic by Deborah Iaria and Luca Portik

to compose and think music, because previously they were all in separate boxes: harmony and melody in one box, rhythm and metre in another, then periods, phrasing, larger formal entities in another, while in the timbre field we had only the names of the instruments, no unit of reference at all" (Stockhausen in Maconie, 1989, p. 95).

If sounds could be composed and built from individual components, then they could also be decomposed and divided. "You can hear a sound gradually revealing itself to be made up of many components, and each component in turn is decomposing before our ears into its individual rhythm of pulses" (Stockhausen in Maconie, 1989, p. 97). Thus, elements of a sound might be decomposed, separated and later reassembled with the component parts of other sounds in order to construct entirely new sounds.

Not only can pitches become pulses and pulses become pitches, but the timbral balance between pitch and noise can also be manipulated. "If the degree of aperiodicity of any given sound can be controlled, and controlled in a particular way, then any constant sound can be transformed into a noise" (Stockhausen in Maconie, 1989, p. 108). In Stockhausen's case, this meant introducing noise of varying types into the synthesis process; by introducing random fluctuations into the signal chain, the sounds were modulated with noise. These days, with the technology of FFT processing, the balance of noise and pitch content can be manipulated in great detail either highlighting or isolating the harmonic components of a sound. The blurring of distinctions between noisy and musical sounds has been in part thanks to the pioneering work of composers at the WDR, including Stockhausen, and this development allows creative sound designers and musicians to take advantage of working with the widest range of sonic materials and in morphing between them.

Nowadays any noise is musical material, and it is possible to select a scale of degrees from sound to noise for a given composition, or choose an arbitrary scale, from the complete range.

(Stockhausen in Maconie, 1989, p. 108)

Stockhausen was also an advocate of the development of space as a compositional parameter (by no means was Stockhausen alone in this; Schaeffer and colleagues at the GRM had invented a system that could spatialize sounds, the *pupitre d'espace*, back in 1951, and there are a number of significant historical examples of the use of space in musical composition, such as Giovanni Gabrieli's use of multiple choirs within the basilica of St. Mark in Venice). However, Stockhausen argued that a majority of Classical music was largely monophonic and that it had lost the potential that the spatial dimension in music could provide. Audiences sat facing forward, listening to a sound emanating from a fixed point on the stage in front of them. This completely negated an entire parameter of sound. To explore the parameter of space, Stockhausen pioneered composition in multichannel formats developing, in the process, new techniques for spatial recording and playback. Why shouldn't sound sources emanate from anywhere around us, and why shouldn't they move?

In 1959, for my work *Kontakte* [Contacts] I had a special table made to my own design. A loudspeaker was placed in the middle, attached to a cable vertically above, and there were four microphones at the points of the compass around the table. I moved the table by hand at very different speeds from zero to six revolutions per second—clockwise and anti-clockwise—and recorded this on four tracks via the microphones. It was played back on four Loudspeakers. For the first time you could hear rotation movements in space.

(Stockhausen in Obrist, 2013, p. 33)

Working with sounds in space Stockhausen identified, "If I have a sound of constant spectrum, and the sound moves in a curve, then the movement gives the sound a particular character compared to another sound which moves just in a straight line" (Stockhausen in Maconie, 1989, p. 102). Therefore, the parameter of spatial motion becomes just as important a compositional parameter as pitch, duration and timbre. This also includes the audience's perception of spaces: "[N]ot only does the sound move around the listener at a constant distance, but it can also move as far away as we can imagine, and also come extremely close" (Maconie, 1989, p. 105). He wanted the audience to liberate themselves from the

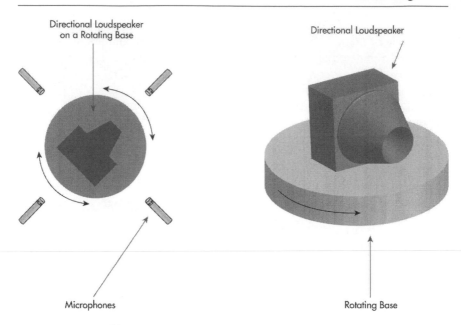

Figure 12.5 Stockhausen's Rotating Speaker Table.

Source: Graphic by Deborah Iaria and Luca Portik

traditional concert hall through their listening, for the sounds to dissolve the walls of the auditorium by immersing the audience within a spatial reality constructed by sound itself. Similar effects are frequently applied in contemporary cinema, extending the frame of the image into off-screen space and immersing the audience in the world of the film through sound. Utilizing spatial trajectories as a core compositional idea and exploring the full spatial potentialities of surround sound systems remain exciting opportunities for future innovation.

12.2.6. *Daphne Oram and the Radiophonic Workshop*

In the UK, Daphne Oram convinced the BBC to found the Radiophonic Workshop in 1958 with a remit to compose music and sound effects for radio and television programs. Unlike their European counterparts, the Radiophonic Workshop remained largely tied to the utilitarian task of delivering sounds for existing programs. Stifled by the lack of opportunity to experiment and innovate, Oram left the studio in 1959 to set up her own studio in an oast house in Kent. From there she composed soundtracks for adverts and programs and developed an innovative synthesizer with a powerful graphical interface, the ORAMICS machine.

Figure 12.6 Publicity Photograph of Daphne Oram at Her Tower Folly Studio.

Source: Photograph: Brian Worth. Image courtesy of the Daphne Oram Trust and Goldsmiths Special Collections & Archives. Catalogue number: ORAM/7/4/016

In 1972, Oram published a book on electronics and music, *An Individual Note*, designed very much to inspire creative practice and to make electronic music practices more widely accessible. Within this text, she set out her ideas for a new type of control interface. Not knobs and switches but intuitive and graphic interfaces to mold and shape sound synthesis: "We wish to design this machine-with-humanising-factors so that the composer can instruct it by means of a direct and simple language. [S]he will want to transduce [her/her] thoughts as quickly as possible via a channel which is logical" (Oram, 1972, p. 97). These ideas lay at the heart of the ORAMICS machine, which was conceptually far ahead of contemporary musical control interfaces. Such ideas are only just being capitalized upon within the digital domain as developers abandon the skeuomorphic obsession with analogue gear and instead embrace the possibilities of alternative design models for intuitive control of parameters and the possibilities of alternative control surfaces.

The Radiophonic Workshop continued to develop sound design and music for the BBC, including some of the most iconic BBC programs on both radio and television, such as *Hitchhiker's Guide to the Galaxy* (1978–1980) and *Doctor Who* (1963–1989). Therefore, while the space and freedom to innovate within purely musical terms was perhaps not so prevalent as their European counterparts, their creative work was disseminated to the whole UK population through national public service broadcasting.

Combining synthesis approaches from *elektronische Musik* and techniques from *musique concrète*, the Radiophonic Workshop developed both conventional musical jingles, electronic orchestration and abstract radio plays. Many future composers and sound designers were inspired by hearing these strange sounds emanating from the television and radio, and the whole profile of sonic creativity was raised as a result of introducing the wider population to electroacoustic sounds.

12.2.7. Computer Music in North America

The United States led the way in developing compositional systems for work with computers. Max Matthews, a researcher at Bell Telephone Laboratories in New Jersey, began to experiment with using the computer to synthesize music, developing his first program in 1957. This program, MUSIC I, was a basic synthesis program, limited to only the generation of triangle waveforms. But with the rapid development of computers and processing systems, the software was soon upgraded, and by 1962 existed in its fourth iteration. MUSIC IV offered a wide array of synthesis and signal processing functions and has been the progenitor of a whole family of audio programs, some of which are still widely available and used today, including the open source CSOUND.[1]

On the other side of the country at Stanford University, John Chowning began experimenting with the use of frequency modulation as a form of synthesis, adapting the MUSIC program to accommodate the particulars of this new approach. FM synthesis could generate a wide variety of sound types with particularly successful results in synthesizing metallic timbres such as percussion, bells and brass sounds. The ability to adjust the modulation process in a continuous fashion also allowed the first experiments in timbre transformation, whereby sounds were able to morph smoothly from one form into another.

In 1974, Canadian composer Barry Truax developed the first tools that applied real-time granular synthesis methods, demonstrating the creative potential of working with grains of sound in contrast to standard continuous waveforms. Through the application of stochastic procedures that Truax had developed, input sounds were spliced into many tiny grains before being transformed and reassembled into new structures and forms.

Figure 12.7   Barry Truax Running the POD System on a PDP-15 at the Institute of Sonology, Utrecht.

Source: Used by permission of Barry Truax. Photo credit: Theo Coolsma

Granular synthesis can therefore be used to highlight the internal spectral details of the input sounds by extending or duplicating individual grains into long streams of sound. The principle of granular synthesis underpins many contemporary time-stretching algorithms, especially those where the original pitch is maintained. Granular synthesis can also be used to create spatial clouds of sound of varying densities, which can be vitally useful for manufacturing ambiences and textures.

Digital techniques and processes continue to evolve, but many of the tools and techniques that we use today can be traced back to the pioneering work of the early electroacoustic music composers whether from early computing or analogue practices.

12.2.8. Soundscape

Alongside technological innovation and the push to diversify musical practices, a new approach to sound emerged. R. Murray Schafer (not to be confused with Pierre Schaeffer) was inspired by the sounds of nature and the Pythagorean idea that the cosmos itself is a musical composition unfolding all around us. "The soundscape is any acoustic field of study.

We may speak of a musical composition as a soundscape, or a radio program as a soundscape or an acoustic environment as a soundscape" (Schafer, 1977, p. 7).

With a team of composers and researchers, he founded the World Soundscape Project (WPS) at Simon Fraser University, Vancouver, seeking to explore soundscapes and to highlight the sonic impact of urbanization and rapid technological change: "The soundscape of the world is changing. Modern [humans are] beginning to inhabit a world with an acoustical environment radically different from any [they have] hitherto known" (Schafer, 1977, p. 3). Their written work and creative projects sought to draw attention to issues of noise pollution and to document how soundscapes around us were changing and to highlight the importance of listening.

Soundscapes and location recordings later evolved into material for compositional practice that could be combined in montage with other soundscapes or combinations of other materials to explore the nature of listening. Hildegard Westerkamp's seminal composition *Kits Beach Soundwalk* (1989) is a prime example of this practice. Soundscape composition has continued to evolve and is sometimes now referred to as context-based composition, composition in which the referential nature of the sound source is a key driving factor within the organization of materials. The combination of traditional musical instruments and soundscapes is another

Figure 12.8   The WSP Group at Simon Fraser University, 1973: From left to right, R. M. Schafer, Bruce Davis, Peter Huse, Barry Truax and Howard Broomfield.

Source: World Soundscape Project and Simon Fraser University

widely popular area of soundscape, for example Bethan Kellough's album *Aven* combines field recordings with strings and electronics.

The increasing portability and miniaturization of recording equipment make the capture of soundscapes ever more accessible, and they are to be found increasingly within gallery installations, commercial recording and narrative film.

12.2.9. Gesture/Texture

For a practice that engages directly with the materiality of sounds, traditional notions and notation systems of music are inappropriate. This new sonic practice requires new forms of representation in both graphic and conceptual forms. From the earliest experiments in sound, there were very strong compositional methodologies, and in the late 1970s and 1980s, new languages to describe these practices began to develop. These discussions began to develop terminologies for conceptualizing sound practice with one of the key composers and theorists being Denis Smalley. Smalley differentiated between two sonic characteristics, that of texture and that of gesture. An ideal texture has no shape over time, it is just a texture, while an ideal gesture is pure trajectory with no textural content. Such pure manifestations of texture and gesture could never exist, but they represent important archetypal ideals within a continuum of sound that tends toward being either more gestural or more textural.

The texture is the timbre of a sound, defined by its spectral characteristics and makeup. As touched upon previously in our discussion of granular synthesis, by extending the sound, attention becomes shifted toward the inner workings of the sounds spectrum. Smalley describes textural music as therefore introspective and concentrated upon internal activity at the expense of forward impetus. It directs attention to the inner world of the sound and can create the impression of suspended time.

Gesture, by contrast, is all about action and trajectory. The notion of gesture is derived from musical performance where a physical gesture results in a sonic outcome. Actions and properties of the gesture are often reflected within the resulting sound. "Sound-making gesture is concerned with human, physical activity [in which a] chain of activity links a cause to a source" (Smalley, 1997, p. 111). However, in an electroacoustic context, both the performer and the source object are removed, leaving only the abstracted sound. Therefore, within these purely sonic contexts we can think of gestures as follows:

- Gesture is concerned with action directed away from a previous goal or toward a new goal;
- it is concerned with the application of energy and its consequences;
- it is synonymous with intervention, growth and progress.

(Smalley, 1986, p. 82)

Therefore, gesture becomes a key compositional ingredient. It signifies movement, motion and communication by providing a trajectory from one point to another. It is concerned with propelling time forward, with moving away from a previous state toward the next state in the structure. Gestural music, then, is governed by a sense of forward motion, of linearity, of narrativity. The British composer Trevor Wishart outlined a series of archetypal gestures within his treatise *On Sonic Art* (1996) using the term "morphology" to refer to the shape of sounds.

These archetypes represent a wide range of gestural forms that might be assembled and combined as building blocks. Once individual morphologies are combined, they begin to possess a higher-level shape (second-order morphology) such as an increase in speed. In turn, these second-order morphologies can be combined to create even larger forms and structures, just as phrases might be combined to create melodies within traditional music.

This construction of larger gestural forms from individual gestures builds upon an idea by Schaeffer who identified that each sound exists as both an object itself and as a structure of smaller component parts. For example, a note on a guitar can be considered as an individual sonic object, but it is also made up of attack, sustain and decay portions that combine to structure the sound and give it its character. Similarly, the sound of someone hitting a ball with a bat might be described as an individual sound event. But we can also break this down into the swish of the bat passing through the air, the thwack of the ball being hit and the ball whooshing away afterward. Each of these component parts might be considered as an individual sound object itself or as part of a larger structure.

- Every object is perceived as an object only in a context, a structure that includes it.
- Every structure is perceived only as a structure of objects that compose it.
- Every object of perception is at the same time an *object* insofar as it is perceived as a unit locatable in a context, a *structure* insofar as it is itself composed of several objects.

(Chion, 2009, p. 58)

What this means is that the notion of gestures and texture can be applied to all levels of composition, from individual sounds to the overarching structures of a composition or a soundtrack. Gesture not only is therefore the history of an individual event but can also be an approach to the psychology of time and its articulation within music and sound design. A new way of thinking about sound.

Smalley describes gesture as an energy-motion trajectory. We can think of the level of energy at a particular point (high or low), the concentration

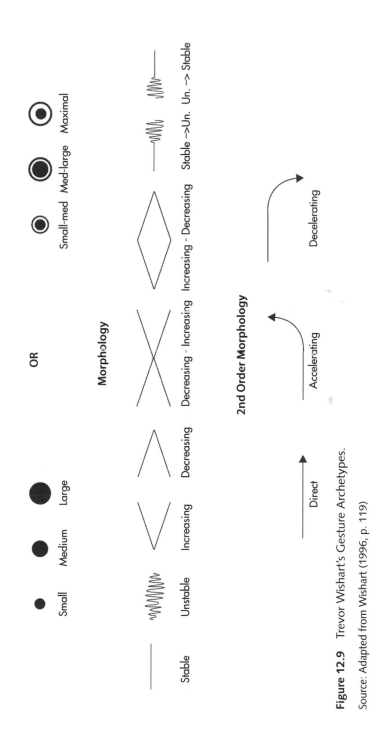

Figure 12.9 Trevor Wishart's Gesture Archetypes.

Source: Adapted from Wishart (1996, p. 119)

of that energy (focused at one point or dissipated) and the development or change of these characteristics over time (increasing or decreasing). These parameters might describe the character of an individual sound or the structure of a whole piece. As such, Smalley developed the term "spectromorphology" to describe the nature of sounds (Smalley, 1986, pp. 61–93). This term reflects the interrelated nature of spectrum and shape that is manifest in time and in space. The spectrum refers to the makeup of frequencies within the sound, while the morphology relates to the shape of the sound over time and how it moves:

> Spectromorphology is concerned with perceiving and thinking in terms of spectral energies and shapes in space, their behaviour, their motion and growth processes, and their relative functions in a musical context.
>
> (Smalley, 1997, pp. 124–125)

> A gesture is therefore an *energy-motion trajectory* which excites the sounding body, creating spectromorphological life.
>
> (Smalley, 1997, p. 111)

Developed within the practices of electroacoustic music, spectromorphology (and the notions of thinking about sounds as either a texture or gesture) can be applied as a framework for thinking about sounds in a wide range of contexts and within many different applications. Gestures can operate within parameters of sound (pitch, loudness, spatial movement etc.) or between them. Thinking of sound in terms of energy-motion trajectories can inform creative practice in both linear and nonlinear narrative contexts providing a fluid way of designing and imagining sounds, ranging from individual events to larger structures of sound.

12.3. Contemporary Electroacoustic Music Practices

In the 21st century, the diversity of electroacoustic music practices and their application are enormous. Many of the techniques, approaches and tools that underpin everyday audio processes across music production, sound design and composition have been developed from electroacoustic music practices, with students of pioneers applying their knowledge and ideas to many different areas and fields of sound practice, all embracing Edgar Varèse's dream of accessing the whole mysterious world of sound (Varèse in Cox and Warner, 2004, p. 20).

The distinct and segregated schools of compositional thought that emerged from the WDR, GRM and others have slowly dissolved through the interchange of the Internet and the combination of ideas that the post-modern world champions into a unified electroacoustic music. With the ubiquity of digital tools providing increasing access to techniques and technologies, electroacoustic music practices are more accessible than ever.

But the ideas of electroacoustic music, as well as the new paradigms of musical creativity that they afford, are the most significant aspects of this art form. You can use a computer to imitate a violin or a piano, but you can also use it to create entirely new sounds, to morph timbres and construct novel sonic spaces, to discover new forms and ways of expression. Electroacoustic music practices are highly applicable to sound design contexts, and engagement with the diversity of creative works that have developed over the past century can provide an invaluable source of inspiration and instruction. This chapter has provided a brief overview to electroacoustic music practices and the new ways of thinking about sound that have emerged from them. The bibliography of recommended readings and discography of seminal works at the end of the chapter provides further materials for exploration, but you will also find many examples of electroacoustic music online and in concerts and performances all across the globe. The Canadian Electroacoustic Community (CEC) runs an online mailing list that has become a vibrant international resource for practitioners to discuss and debate the various merits of sound practice and to disseminate information about upcoming events and opportunities such as CEC-Conference,[2] and there are many other organizations around the globe, such as the Society for Electroacoustic Music in the United States (SEAMUS),[3] the Electroacoustic Music Studies Network,[4] the Australasian Computer Music Association,[5] the International Computer Music Association[6] and the International Confederation of Electroacoustic Music,[7] that provide links to many affiliated member organizations in many other countries.

Electroacoustic music in the early 21st century has sought to redefine its own history, investigating marginalized or forgotten practitioners who have been neglected from the canonical history of electroacoustic music, which tends to focus on the large studios. The record label Sub Rosa is one such example; through its series of releases, *An Anthology of Noise and Electronic Music 1–7*, it has explored and released a wide range of previously unpublished compositions and has since made the move toward compilations of non-Western electronic music, through output such as *Anthology of Chinese Experimental Music 1992–2008*, *Anthology of Turkish Experimental Music 1961–2014*, and *Anthology of Persian Electronic Music: Yesterday and Today*.

Electroacoustic music practitioners continue to innovate and develop new practices, new tools and new interfaces for controlling, shaping and exploring this art of sound.

Figure 12.10   A Concert at the GRM's PRESENCES Électronique Festival in 2012.

Source: Ina GRM. Photo by Didier Allard

Acknowledgments

Very special thanks to Simon Emmerson for his comments and support on an earlier draft of this chapter. Special thanks to Luca Portik and Deborah Iaria for their graphic renderings within this chapter, to Jean-Baptiste Garcia and Ina GRM for images of Pierre Schaeffer et al. and the GRM concert, to the Daphne Oram Trust and Goldsmiths Special Collections & Archives for the image of Daphne Oram, and to Barry Truax for his help in sourcing the images of the World Soundscape Project and the POD machine. Thank you also to Michael Filimowicz for his constructive comments throughout the editorial process.

Notes

1. https://csound.com/
2. https://cec.sonus.ca/cec-conference/index.html
3. https://seamusonline.org
4. http:// www.ems-network.org/

5. http://acma.asn.au/about/
6. www.computermusic.org/
7. www.cime-icem.net/cime/

Bibliography of Recommended Reading

Cage, J. (2013). *Silence: Lectures and writings*. New York: Marion Boyars.
Emmerson, S., ed. (2018). *The Routledge research companion to electronic music: Reaching out with technology*. London: Routledge.
Landy, L. (2007). *Understanding the art of sound organisation*. Cambridge, MA: MIT Press.
Manning, P. (2004). *Electronic and computer music*. Oxford: Oxford University Press.
Kahn, D. (1999). *Noise water meat*. Cambridge, MA: MIT Press.
Oliveros, P. (2005). *Deep listening: A composer's sound practice*. New York: iUniverse.
Rodgers, T. (2010). *Pink noise: Women on electronic music and sound*. Durham, NC: Duke University Press.
Schaeffer, P. (2012). *In search of a concrete music*. Translated by Dack, J. and North, C. Berkeley: University of California Press.
Voegelin, S. (2010). *Listening to noise and silence: Towards a philosophy of sound art*. London: Bloomsbury.

Discography of Seminal Works

Bayle, F. *Toupie Dans Le Ciel*. INA-GRM NIA C3002.
Derbyshire, D. *"Falling" the dreams*. BBC Worldwide Music.
Kellough, B. (2016). *Aven*. [CD]. Touch Music.
Matthews, M. *Bicycle built for two*. Decca DL 79103.
Oliveros, P. *A little noise in the system. An anthology of noise and electronic music #1*. Sub Rosa. SR190.
Oram, D. *Four aspects. An anthology of noise and electronic music #2*. Sub Rosa. SR200.
Parmegiani, B. *De natura sonorum*. INA-GRM—INA C 3001.
Radigue, É. *Trilogie de la Mort*. Experimental Intermedia Foundation. XI 119.
Russolo, L. and Corale, A. *An anthology of noise and electronic music #1*. Sub Rosa. SR190.
Ruttmann, W. W. *An anthology of noise and electronic music #1*. Sub Rosa. SR190.
Schaeffer, P. *Cinq études de bruits: Étude Violette. An Anthology of Noise and Electronic Music #1*. Sub Rosa. SR190.
Stockhausen, K. *Kontakte*. Stockhausen-Verlage CD3.
Smalley, D. *Valley Flow empreintes*. DIGITALes IMED9209.
Tenney, J. *Five stochastic studies*. DECCA 71080.
Truax, B. *Nautilus*. Imperial Record Corporation, Vancouver SMLP 4033.
Varèse, E. *Poème électronique. An anthology of noise and electronic music #1*. Sub Rosa. SR190.
Varèse, E. *Tuning up. Varèse the complete works*. DECCA 4602082.
Vertov, D. *Radio ear Radio Pravda. An anthology of noise and electronic music #7*. Sub Rosa. SR300.

Westerkamp, Hildegard. *Kits beach soundwalk*. Empreintes DIGITALes—IMED 9631.
Wishart, Trevor. *Red bird*. EMF CD022.

References

British Broadcasting Corporation. (1963–1989). *Doctor Who*. [TV program].
British Broadcasting Corporation. (1978–1980). *Hitchhiker's guide to the galaxy: Primary and secondary phases*. [Radio].
Chion, M. (1994). *Audio-vision: Sound on Screen*. Translated by C. Gorbman. New York: Columbia University Press.
Chion, M. (2009). *Guide to sound objects*. Translated by J. Dack and C. North. Available at: https://monoskop.org/images/0/01/Chion_Michel_Guide_To_Sound_Objects_Pierre_ [Accessed March 2019]. Schaeffer_and_Musical_Research.pdf [Accessed May 2018].
Cox, C. and Warner, D. (2004). *Audio cultures*. New York: Continuum Press.
D'Escrivan, J. (2007). Imaginary listening. In: *Proceedings of EMS2007*. Leicester. Available at: www.ems-network.org/IMG/pdf_EscrivanRuskinEMS07.pdf [Accessed August 2018].
King Kong (1933). [Film]. RKO Pictures.
Maconie, R. (1989). *Stockhausen: On music*. London: Marion Boyars.
Mamoulian, R. (1932). Jekyll and Hyde. [Film]. Paramount Pictures.
Obrist, H. U. (2013). *A brief history of new music*. Zürich: Ringier & Les Presses Du Réel.
Oram, D. (1972). *An individual note of music, sound and electronics*. London: Galliard.
Russolo, L. (1913). *L'Arte dei rumori*. Poesia: Edizioni Futuriste. Milano: Movimento Futurista.
Schafer, R. M. (1977). *Soundscape: Our sonic environment and the tuning of the world*. New York: Destiny Books.
Schaeffer, P. (2012). *In search of a concrete music*. Translated by C. North and J. Dack. Berkeley: University of California Press.
Smalley, D. (1986). Spectro-morphology and structuring processes. In: S. Emmerson, ed., *The language of electroacoustic music*. Basingstoke: Palgrave Macmillan, pp 61–93.
Smalley, D. (1997). Spectromorphology: Explaining sound shapes. *Organised Sound*, 2(2), pp. 107–126.
Varèse, E. (1916). Composer Varèse to give New York abundance of futurist music. *New York Review*. In: Olivia Mattis, op. cit., p. 57.
Vertov, D. (1929). *Man with a movie camera*. [Film]. VUFKU.
Vertov, D. (1931). *Enthusiasm: The Symphony of Donbass*. [Film]. Ukrainfilm.
Vertov, D. (2008). 'Kak eto nachalos?' (How has it begun?). *'Iz Naslediia' (*From the Heritage*)*, Translated by A. Smirnov. Vol. 2. Moscow: Eisenstein Centre, p. 557.
Westerkamp, H. (1989). *Kits beach soundwalk*. Empreintes DIGITALes–IMED 9631.
Wishart, T. (1996). *On sonic art*. London: Routledge.

Electronic Dance Music in Narrative Film

Roberto Filoseta

13.1. Introduction

Electronic dance music (EDM) is the latest addition to the sonic resources of narrative film.

As a loose umbrella term, "EDM" refers to a form of popular music that owes its existence to recent developments in electronic sound production technologies, its aesthetics tightly bound up with the tools, procedures and processes on which it relies heavily for its realization. It includes the more common styles of house, techno, drum 'n' bass and trance, as well as a plethora of other genres and subgenres that, confusingly, are not necessarily conceived for dancing (see McLeod, 2001; Butler, 2003, 2006). Nevertheless, as the focus of this writing is largely on audiovisual rhythm, I discuss EDM as a genre specifically concerned, at its core, with beat-driven, groove-based sonic structures.[1]

Some authors use the term "electronica" to refer to the same musical territory. But other authors (McLeod, 2001; Butler, 2003, 2006) have pointed out that the "electronica" label, used as a catchall term, is in effect a marketing gimmick popularized in the 1990s by the music industry. In its more current sense, instead, the term "electronica" denotes a genre *within* the EDM umbrella where, confusingly, it may either indicate a beatless style or refer to the earlier electronic productions of the 1970s and early 1980s, sometimes also called synth-electronica. These latter are characterized by luscious synthesizer parts articulating more conventional melodic and harmonic material, with less emphasis (and often less sophistication) on the rhythmic elements. Examples of (synth-)electronica in film include John Carpenter's *Assault on Precinct 13* (1976), Alan Parker's *Midnight Express* (1978) and Ridley Scott's *Blade Runner* (1982). These films represented something of an exception in an industry dominated by the conventional orchestral scoring practice. But by the late

1990s, a surge can be observed in the number of films featuring EDM in their soundtracks.

Certainly, technological developments have played a role in this phenomenon—the surge coincides with the transition from analogue to digital technologies across audio and video production processes, which brought unprecedented ease to the manipulation and structuring of audio and visual material. By that time, EDM had reached the status of a mature and recognized field, encompassing a rich variety of stylistic approaches and a growing number of practitioners and audiences. But none of those factors, in themselves, could have started the trend if there had not been, fundamentally, a genuine search for an alternative language eschewing the worn devices of orchestral underscoring. As Danijela Kulezic-Wilson remarks, "Cinema overexploitation of conventional scoring practices has reached the point when music's impact in film is either devalued by its overuse, or can produce an effect opposite of that desired, especially when employed with the intention of augmenting audiences' affective responses" (Kulezic-Wilson, 2017, p. 134). The language of cinema itself has evolved considerably, and the conventional approach to the soundtrack is becoming inadequate to the needs of contemporary filmmaking. Thus we see a growing number of directors looking at alternative models, and the structures of EDM have a significant role to play in this search.

All this is good news for the electronic composer/producer. But what does EDM have to offer specifically to narrative film? What are its distinctive features, and how are these deployed by filmmakers to construct more dynamic and better integrated audiovisual structures? To answer those questions, in this essay I will first provide a concise account of EDM's main features, followed by a short section reviewing the fundamentals of film rhythm and its relation to music rhythm. I will then proceed to illustrate how EDM can function in narrative cinema, focusing on two films that can be regarded as seminal works in this trend: Tom Tykwer's *Run Lola Run* (1998) and Darren Aronofsky's *Pi* (1998). These films have been selected for their remarkably integrated use of EDM in their structures.

To clarify this further, I start by distinguishing between two broad lines of films. On one line we have works that are in some way *about* EDM cultures. In such films, EDM is, predictably, what we expect to hear, given their specific focus on characterizing the club "scene," the DJs, the rave, and the associated sex and drugs narratives. Danny Boyle's *Trainspotting* (1996) was perhaps the initiator in this trend, soon to be followed by a few other worthy successors like Justin Kerrigan's *Human Traffic* (1999), Doug Liman's *Go* (1999) and Hannes Stöhr's *Berlin Calling* (2008), to name but a few. In an article comparing *Run Lola Run* and *Berlin Calling*, Sean Nye places the latter within a "tradition of *techno scene film*" (Nye, 2010, emphasis in original), stating, "This type thus acts like a sociological study of club scenes in film form" (Nye, 2010, p. 122). These films will *necessarily* include EDM to construct a specific diegetic reality reproducing "real-life" experience. At their best, they may exhibit an integrated use of EDM and would very much deserve detailed analysis. But my aim in this essay is to tease out what EDM has to offer to narrative film in general, as a deliberate stylistic choice alternative to more conventional orchestral scoring, not dictated by diegetic necessity.

This brings us to the second line of films—works that are not so specifically focused on narrating the EDM world as such. Many soundtracks of the past three decades include elements of EDM as a deliberate choice, but it is difficult to find works that achieve the degree of stylistic consistency observable in *Run Lola Run* and *Pi*. Instead, it is more common to find EDM used within or alongside more conventional orchestral parts. In films like *The Matrix* (Wachowski Siblings, 1999) and *Lara Croft: Tomb Raider* (Simon West, 2001), for example, the EDM elements are overwhelmed by some less imaginative orchestral material of an epic quality. As Mark Minett puts it, "Revell's work on *Tomb Raider* can be seen as typical of the bombastic Hollywood blockbuster approach to EDM" (Minett, 2013).

It is the same for films like *The Saint* (Phillip Noyce, 1997), *The Beach* (Danny Boyle, 2000), *Spy Game* (Tony Scott, 2001), *Man on Fire* (Tony Scott, 2004), *Layer Cake* (Matthew Vaughn, 2004), *The Sentinel* (Clark Johnson, 2006), and many others. There are, no doubt, many interesting aspects to these films' soundtracks, some remarkable cues and a sophisticated use of sound design, but ultimately the inclusion of some interjected or intermixed EDM elements does not change the fact that these works remain very much rooted in more conventional scoring practice, featuring well established orchestral gestures that range from epic to clichéd sentimental.

Much more stylistically consistent, instead, are films like *Sexy Beast* (Jonathan Glazer, 2000) and *Amores Perros* (Alejandro González Iñárritu, 2000). These films, too, include both EDM and non-EDM music as part of their soundtracks, but they neatly avoid falling on the conventional orchestral model, resorting instead to other forms of popular music for their non-EDM cues. And finally we have films like *Fight Club* (David Fincher, 1999) and *Hanna* (Joe Wright, 2011), which have entirely EDM-based soundtracks (though *Hanna* does include some narratively relevant North African diegetic music), remarkably integrated with elaborate sound design.

Run Lola Run and *Pi* are part of this latter group[2] and remain prime examples of a more radical use of EDM in film. As will be shown in their respective analyses, these films do not simply employ some EDM as part of their soundtrack; their audiovisual strategy consistently exploits specific formal features of EDM, particularly its rhythmic structures and its distinctive sound quality, for specific narrative ends. They therefore offer a particularly rich terrain for analysis in the context of this writing.

13.2.   Electronic Dance Music: Main Features

Rhythm is the essence and main principle of organization in EDM, the parameter on which all other parameters depend. Unlike other styles of popular music, in which the drums function mainly in a supporting role to melody, harmony and vocals, in EDM the percussive elements are literally foregrounded, dominating both the mix and the musical discourse. Melodic lines or vocal parts (when present) are themselves mostly structured as looped patterns contributing to the articulation of rhythmic relationships (except for some drones and pads filling the texture). The bass frequencies are particularly emphasized; the role of the bass drum and bass line is primary in establishing the music's pulse.

Vijay Iyer stresses "EDM's connections to musics of the African diaspora" (Iyer, 2008). And indeed in EDM the logic of African percussion music can be seen in the way in which several short repeating patterns, individually simple, combine to create complexity at the higher structural level. Nevertheless, the electronic nature of EDM's production processes takes groove-based music to a new terrain, with its own distinct aesthetics and potentialities, as well as challenges. In terms of its rhythmic structures, for example, electronic music production offers two compositional possibilities specific to studio realization. The first is the possibility of structuring events on a consistent, machine-accurate alignment of beat and chronometric time; and indeed EDM practitioners do make regular use of the "quantize" function offered by sequencers (at least for some of the music's layers). The second possibility lies at the opposite end of the spectrum: creating irregular patterns, independent lines of nonaligned rhythm, loops that produce metrical ambiguity, go in and out of phase etc.—but even this device can only function (can only be "out") in *relation* to a strictly regulated beat structure.[3]

Regular, cyclic repetition and a strong sense of meter are the fundamental characteristics of EDM as a "dance" music.[4] This does not mean that EDM cannot articulate sophisticated relationships; Mark J. Butler (2003, 2006), in particular, has written extensively to highlight EDM's rhythmic complexities and ambiguities. But all such nuances occur within a well-defined, predictable temporal framework; as he states: "Though individual rhythms within the network tend to be highly syncopated, as a whole they usually express the meter clearly" (Butler, 2003, p. 103). This *has* to be the case if EDM is to perform its intended communal function; predictability in the music's timing and structure is essential to free-form, noncodified dancing, as dancers have to coordinate their movements in *anticipation* of the beat. The notion of "anticipation" points to the listener's active role in the construal of meter. Recent studies in the psychology of perception usefully conceptualize the perception of rhythmic structures as "an inter-action between what is heard ('rhythm') and the brain's anticipatory struc-turing of music ('meter')" (Vuust and Witek, 2014). In other words, meter is *inferred* by the listener; it is the pulse *implied* by a certain rhythm, but the pulse itself may or may not be marked by aural events. The "four-on-the-floor" bass drum pattern found in techno, for example, articulates the meter fully—rhythm and meter coincide; whereas in "breakbeat"-based genres like jungle or drum 'n' bass the bass-drum pattern does not fully duplicate the meter: some of the beats are omitted or displaced to the off-beats (except for the obligatory downbeat at the beginning of each bar).

Within this logic of cyclical repetition, forward movement is generated by a process of addition and subtraction whereby several interacting rhyth-mic layers, constituted by loops of differing lengths, are brought in and out at specific points, typically marking musical statements lasting some multiple of four bars. As Butler explains, "EDM's structure is modular, consisting of relatively generic patterns that can be combined both verti-cally and horizontally in seemingly endless ways" (Butler, 2003, p. 70). Such a modular approach results in an open form, lacking the kind of development that characterizes, for example, the Western music canon and displaying instead a circular conception of musical time.

This ability to articulate both forward drive and stasis makes EDM an attractive option as a film soundtrack, where it can function to inflect the temporal perception of the action and narrative. It is true that similar rhyth-mic structures can also be found in other groove-based musics, e.g. Afro, Latin, samba, funk, etc., but other features set EDM apart from its musical siblings. One of these was already pointed out: EDM reliance on studio tools and techniques for its realization, which enables precisely controlled relationships between sound events and timeline in ways beyond the capa-bilities of human performance. I will now highlight a few other aspects that make EDM a distinctive option as film music.

EDM is further characterized by its particular *sound*, either constructed through electronic synthesis or derived from acoustic sources recorded and manipulated into individual sound qualities. The characteristic timbres of EDM often include much noise-based material, resulting in musical parts that integrate better with a film's sound design and Foley or that even function as sound design or as a cross between music and sound design roles—these latter possibilities, in particular, represent a fertile trend in recent soundtrack practice.

Given its predominantly percussive and noise-based nature, EDM relegates pitch-based material to a secondary role, often reduced to not much more than a fundamental drone. When a harmonic universe can be detected at all, it tends to be based on modal relationships and mostly on minor modes (Dorian, Aeolian, Phrygian); this modal character further accounts for the nondevelopmental quality of EDM. An alternative approach to the analysis of EDM's tonality is given by Rene Wooller and Andrew Brown (2008), who question the applicability of concepts from traditional musicology, and propose instead four "descriptive continuums" (Wooller and Brown, 2008). These are identified as: Rate of Tonal Change, Tonal Stability, Pitch/Noise Ratio, and Number of Independent Pitched Streams. The authors' findings for Rate of Tonal Change in EDM confirm my previous assessment of a low rate in general: "Overall, EDM is skewed more toward the 'drone' end of the spectrum" (Wooller and Brown, 2008). The values for Tonal Stability, as "an estimate of how strong the sense of tonality (as tonicity) is" (Wooller and Brown, 2008), were found to be in midrange, reflecting a broadly modal language and a prevalence of minor modes, to which the authors add the Mixolydian as the next likely structure: "in the vast majority of cases, there is a clear tonic, regular scales are used; most commonly pentatonic minor, followed by minor and Mixolydian" (Wooller and Brown, 2008). Pitch/Noise Ratio was again found to be in midrange, though, the authors add, "varying substantially between subgenres and individual tracks" (Wooller and Brown, 2008). The Number of Independent Pitched Streams, essentially the relative density of a texture in terms of pitched voices or the degree of polyphony, was found to be in midrange, though again with frequent deviations.

In terms of vocals, many tracks dispense with them altogether. When vocals are present, EDM tends to avoid sung lyrics; the voice is mostly used to articulate spoken lines or wordless sound; both are often manipulated and integrated within the textural design (rather than foregrounded, as is the case in conventional song structures).

Finally, we need to consider EDM's distinctiveness in its functioning as a cultural code when inserted in narrative film. Its electronic nature may connote the synthetic, the technological, the industrial. Its roots in youth culture may connote the urban, the hip, the rebellious. As a dance music, EDM may connote body rhythm, physical energy, sexual drive. As a club music, it may connote rave culture, drug taking, excess, abuse, sexual overdrive and so on. Robynn Stilwell offers a more colorful expression to characterize the cultural connotations of techno soundtracks, stating that "films that choose hard-core techno have already constructed a certain 'badassness' into the film" (Editors, The 2003). Nevertheless, these are only potentialities that need to be activated by a specific narrative context; they should not be intended as handy tokens for signaling unequivocally any simple meaning. Effective narration relies on complex contextual and transtextual relations that cannot be generalized.

Interestingly, despite these many possible associations, EDM seems to lack a connection to a specific geographical area. As Caryl Flinn notes, "Techno paradoxically is grounded to region and place—for example, local rave cultures, clubs, and genres of techno of various cities—at the same time that it evades or evacuates geographical specificity" (Flinn, 2004, p. 201). This is another reason why a filmmaker might choose EDM over other groove-based music: in order to avoid too specific geocultural references that may not be appropriate to a certain narrative situation.

13.3.   Rhythm in Relation to Audio and Visuals

The idea of applying a musical logic to the structuring of film rhythm arises very early in the history of cinema. Writing in 1925, historian and critic Léon Moussinac remarked, "If we attempt to study cinegraphic rhythm, we can see that it has an obvious counterpart in musical rhythm" (cited in Mitry, 1997, p. 111). As early as 1915, with *The Birth of a Nation*, Griffith was already cutting his films to create defined rhythmic patterns in the image track, soon to be followed by Abel Gance with *La Roue* (1923). Such techniques were eventually theorized and further refined by the directors of the Soviet Formalist school, Lev Kuleshov, Dziga Vertov, Vsevolod Pudovkin and Eisenstein, among others. The term "montage" (from the French, "to assemble") was adopted to indicate a particular way of cutting and splicing film material, an approach not bound by the rules of "continuity editing" typical of classical filmmaking, which relies on devices like cutting on action, eyeline match and 180° staging to disguise the cinematic artifice and preserve the diegetic illusion. Discontinuity of action, time and space may, instead, be deliberately emphasized. In fact, montage is more than merely an editing technique—it is a sophisticated principle concerned with juxtaposing shots in ways that, by collision and interaction, generate an idea in the mind of the perceiver. In *Strike* (1925), for example, Eisenstein intercuts shots of soldiers firing at the defenseless workers with nondiegetic shots of a bull being slaughtered.[5]

As a technique of editing focused on shot-to-shot relations, montage privileges rapid cutting of film material, and this makes it particularly amenable to rhythmic patterning, a strategy that Eisenstein pursued consistently in the celebrated Odessa Steps sequence in *Battleship Potemkin* (1925). When sound was finally introduced in the late 1920s, Eisenstein simply expanded his earlier typology of montage methods, since these were already largely based on musical analogies, and theorized five main versions of *vertical* montage: metric, rhythmic, melodic, tonal and overtonal. Especially relevant to this discussion are the metric and rhythmic versions of vertical montage. Metric montage is simply matching shot lengths to specific musical features, e.g. beats, bars, phrases and so on. Rhythmic montage, however, is more complex, as it takes into account what goes on within shots. To appreciate exactly what is involved, we need to examine how the notion of rhythm applies to music and to film.

In music, rhythm and meter are well-defined and objectively measurable parameters; we can safely identify pulse, calculate tempo as beats per minute, determine groupings and so on. But film rhythm is a much more complex notion. Rhythm in film is generated at three levels, often operating simultaneously: (1) *editing* rhythm (or *cutting* rhythm), i.e. the rate at which the individual shots succeed one another; (2) *figure movement*, i.e. the movement of the objects and characters represented within a shot; and (3) *camera movement*, i.e. panning, tracking, craning, zooming etc. Of those three levels, editing rhythm is the only chronometrically measurable parameter; it is therefore no surprise that most of the experimentation with film rhythm since the beginning of cinema has concentrated on specifying mathematical relationships in the cutting of film material. But while it is indeed true that, as Jean Mitry observed, "film rhythm is experienced primarily by virtue of the effect of editing" (Mitry, 1997, p. 125), it does not follow that shot durations could be organized into a visual rhythm in the same way that sounds can be organized into musical rhythm. To begin with, the ability of our visual perception in detecting shot length with the required degree of accuracy is questionable; as Mitry noted, "Obviously the audience is aware more or less of relationships of time but is incapable of evaluating *in any precise way* their metrical value" (Mitry, 1997, p. 149, emphasis in original). Further and more importantly, "[R]hythm is not made up of simple relationships of duration. [. . .] Rhythm has more to do with *relationships of intensity*" (Mitry, 1997, p. 125, emphasis in original), themselves contained within relationships of duration. "Intensity" results from the interaction of virtually an infinite number of variables at work in a particular shot: framing (close-up, long shot, etc.), lighting, camera angle, figure movement, camera movement and so on—all these variables affect the perception of intensity in a shot. Furthermore, intensity is a relative concept. A shot does not so much possess, of its own, a certain degree of intensity; the shot acquires its intensity from its relationship to the preceding and following shots, as well as from its role within the whole narrative context. Thus, as a simplified example, a close-up may be perceived as an accent when presented after a series of long shots, whereas the same shot would not necessarily function as an accent when appearing among a series of close-ups. Intensity, in its various degrees, affects our perception of shot duration. As Kulezic-Wilson notes, "Depending on the content, composition, framing, camera movement of the shot and its 'density,' two shots of the same length might be perceived as being different in duration"

(Kulezic-Wilson, 2015, p. 38). That is why the notion of structuring visual rhythm merely by calculating shot length is fallacious.

However, it is at this point that the properties of sound can be harnessed to assist the task of structuring film rhythm. One of the fundamental functions of sound in film can be seen in its ability to temporalize the images. Of music, in particular, Mitry observed that *"it provides the visual impressions with the missing time content by giving them the powers of perceptible rhythm"* (Mitry, 1997, p. 265, emphasis in original). It is by effectively deploying the temporalizing function of music that any metric and rhythmic organization of visuals becomes finally perceivable as such. At this point, the possibilities of creating and controlling *audiovisual* rhythm open up to the filmmaker. With regard to the specific rhythms of EDM, we will see how a repetitive pattern on the soundtrack can provide a reference beat against which the complex visual rhythms could be related and understood. The next sections examine how these possibilities are effectively exploited by the directors of the films analyzed in this chapter.

13.4.   Techno and Kinesis in Tom Tykwer's *Run Lola Run*

Tom Tykwer's *Run Lola Run* (1998) offers a particularly good example of the potentiality of EDM in the structuring of audiovisual rhythm for kinetic effect. In its exuberant visual style, *Lola* epitomizes David Bordwell's concept of "intensified continuity" (Bordwell, 2002a). Bordwell rejects the notion that cinema has entered a "post-classical" phase in the 1960s, maintaining instead that, while much of contemporary filmmaking (post-1960s) has tended to emphasize certain techniques to extremes, the fundamental principles of classical continuity have continued to remain the backbone of mainstream (and some nonmainstream) filmmaking. Four main devices are identified as salient marks of intensified continuity: "rapid editing, bipolar extremes of lens lengths, a reliance on close shots, and wide-ranging camera movements" (Bordwell, 2006, p. 121). All these techniques have a direct impact on film rhythm: cutting rate and camera movement affect meter, speed and flow, while extremes in framing and lens lengths produce a range of intensities and accents that can be structured into defined patterns by means of editing. They are therefore ideal for creating highly kinetic audiovisual structures, and *Lola* exploits such techniques consistently to that end.

The film is organized as a forking-path plot (Bordwell, 2002b): the same scenario is re-proposed three times, each time developed with slight but significant variations in the sequence of events, each leading to a different outcome. Lola has 20 minutes to find 100,000 marks and rescue her boyfriend Manni who, having lost a bag containing payment from a dodgy deal, fears he is going to be killed once the gang's boss turns up at the agreed time—20 minutes from the time of his desperate call from a phone booth to Lola. This reversal of conventional roles, in which the female is cast as the lead character saving the powerless male companion, has prompted a persistent reading of the film in terms of gender relations, though writers differ in their assessment of the film's effectiveness in challenging stereotypical values. While Maree Macmillan, for example, considers *Run Lola Run* as "perhaps one of the first examples of a new breed of truly post-modern explorations of Woman" (MacMillan, 2009), Ingeborg O'Sickey sees Lola as a "deflated heroine" (O'Sickey, 2002).

As the title promises, a significant amount of footage is devoted to portraying Lola dashing through the streets of Berlin. These sections are characterized by extended montage sequences and feature a nondiegetic techno soundtrack mixed in the foreground; diegetic sound is largely suppressed, except for rare occurrences in which selected ambient sounds are themselves integrated into the musical-sonic discourse. This techno music acts as a "main theme" to the film, with its driving pulse lending physical substance to the urgency of the action, and providing a perceptible time grid over which the narratively significant manipulations of the visual/narrative flow can be measured. As Mitry observed, "The time experienced by the characters in the drama [. . .] may be perfectly well recognized—but it is *understood*, not *experienced*. Film needed a kind of rhythmic beat to enable the audience to measure internally the psychological time of the drama, relating it to the basic sensation of real time" (Mitry, 1997, p. 248, emphases in original). Not only does this music supply the chronometric reference that enables the audience actually to *experience*—not just *understand* at the intellectual level—the pressure of the clock ticking inexorably toward the deadline while the protagonists are trying to avert disaster; as a "dance" music, techno has the effect of translating the kinesis of the fast-paced images into the kinesthetic—the visceral sensation of movement itself, felt by the audience at the physical level, thanks also to the emphasis on bass frequency content from the prominent kick drum. This view is supported by the work of psychomusicologist Paul Fraisse, who wrote, "When a piece of music contains recurring isochronous patterns of strong, i.e., stressed, beats, it gives rise to a *motor* activity which develops in synchronization with the strong beats of a musical performance" (cited in Mitry, 1997, p. 274, emphasis in original). In essence, this strategy is aimed at fostering character empathy by stimulating in the audience the psychological pressure of time passing and the dynamic sensation of a motor activity. Thus, rather than merely *witnessing* the protagonists' ordeal, the audience can actually *experience* their mental and physical strain.

A few writers have remarked on the "musicality" of *Run Lola Run*, referring to the way its audiovisual strategy exploits particular features of its techno soundtrack. In this vein, Katherine Spring (2010) has high-lighted sections of the film characterized by congruency in the accent structure across audio and visual channels. In her analysis, Spring draws from Marshall and Cohen's (1998) Congruence-Associationist model, which focuses on the relationship between audio and visual accents to measure audience responses to audiovisual composites. Simply put, the model predicts that "the greater temporal congruence the greater the focus of visual attention to which the meaning of the music consequently can be ascribed" (Bolivar, Cohen and Fentress, 1994).

It is clear that the regular structures of EDM provide ample scope for organizing audiovisual material into rational patterns implying a "musi-cal" logic. However, I am concerned that an overemphasis on structural congruence may perpetuate a view of EDM as only good for "MTV-style" filmmaking,[6] which would not do any justice to either of the two films analyzed in this writing. Besides, the effectiveness of temporal audiovi-sual congruency should not be taken for granted. Marshall and Cohen's original study (1998) relied on very basic animated graphics. But as Scott D. Lipscomb reports, "[A]s the stimuli become more complex, the impor-tance of accent structure alignment appears to diminish" (Lipscomb, 1999). Lipscomb conducted a study comparing audience responses to three conditions of temporal alignment between audio and visual accents: *consonant*, *out-of-phase*, and *dissonant*. The experiment involved two sets of audiovisual composites, animation and motion picture excerpts, and found that, in contrast to ratings for animations, "in actual motion pic-ture excerpts, subject ratings of effectiveness are no longer consistently highest for consonant alignment conditions and lowest for dissonant align-ment conditions" (Lipscomb, 1999). In this chapter, therefore, I intend to go beyond the obvious instances of structural congruence and show that EDM can offer a wider range of possibilities for meaningful audiovisual alignment. In effect, a strategy too reliant on congruency of audiovisual accents would soon become obtrusive, and while this may be exploited as a deliberate effect, it would soon defeat its purpose if overused. Indeed, even in a film like *Lola*, which revels in overt narration, we find that met-rically organized audiovisual structures are limited to selected moments.

To understand *Lola*'s musical strategy, it is necessary to look at how the film conceptualizes time; instances of audiovisual congruency need to be seen in relation to the overall strategy of temporal organization. As Caracciolo puts it, "*Run Lola Run* is a film obsessed with time, not only diegetically, because of the twenty-minute time window Lola has to reach her boyfriend, but also thematically" (Caracciolo, 2014, p. 65). Being simultaneously *subject* and *object*, time in *Run Lola Run* is purposefully manipulated at several interacting levels. Michael Wedel (2009) offers a good analysis of such levels, which span the whole spectrum from the macro-interval, i.e., the film itself, with its cyclical macro-structure, to the micro-interval: the flash-forward of still frames as the smallest unit of cinematic time. As Wedel observes, "In between those two extremes the kinetic experience of *Run Lola Run* can be reconsidered to unfold by a permanent modulation, alternation, and interference of different speeds and frequencies" (Wedel, 2009, p. 140). In other words, moments of strict audiovisual temporal congruence are just one mode of synchronization among several at work in the film. They function as one stage on the continuum of possible time relationships and time representations, but the overall synchronization strategy in *Lola* is far more complex and the audio-to-visual relationship, overall, is not reducible to a *conformance* model[7] (cf. Spring, 2010).

Take, for example, the segment starting at 00.25.20, the climactic moment when Lola, in her first life, is about to reach Manni at the meeting spot. The screen splits in the middle; we see Manni on the left half and Lola on the right, in different spaces. Over the steady four-on-the-floor time reference of the techno track, playing at full volume and with all diegetic sound suppressed, the images articulate contrasting speeds and tempi. On the left half we have slowed down footage, but even here there is a contrast between the rhythm of the busy shoppers coming and going in the background and the almost immobility of Manni in the foreground. On the other half of the screen we have a tracking shot, involving two independent speeds of figure and camera movement, showing Lola's running action in slow motion, while the wall behind her seems to move at an accelerated speed (a combined effect of tracking, soft focus and close-up framing). There is hardly any cutting from 00.25.20 to 00.26.00 (only two barely perceptible cuts in Lola's half screen); it is practically all *internal* rhythm (figure and camera movement); nothing is synchronized to the music's beat. As the philosopher Henri Lefebvre noted, "We know that a rhythm is slow or lively only in relation to other rhythms" (cited in Wedel, 2009, p. 139). And indeed this sequence capitalizes on Lefebvre's observation by having four different speeds on the image track (actually five, once the clock face appears in the lower part of the screen) playing over the steady pulse of the techno track, this latter functioning as the central time reference against which the visual rhythms can be related and understood. We could appreciate this sequence for its formal design, but there is more to it: this strategy is narratively significant. The juxtaposition of isochronous pulse and multiple speeds in the images reflects the tension between chronometric time and experienced time. The driving beat emphasizes, by contrast, the slower speed in the footage; the synergetic result is stillness and great kinetic energy simultaneously: Lola runs; Manni waits; the clock ticks away, indifferent and unstoppable. The suppressed diegetic sound is a cue for subjectivity: against the objective time marked by the music's steady pulse, the protagonists' state of mind is evoked as the subjective experiencing of a warped temporal dimension—the interminable strain of "the last minute," the dilated time-space of "the last mile," the desperate yearning for a "last chance."

I suggest that a significant portion of *Lola* is characterized by *this* type of audiovisual relationships, this more sophisticated concept of "musicality." Complexity, multiplicity and simultaneity are integral to this film's fundamental preoccupation with time and chance, and it is precisely by contrast with the asynchronous elements that instances of audiovisual structural congruency gain their meaning as "encounter," or convergence.

So far I have been characterizing the music in the "running" sequences as techno music; it is now time to refine my account of it. A relatively minor change occurs in the first part of the second run: the four-on-the-floor pattern of the bass drum is replaced by a sparser and constantly varied pattern emphasizing the off-beats, while the snare drum and electric guitar articulate a jungle/breakbeat pattern that foregrounds the pulse of the 16th notes. By having this round begin with an alternative version of the music, the film makes room for the meaningful return of the four-on-the-floor pattern. This is reintroduced when Lola enters her father's bank (00.38.10), and by the time Lola marches her father at gunpoint to the cash office, there is indeed a strong sense of synchrony between Lola's pacing and the bass drum pattern (accent structure alignment). This device effectively functions to signify Lola's chilling determination in carrying out a crazy act, as the building is monitored and soon to be surrounded by police: the stomping walk, expressed as a pounding bass drum, tells that Lola, with that gun in her hand, means "business."

The third iteration (00.52.46) starts with a jungle/breakbeat pattern (revealed at 00.53.39), similar to the beginning of the previous run. But by 01.00.00 we find a significant change of style with the start of a new music cue (separated by 24 seconds of musical silence from the previous cue). This music follows the same groove-based logic of the previous cues, but it is neither techno nor electronic-sounding—at least not in the rhythm section; there may be synthetic drones and processes, but the overall sound quality is mainly acoustic. Ethnic-sounding drums articulate an Afro-inspired pattern, combined with flute sound and wordless vocals delivering Indian-flavored lines at selected points. Here the acoustic timbres seem to suggest a more "organic" and humanized diegetic time-space, as opposed to the more synthetic and mechanized universe of the techno sections. This change of music style marks a significant turning point in the narrative. The cue starts when Manni spots the tramp who has got his money—this is the beginning of a turn of luck for Lola and Manni. From then on, fate will be kinder to the couple: Manni chases the tramp and recovers his money; Lola goes on to a spectacular win at the casino.

With regard to its rhythmic structure, this music functions in the same way as the previous techno: to provide a temporal reference by its regular pulse against which the rhythms of the images and narrative can be related and felt. A good example of effective design in the audiovisual rhythm can be seen between 01.00.52 and 01.01.45. These 53 seconds are composed of only two shots: the first approximately 15 seconds long and the second approximately 38, joined by a soft cross-fade; editing rhythm is therefore totally absent from this sequence—it is all figure and camera movement. All diegetic sound is suppressed. In relation to the steady pulse of the Afro drums, the bouncy rhythm of the running Lola gradually shifts from out of phase at the beginning to a strong sense of audiovisual synchrony around the sequence's midpoint and gradually out again. The effect is emphasized by framing, which becomes gradually narrower until Lola's face is shown in extreme close-up. The image background adds a third independent line of rhythm, appearing to speed up as it gradually goes out of focus. In sequences such as these, audiovisual relationships are articulated dynamically and can only be understood in their diachronic unfolding as a complex gesture cutting across all stages of the *conformance-complementation-contest* continuum.[8] In terms of narrative function, this is another moment of extreme subjectivity; we hear Lola's voice-over saying, "[W]hat shall I do? [. . .] help me, please [. . .] I'm waiting," as she runs, eyes closed, into the "zone." The interplay of temporal dimensions reflects Lola's trancelike state, while the shifting audiovisual alignment could be read as a metaphor for Lola's "tuning-in" to her inner powers. And indeed her "powers" do deliver: Lola's meditative state is abruptly interrupted by the diegetic sound of screeching tires from a lorry that almost runs her over—and that is how she finds herself "by chance" in front of the casino. Here we find a strategy quite typical of montage sequences: the diegetic sound of screeching tires pierces the nondiegetic bubble that had been in place for the previous 53 seconds, producing an abrupt switch from floating subjectivity to anchored objectivity.

To summarize, the regular structures of EDM may be effectively utilized to shape and control audiovisual rhythm, not only by establishing simple metric congruency of music and images but also by exploiting the reliable temporalization afforded by the isochronous pulse to establish complex, polyrhythmic and phasing audiovisual relationships. As Claudia Widgery remarks, with specific reference to music's temporal function in film, "The key is to maintain a balance between integration and autonomy, for both are necessary if the temporal potential of music for film is to be realized" (Widgery, 1990, p. 381).

An analysis of *Lola*'s musical strategy could not be complete without examining the relationship between the techno music and the other musics included on the film's soundtrack. In the latter part of the first run, the entrance of Dinah Washington's "What a Difference a Day Makes"—immediately after the techno music stops (00.28.30)—creates a strong unexpected contrast of great effect. This cue starts when Lola and Manni flee the supermarket they have just robbed. All diegetic sound is initially suppressed, until the first police car appears (00.28.58). The cue is abruptly cut off by the gunshot-sound that kills Lola (00.29.29), and from the long reverb tail of the enhanced gunshot-sound another contrasting music emerges: this is Charles Ives's *Unanswered Question*, an apt quotation in relation to the interrogatives posed at the film's opening. Thus, we have three moments of strong contrast that take the audience for an affective roller coaster ride: in the robbery segment, we find a techno soundtrack fully involved in the action; the framing gets progressively narrower until, in the last shot, Lola's face is shown in extra close-up—we are *with* the characters. As the camera cuts to Lola and Manni fleeing the supermarket, the framing widens considerably into a decidedly objective series of shots; in complete contrast with the involved role of the previous techno, Washington's gentle jazz ballade in 12/8 could not be any more detached from the action, its only connection with the narrative being its lyrics, commenting on time, love and chance in a manner resembling the chorus of ancient Greek theater. The tension is defused, and the grip on the audience is significantly relaxed—we are now mere witnesses, spatially distant and emotionally disengaged. The gunshot event comes as yet another jolt, abruptly stopping the song, and after a few seconds of almost total silence, the music track reconnects to the action by introducing the empathetically aligned Ives's music—we are now back into character subjectivity (slowed-down footage, time-stretched sound design, close-up framing), emotionally involved in the couple's muted shock and grief.

Finally, we come to another important element of *Lola*'s soundtrack: the lyrics within the techno music. In line with my account of EDM's features previously given, it can be seen that lyrics in *Lola* are largely spoken (except for the "Love is" verses, and a stray line of "Help me," which are sung) and often sonically processed in various ways. Their delivery is monotonic and percussive, adding a further layer of rhythmic emphasis to the mix. Both female and male vocals are included, and, significantly, these are heard only in conjunction with Lola's action and never heard when Manni is in the picture. The female lyrics are intended to be understood as Lola's own "voice" and are in fact delivered (mostly) by Franka Potente herself, contributing to an effective characterization of the main protagonist and thus fostering an emotional bond with the audience. As MacMillan observes, "This non-diegetic overlay of vocals works to provide insight into Lola's sense of identity by representing the character's inner thoughts, fantasies and desires" (MacMillan, 2009, p. 111). Indeed, themes of fantasy and desire are conspicuously articulated in the "I wish"

lines, which constitute a sizable portion of the lyrics in *Lola*, and can perhaps be taken as the central theme among the lyrics' sections.[9] These verses
include a few cross-references to key motifs in the narrative, specifically
the heartbeat, which refers to both the pulse of the running clock time,
as well as the actual heart, that of the racing Lola and that of the security
guard (Lola's real father), which stops and is reactivated by Lola's lov-
ing powers. A second important cross-reference recurring in these verses
is that of the "princess" (with armies at her hand) and the "ruler" that
Lola wishes to be, a clear desire for power or at least empowerment. And
indeed, as "princess" and "royalty" is precisely how she is greeted, sarcas-
tically, by the security guard on her arrival at the bank. These lines seem
at odds with the image, commonly encountered in writings, of Lola as a
subversive heroine; one might ask whether Lola is genuinely critical of the
status quo, or is she merely wishing for more power for herself within a
social order that, whether patriarchal or matriarchal, is still based on hierarchical power relations. Alternatively, one may, following Tom Whalen,
understand these lines as quotations, as part of the numerous fairy tale
signs diffused through the work. "For what is *Run Lola Run*," asks Whalen
rhetorically, "if not a fairy tale, albeit of the self-conscious, philosophical
variety" (Whalen, 2000).

In "Running Two," we observe a change toward a more assertive attitude. The "I wish" lines temporarily disappear, giving way to three new
sets of verses. Two of them are again, supposedly, Lola's "voice": the
first set constructs Lola as a character determined to rescue her partner
("I want to fight"/"I want to see you again," rhythmically spoken); the second expresses Lola's infatuation with more "lyrical" terms ("I can't think
of anything but you," sung). As they appear interwoven on the soundtrack,
these two sets reaffirm Lola's belief that "love can do everything" (revealed
early in the film), a fundamental trait of Lola as a Romantic character. As
Grant McAllister puts it, "If there is indeed another discourse in this film's
deceptively postmodern text, it speaks with a Romantic twang" (McAllister, 2007, p. 331). Finally, as the third new set of verses in this section, we
find male (spoken) vocals exhorting Lola to "never, never, give up" and
"do, do the right thing." It should be noted that these verses are first briefly
introduced on female voice (00.34.08–00.34.22), before being promptly
taken over by a male, which makes attributing that "voice" to any specific
character or entity problematic. If, however, we view Lola and Manni as
in a Jungian anima/animus relationship, as proposed by Vadim Rudnev
(2003), the male voice in Lola's space may be understood as Lola's own
mythical *imago* of Manni.

Nevertheless, in "Running Three," the "I wish" and "Never, never"
verses come together, interwoven, as in a synthesis of the previous two
runs,[10] signaling the reunion of the two lovers and perhaps, as some

authors have suggested (e.g. Rubenstein, 2010), a reconciliation of the genders. But then again there are writers (e.g. O'Sickey, 2002) who see a refutation of the reconciliation thesis in the track that plays last over the film's closing credits. Rapped on male voice and all in German, "*Komm zu mir*" has generated some debate for its closing line that states (translated): "I don't need you any more" (cf. O'Sickey, 2002). If this track really is that much of a key to understanding *Lola*, I am wondering whether non-German-speaking audiences are effectively missing something from this film. Consider, also, that one would need to sit through 3.42 minutes of rolling credits to finally catch that very last line.

But why worry about the film's last line in the closing credits, when *Lola*'s key is so conspicuously on display in the first lines of the opening titles: "after the game is before the game" reads one of the two quotations at the very start. "The ball is round, the game lasts 90 minutes," states Schuster, the security guard, kicking a ball in the air to start the game. Plus: ticking clocks, animated sequences, video game references, croupier, roulette, multiple-lives characters. These are strong cues in a privileged position within the film's structure. And considering that Schuster, having stated the basic rules of the game, immediately adds: "anything else is pure theory," it seems clear that the film is suggesting a particular framework for interpretation. In Whalen's words, "What is important while viewing *Run Lola Run* is to acknowledge the director's central conceit of life/art/film as a game" (Whalen, 2000).

13.5. Breakbeat, Noise and Glitch in Darren Aronofsky's *Pi*

Darren Aronofsky's *Pi* (1998) offers a good example of a soundtrack in which the diegetic sounds of the action and the nondiegetic music appear to coexist on a continuum rather than as neatly separate strands. As Kulezic-Wilson notes, "diegetic and non-diegetic sound effects, as well as parts of Max's inner monologues, are often perceived as part of the techno-score" (Kulezic-Wilson, 2015, p. 157). Being remarkably alike in their distinctly noise-based quality, the music, sound design and Foley in this film perform as tightly interrelated parts, merging into a unified sonic discourse that renders their distinction often meaningless.

Pi's plot follows a gifted mathematician, Max Cohen, in his obsessive search for the fundamental master pattern underlying all phenomena. If a pattern could be found within the seeming irrationality of π, whose decimals extend to infinity without, apparently, ever falling into a repeating sequence, then everything has a pattern—it is just a matter of finding the key. In this sense, *Pi* is a tale about the human struggle to understand the world by searching for an order in the chaotic nature of phenomena. This conflict between order and chaos, as painfully lived by the protagonist, is the essence of *Pi*, and the film constructs a robust audiovisual strategy to articulate this conflict convincingly.

Max pursues his quest from his modest flat, kitted out as a rather anachronistically fashioned computer lab—not some tidy desktop machine but a messy contraption of hacked and precariously put together parts taking up most of his living space. This idiosyncratically low-tech aspect of the mise-en-scène is matched by the film's black-and-white photography, characterized by a grainy and overexposed picture quality. Max's life is punctuated by debilitating headaches; these are dealt by a subcutaneous injection that he administers to himself with a vaccination gun, the effects of which are only felt after a protracted crisis that sees the protagonist contorting in excruciating pain and experiencing hallucinations. The cinematography is characterized by fast-paced editing, at times becoming a montage of flashing images of disorientating effect; longer takes are themselves often dominated by shaky camera movement. In short, everything in *Pi*'s narrative and visuals screams "noise"—and the soundtrack obliges.

It is quite common to come across writings that characterize *Pi*'s soundtrack as "techno," but I would like to suggest a sharper characterization, in line with my account of EDM's features, as given in the relevant section of this writing. The main theme, which Aronofsky (1998) designates as the "Max theme," is actually a most typical version of the "Amen break."[11] This epitomizes the "breakbeat"-based styles, whereas techno actually belongs to the opposite camp of the basic binary distinction recognized in EDM, the camp of the "four-on-the-floor" styles. Since breakbeat-based patterns originate from real sampled drums from old records, as opposed to the drum-machine-based constructions of techno, they tend toward higher syncopation and irregularity and are therefore particularly suited to articulate the order/chaos conflict underlying this film's concept. This main theme, which is heard right at the film's opening, sets the tone for the whole soundtrack, which is indeed characterized by various sorts of "breaks." Aronofsky (1998) mentions hip-hop as an influence to his audiovisual style, probably with reference to the common techniques of scratching, sample triggering and punch-phrasing used by DJs. But another equally relevant connection could be made with reference to the "lo-fi" and "glitch" aesthetics, as a distinctive quality pervading both the audio and the visual channels in *Pi*. And this would fit neatly with the film's narrative intent, as the glitch procedure is itself, fundamentally, concerned with reconciling chaos and order, randomness and structure. Moreover, glitch is concerned specifically with the noises of the computer itself: the mechanical whirring and buzzing of printers and peripherals, the electronic beeping of screens and, most significantly, the digital "noise" of information processing: unwanted data, errors, debris in the channel, the noises of malfunction and system failure. We may therefore conceptualize *Pi*'s soundtrack as the sonic manifestation of computation itself as raw, swarming digits. When Max is confident about his latest assumptions, these data coalesce into the positive driving beat of the Amen-break theme—the ordered "pattern" Max is yearning for. When his assumptions are confronted with the chaotic reality of π, however, a critical limit is reached: the computation crashes, spitting out random digits that refuse to arrange themselves into anything other than a disorderly state that translates on the soundtrack as screaming sonic matter.

The same processes that drive Max's computer, Euclid, apply to Max himself, since the two operate in a symbiotic relationship. It is significant that Max sits at the center of a computer system that completely surrounds him and almost engulfs him. As Tarja Laine has pertinently observed, "The relationship between Max and Euclid can be characterized as prosthetic, which Marshall McLuhan defined as a physical extension of the self by means of media" (Laine, 2017, p. 28). In this symbiotic relationship, Max and Euclid crash and melt together as they struggle to find the key that could unlock the mystery of life.

For Euclid, meltdown is a blown microchip; for Max, though, it is acute physical and existential pain. But the cause of the breakdown is the same—their preoccupation with π. Effectively, *Pi*'s narrative is predicated on the "forbidden knowledge" trope—that which will blind you/maim you/kill you. This is implied quite literally at the film's opening: already as a child Max had to endure temporary blindness because he wanted to "see the Light." This episode from Max's childhood sets the excuse for the subsequent attacks, while other elements intervene to add a layer of ambiguity to the narrative, i.e. Max's compulsive behavior and psychological state, which throws into question whether what we see is actually happening to Max or whether it is simply imagined. But essentially, Max is (or becomes, following his sun staring) a prodigious child ("published at 16, PhD at 20," as we learn from his teacher) whose inquisitive nature keeps pushing him ever further toward "dangerous" knowledge, and his suffering gets increasingly worse as he seems to be getting closer to discovering *the* pattern. "You fly too high. You'll get burnt," warns Sol, his teacher who, on his part, has in fact already suffered a stroke *while* working on π and apparently getting close—too close, obviously—to the answer: he knows of the 216-digit number and of the meltdown it causes, in which supposedly the computer becomes "aware of its own structure." After his first stroke, Sol had given up his quest, and the moment he resumes his work, spurred by Max's findings, he suffers the second, fatal stroke. When Max enters his mentor's apartment, he finds a handwritten paper showing a series of numbers, which is to be understood as the key to π—that which has killed the old man. Max takes the paper home and there, while contemplating the numbers, he has his final, most violent attack, after which he decides to eradicate his knowledge of the code by burning the paper and drilling a hole in his skull. Supporting and corroborating the "forbidden knowledge" trope is a related trope: only the "pure," or the "chosen," can be admitted to the "ultimate truth," can bear with impunity the vision of the "Light" etc. And indeed Max eventually declares himself to be the "chosen one" to the Hasidic Jews who are trying to coax his services. Learning from Rabbi Cohen that the 216-digit number is the "true name of God," Max mutters: "That's what happened: I saw God!" That is why the migraine attacks are treated so prominently in the narration, with both the cinematography and the soundtrack richly elaborating on the symptoms and the devastating effects experienced by the protagonist. Understood in this context, the migraine sections are no longer a question of the narration dispensing a dose of discomfort to the audioviewer simply as a way of characterizing Max as a troubled paranoid but a way of articulating a more fundamental tenet of *Pi*'s construct: the agony of pursuing, or beholding, that which is beyond human comprehension.[12]

In *Pi* this ancient trope is tastefully reworked into a more contemporary version that adds a strong psychological dimension to the drama; this is achieved particularly through a highly subjective filmmaking style. The range of knowledge made available by the narration is severely restricted to the protagonist's experiencing of the events, and the cinematography is replete with Max's POV shots. We perceive largely through Max's senses, and this fact generates a grey area of film space in which sounds can never be fully diegetic or nondiegetic. In *Pi*, sounds exist in the liminal, "in-between" space opened up by the subjective narration. In this osmotically structured space, sounds float rather than being firmly anchored to their supposed origin and can therefore perform as a unified, cohesive sonicscape signifying at more than one level simultaneously. Thus, for example, a sound could be referencing diegetically a certain detail of the action (e.g. computer beep, printer noise, keyboard tapping etc.) while functioning at the same time as a "musical" element. Cast into this dual role, sound acquires a more abstract, symbolic significance beyond its mimetic function—it is no longer merely indicating "the computer is beeping" or "the printer is buzzing" but is conveying to the audioviewer the full dramatic implications of such sonic outputs.

As an example, we could look at the sequence starting at 00.04.00. The image cross-fades from an exterior shot showing a treetop to an interior shot framing the scrolling LED display of Max's computer setup. Over a nondiegetic synth sound continuing from the previous shot (though with a slight variation in its pitch pattern, now reiterating *C–C–F–Bb*), we hear a complex of other sounds with a supposedly diegetic origin, all connected to the machine's workings, among which are an indistinct low hum, a touch of white noise peaking at certain points and a rather dense series of faint, random beeps. When the camera frames the blinking cursor on the computer screen, a more prominent, regular beeping appears on the soundtrack, initially beating roughly at half-note intervals. Max presses the return key (key close-up) and the beeping doubles in speed (in synch with data appearing on the screen), metrical for a short time but soon refusing to settle into a regular pattern. Instead, the groove promised but not realized by the beeping is promptly taken over by the printer's buzzing sound articulating a ternary pattern. A telephone ringing joins the soundscape, and the synth motif is replaced by a bass drone, signaling the ominous nature of the call. In quick rhythmic succession, the printer's sound pattern ends on the briefly sustained tone of the paper ejecting (printer close-up), Max slams the handset down (Max close-up), and tears off, with conspicuous sound, the paper printout (printer close-up); on which shot the sequence sharply ends (00.04.57).

In this short excerpt, the diegetic sounds, heard over the minimal harmonic support of the nondiegetic track, acquire a "musical" sense—not necessarily as metrical patterns but as freer motifs interacting with the images to create a more complex audiovisual rhythm: fragments of patterns articulating the order/chaos tension underlying the narrative and the elusiveness of the order Max is striving for. These are not just mimetic sounds reproducing a supposed "reality" but narratively significant structures conveying a bundle of affects, e.g. the anxious anticipation of the results following Max's pressing of the return key, the implications of Max committing himself further to the perilous quest and so forth.

Two factors are at work to enable this strategy. One is the consistency of the sonic palette: it is thanks to the well focused strategy previously outlined—the idea of an overall sonic print meaningfully based on digits and breaks—that integration is achieved. The second is the purposeful structuring of events into rhythmic patterns across audio and visuals; this is indeed a fundamental characteristic of the whole film. As Kulezic-Wilson remarks, "In the same way Max believes everything we touch is infused with the pattern of a spiral, so is π permeated with repetitions and patterns in all elements of its narrative and audio-visual structure on both the micro- and macro-levels" (Kulezic-Wilson, 2015, p. 140).

Thus we have patterns of pill taking, door locking/unlocking, migraine attacks, recurring verbal statements and so on. Of these, the shorter motifs, like the pill taking or the lock/unlock action, are of particular interest as audiovisual structures. These are miniature sequences of rhythmically organized audiovisual montage, striking in their speed and brevity—the pill-taking sequence lasts only two seconds, cut into four shots, themselves filled with fast movement within the frame (extreme close-ups). Here the role of sound is significant in achieving the desired effect. Besides providing an acoustic rendering to the visually represented, sound here performs two important (overlapping) functions. The first is what Chion (1994) labeled the *spotting* function. Chion noted this crucial role with regard to rapid visual movements in fight scenes: "The ultrabrief image of the punch all by itself would not become engraved into the memory, would tend to get lost. But an ultrabrief but clearly delineated sound has the advantage of etching its form and tone directly into consciousness, where it can repeat as an echo" (Chion, 1994, p. 61). It is sound that makes us register (audio) visually such a fast succession of images; this is particularly important in the context of *Pi*, in which this sequence needs to be recognized by the audioviewer as a repetition at various points in the film. The second function accomplished by sound is to make the sequence perceptible as a well-defined rhythmic structure, thus turning it into a self-contained symbolic unit predisposed to be redeployed to form larger patterns at a higher structural level. This device has the effect of transposing a simple action into a more complex idea—a sharp, stylized audiovisual gesture with a precise narrative function, compressing time and meaning to make its point with maximum efficiency. As Kulezic-Wilson elaborates, "In its first appearance the pattern indicates the repetitiveness of the action, which for Max is an unavoidable routine, but later suggests also the urgency of it, the anticipation of the pain and the fear that follows it" (Kulezic-Wilson, 2015, p. 144).

The fact that Aronofsky has referred to these punchy structures as "hip hop sequences" (Aronofsky, 1998) is indicative of the musical thinking behind the director's individual style. Indeed, one of the interesting aspects about the phenomenon discussed in this writing has to do with the influence exerted by the particular structures of EDM on filmmaking style—it is not simply a case of swapping an orchestral score for an EDM track but a significantly different approach to filmmaking, with a strong commitment to an integrated audiovisual strategy and keen to exploring and exploiting the features of EDM structures to such ends.

Nevertheless, musically aware filmmaking, and EDM-influenced film structures in particular, need not be all about metrically organized material and audiovisual structural congruence. Indeed, I would suggest that it is precisely because EDM structures, as Katherine Spring correctly notes, "facilitate accent structure alignment" (Spring, 2010) that special care needs to be exercised in organizing audiovisual material, ensuring an appropriate balance between synchrony and asynchrony across all elements of audio and visuals. Devices like Aronofsky's hip-hop sequences, for example, are effective as "breaks" only as long as there is a "flow" to contain them. Overused, the device would soon lose its power. Similarly, a strategy based predominantly on congruency between audio and visual

accent structures would soon defeat its purpose: if everything is emphasized, nothing is significant. In this respect, the audiovisual team who
worked with Aronofsky on *Pi* did a very good job at exploiting the full
range of relationships. And here it is not just a question of aesthetic refine-
ment: the film relies on contrasting modes of synchronization to articulate
the chaos/order conflict underlying its narrative.

Evidence of a strategy that carefully avoids beat matching in favor
of more sophisticated audiovisual relationships can be found practically
throughout the film; already in the short excerpt analyzed here, a careful
balance can be seen between metrical and free-flowing relationships. The
effect can be quite subtle at times and yet significant, as in my next example, a case of composed asynchrony within a groovy section. At 00.16.40,
we find the first occurrence (after the opening titles) of the main theme.
As I have said, this theme is triggered at points where Max is confident
about his assumptions, hence the reassuringly ordered metrical pattern.
First we hear the intro segment of the theme over Max's stated assumptions; then the full Amen-break pattern kicks in (00.17.09), in response
to Max's question, "So what about the stock market?" And finally, when
Max states "press return," the pattern stops on the first beat of the 4/4 bar
(00.17.32). These are the "rational" relationships on which the sequence
is organized. Nevertheless, first it can be observed that film editing over
the above segment is consistently avoiding cutting on the beat, articulating instead its own independent rhythm—vertical (audiovisual) relationships are nonmetrical. Second, asynchronous relationships can be
observed within elements of the soundtrack, too: when Max states "press
return," his fingers hesitating over the keyboard (00.17.32), the Amen-
break pattern stops, and a single-line electronic pulse emerges. Over this
nondiegetic tone that continues to mark beats one and three of the bar, we
hear the supposedly diegetic electronic beep associated with the blinking
cursor on Max's computer screen, apparently pulsing at the same rate as
the nondiegetic tone and placed—initially—on beats two and four. However, this relationship gradually goes out of phase until, after 22 bars, the
computer beep has shifted to beats one and three, and the Amen break
resumes on the first beat of the 23rd bar, in synch with Max's pressing
of the enter key (00.18.04). This may seem a trivial detail, but suppose
the two electronic tones had continued in the same metrical relationship
over 22 bars, backbeat disco style (tuh-beep-tuh-beep . . .)—the tension
would be significantly defused by the reassuringly predictable beat; the
reentering of the Amen break would lose its impact as the first synch point
after 32 seconds of asynchrony; and the complex signification of acute
conflict at that crucial moment (to press or not to press the return key?)
would be lost. In short, there are some metrically organized relationships
at one level of structure, but even here—a section featuring the groovy

main theme—contrasting modes of synchronization can be seen at work. In fact, it is precisely thanks to the asynchronous elements that the two key moments in the sequence marked by the stop and restart of the Amen break obtain their effectiveness *as* synch points.

A much more obviously complex example can be seen starting at 00.47.52. This sequence plays a situation similar to the previous example (the protracted interval between Max stating "press return" and his eventual pressing of the key), though in a further intensified manner, with Max now deep into the snares of π. Here we find no less than four independent rhythms on the soundtrack: one is a diegetic screen cursor sound beeping roughly at quarter-note intervals; three are nondiegetic synth lines: one playing a slow four-note loop in mid-low range (*C–G–Eb–Bb*), another playing a groovy one-bar loop reiterating *G* in mid-high register, varied by dynamic filtering; the third (fading in at around 00.48.30) is the familiar synth tone in low-mid register associated as a leitmotif to the impending migraine attacks, pulsing at roughly quarter-note intervals. These four loops are all playing as independent lines, in continuously shifting relationship to one another and with absolutely no sense of a common meter. To these we may add as a fifth rhythmic element the lovemaking sounds attributed to Max's neighbor, Devi, which also have a carefully constructed flow of peaks and dips. Meanwhile, on the image track, after a pill-taking sequence is punched in, a long take is started (00.48.30 to 00.49.06), showing Max nervously walking in circles around his lab, followed by the camera, creating a complex visual rhythm combining figure and camera movement; this constitutes a sixth independent line of rhythm, accelerating over the course of the shot. The audiovisual result could not be any more sophisticated in its purposeful and carefully composed complexity, generating a multiplicity of time-levels that, while not quite rational in their relationship to one another, can be grasped in their totality as a state of stable instability. This sequence, like many others in the film, shows a carefully controlled balance between synchrony and asynchrony, metrical and free-flowing structures as a deliberate strategy for *Pi*—the rational and the irrational are featured as inextricably linked elements in this drama.

I would finally like to touch on the important migraine sequences. The attacks always start with a twitching of the thumb (close-up) sonified by subtle bursts of white noise, the typical sound indicating interference, corrupted data, malfunction in a digital system. This is soon followed by a pulsating synth tone marking quarter-note intervals; this sound strongly recalls the typical warning tone indicative of system failure, an "alert" sound, often a signal of impending disaster. Here, though, it is realized in a more sophisticated way: it is made of two attacks placed on the first and second 16th notes of a crotchet, the first a low *G*, the second a *G* an octave above. As the loop is subtly manipulated by dynamic filtering, over the course of its duration, it is possible to hear the downbeat shifting to the high G. It remains nevertheless stable enough as an isochronous pulse, and it functions as a unifying backbone to the sequence, temporalizing the images with a reliable chronometric reference and a sense of continuity. Another sound emerges, a rather abrasive timbre alternating Ab and G below (m2) as a slower loop in mid-high register, adding to the tension by the dissonant minor 9th formed with the fundamental G of the "warning" tone. Finally, the more intense sounds directly associated with Max's pain appear. In the first crisis (00.08.25), we find a fast spinning loop alternating two pitches roughly equivalent to quavers, screaming in high register. On the second beat of its fourth bar, this loop appears to trigger a still higher and shriller sound (roughly $F\#$), sustained a few seconds and then morphed into a mellower sustained tone (roughly F). There is, in addition, a mid-low howling sound expanding Max's vocal laments. Most importantly, this "pain-sound" complex is triggered by a punched-in structure lasting half a second and consisting of two intense shots over a noisy tone combining a high-frequency and a low-frequency signal; this is repeated after a while, and when retriggered a third time, it cuts off the pain-sound.

Thus, we have three clear signals of malfunction in logical succession of increasing severity that function to characterize the migraine attacks as a case *analogous* to "system failure": first the glitchy statics over the twitching thumb, the initial symptoms of corrupted data; second the obsessive warning tone; and finally the distinctly "broken" quality of the punched-in micro-events, these latter triggering the spinning pain-sound. Very much like Euclid's chip, Max's brain, faced with the irreducible complexity of π, reaches overload and "crashes." The computation (as explained by Sol) "gets stuck in a particular loop," and the system "spits out a long string of numbers." The spinning pain-sound could thus be understood as a supposed decoding of those numbers into sonic information, as if that data could simply be fed to a D/A converter and output as sound waves—the enigmatic 216-digit code is made audible, and it's not pretty. This idea strongly recalls a typical procedure of glitch, which consists in forcing the computer to convert the raw data from nonaudio files (e.g. ASCII text, pictures, etc.) into audio signal.

In summary, even after 20 years since its release, *Pi* remains one of the finest examples among films exhibiting an "integrated soundtrack," an approach to film sound discussed in current literature as a strategy of "blurring the line between music and sound design" (Kulezic-Wilson, 2017, p. 129). While such an approach is not entirely new, an increasing trend in this direction has been observed in films of the past 20 years. This trend is often portrayed by writers as a "musicalization" of the soundtrack (Donnelly, 2013, p. 366). But, of course, we could equally talk of a "noisyfication" of the music track. The musical revolution started by the likes of Russolo, Varèse, Cage and Schaeffer brought "any and all sounds" (Cage, 1961, p. 4) into music practice. This aesthetics eventually influenced (directly or indirectly) various forms of popular music, among which EDM is perhaps the genre that most consistently embraces the noise-sound approach to music making. When the inclusive logic of such aesthetics is taken into film, the boundaries between soundtrack elements are severely undermined—nondiegetic "music" will tend to absorb the diegetic sounds into its noisy structures, turning them into sonic "gestures," while the diegetic space will tend to attract the "music" into its environment, making it its "ambience." The resulting ambiguity can be purposefully exploited by filmmakers for effective signification, as well illustrated by *Pi*. But in *Pi*, this strategy is particularly effective because its sonic materials are so well connected conceptually to its subject: in this drama of numbers, digits, data processing, information and knowledge, the integrated-soundtrack approach (as a stylized idiom) frees the "noise," the "ruptures," the "overloads" and the "breakdowns" from their indexical function, turning them into powerful signifiers functioning simultaneously at the literal and the symbolic levels.

13.6. Conclusions

The distinctive traits of EDM offer substantial possibilities to narrative cinema, especially with regard to the structuring of audiovisual rhythm and the integration of soundtrack elements (music, sound design, Foley, dialogues) into a synergetic whole.

As a groove-based music, EDM can supply a reliable pulse against which the flow of the images and narrative can be related and experienced by an audience. Its emphasis on rhythmic structures at the expense of melodic and harmonic development can be harnessed to invest film sequences with a marked sense of kinesis, transposing "a purely visual experience of motion into a visceral one as well" (Widgery, 1990, p. 384). Its open, nondevelopmental form, with no clear beginning and ending, makes it particularly flexible for cutting scenes to musical sections, and its metrical organization facilitates the "cross-modal alignment of accent structures" (Lipscomb, 2013, p. 192), including the precise alignment of beats and bars to the film's editing rhythm or to specific visual events, a device that could be exploited to construct calculated patterns of audiovisual rhythm. I have suggested in this chapter that this latter property of EDM calls for considered handling. While the stylized and even musicalized effect produced by metrically organized audiovisual structures can be exploited purposefully for specific narrative ends, the device can easily become ineffective if overused. Audiovisual structural congruence is not the only or even the most effective device for bringing audio and visual into meaningful interaction; the features of EDM offer a wide spectrum of possibilities for structuring audiovisual rhythm, as illustrated by the analyses in this writing.

The synthetic and noise-based sound of EDM facilitates integration with other elements of the soundtrack. This is enabled particularly by the vertical openness of EDM's form, evolving through the constant addition and subtraction of layers and therefore capable of assimilating diegetic sounds as part of its structure. A sonically integrated audio track offers a wider spectrum of possibilities for signification, as all soundtrack elements can be manipulated and structured into symbolic units to articulate a more complex meaning, richer in connotations and capable of making a narrative point with maximum efficacy.

The two films analyzed are prime examples of the possibilities highlighted in the chapter. But what is crucial about both films is that their "integrated" nature is first of all the result of a procedural approach to filmmaking that engages with the soundtrack at an early stage of production rather than leaving it to the last, as is most often the case. Integration starts at the conception stage and implies a filmmaker's awareness of the potentialities of sound in film. In certain cases, we see the directors themselves contributing directly to the soundtrack, as is the case with Tykwer, who composed/produced the music for *Run Lola Run* in collaboration with Johnny Klimek and Reinhold Heil. This approach is also a reflection of a changing practice brought about by technological developments. As Claudia Gorbman notes:

> Over the last twenty years the advent of digital recording and storage of music as well as of digital video editing have made it possible for directors to exert much greater control over the selection and placement of music in their films, and has liberated the music soundtrack from the rarefied province of specialists.
>
> (Gorbman, 2007, p. 151)

To this we may add the changes in industry models over the same period, which saw a decreasing monopoly of the major Hollywood studios and the rise of the independent studios, smaller entities with less rigid divisions of specialism compared to the highly compartmentalized structures of the historic studios. But whether the directors are themselves contributing to the soundtrack is not the essential point; filmmaking is always a collaborative enterprise. What matters is a director's understanding of the possibilities offered by sound in film and of the necessity of involving sound specialists at an early stage of conception and in a collaborative, two-way relationship. A promising trend in this direction can be observed in recent years, with a growing number of directors seeking a deeper engagement with sound in their films. Significantly, this approach is often defined by a particular interest in sound design, and this constitutes a fertile ground for EDM, given its special ability to integrate with, cross into or function as sound design. Such developments in filmmaking practice, coupled with a changing aesthetic sensibility craving alternatives to conventional orchestral scoring, will likely result in an increasing presence of EDM in film and an increasing range of opportunities for EDM practitioners.

Notes

1. This is consistent with the notion expressed by Butler: "Historically, [. . .] EDM has been defined by its relationship to the dance floor" (Butler, 2003, p. 8).
2. *Run Lola Run* includes some non-EDM music as quotation and sharp contrast (effectively a kind of Eisensteinian dialectical montage in the audio domain), but nothing like conventional orchestral film music.
3. Moments of total absence of meter do occur but only temporarily, as intros, outros, interludes. Eventually the meter must be (re)established.
4. I am aware that a few authors consider beatless (sub-)genres as part of the EDM umbrella. As clarified in the introduction, in this essay I am only concerned with the beat-driven genres that constitute the core of the EDM repertoire as a music *for* dancing. Significantly, Mark Butler refers to the beatless within EDM as an anomaly: "'Ambient,' or 'downtempo,' is something of an anomaly within EDM, in that it is slow and often beatless" (Butler, 2003, p. 90).
5. It is important to note that Eisenstein's theories went through two distinct phases. The example from *Strike* reflects his "dialectical" phase, which emphasized tension and shock as a montage principle (montage of *attraction*), whereas in his second phase (from 1930) he regressed to a notion of unity and synthesis of elements more akin to the Wagnerian *Gesamtkunstwerk*, a position he had originally repudiated. See Bordwell (1974, 2005) for a detailed account.
6. See Calavita (2007) for a robust critique of the "MTV aesthetics trope," which Calavita characterizes as a "film criticism fallacy."
7. I am referring to Nicholas Cook's general theory of multimedia, which, very succinctly put, posits three basic models: *conformance*, characterized by a relationship of congruence between media; *complementation*, exhibiting a degree of "undifferentiated difference"; and *contest*, marked by "collision or confrontation" between media (Cook, 1998, p. 102).
8. See note 7.
9. Caryl Flinn maintains that "Believe" constitutes "the central musical theme of the film" (Flinn, 2004, p. 205), even though this track is never actually heard in the film. Flinn seems to rely too much on the audio CD version for her analysis. While such an approach may be perfectly appropriate for Flinn's predominantly cultural analysis of *Lola*, in this writing I am specifically focusing on the experience provided by the film itself.
10. See Whalen (2000) for a detailed account of *Run Lola Run*'s dialectical structure as thesis, antithesis and synthesis.

11. The term "Amen break" refers to a four-bar drum solo found in "Amen, Brother" (1969), performed by The Winstons. This has become the world's most sampled audio segment; its structure forms the basis of many breakbeat-driven styles, e.g., hip-hop, drum 'n' bass, jungle etc.
12. Cf. Anderson (1998), who writes off the migraine scenes as gratuitous: "When the plot sticks with the math, [it's] genuinely exciting and interesting, but when we go to Max's headaches and hallucinations, we get into some seemingly gratuitous nightmare imagery (brains, blood, tumors, etc.) that will turn some people off" (Anderson 1998).

References

Anderson, J. M. (1998). Interview with Darren Aronofsky: Easy as 3.14. *Combustible Celluloid*. Available at: www.combustiblecelluloid.com/daint.shtml [Accessed June 2018].
Aronofsky, D. (1998). Director's and actor's commentary. *Pi: Faith in chaos*. [DVD], Special Features, Region 2, Pathe!
Bolivar, V. J., Cohen, A. J. and Fentress, J. C. (1994). Semantic and formal congruency in music and motion pictures: Effects on the interpretation of visual action. *Psychomusicology*, 13, pp. 28–59.
Bordwell, D. (1974). Eisenstein's epistemological shift. *Screen*, 15(4), pp. 29–46.
Bordwell, D. (2002a). Intensified continuity: Visual style in contemporary American film. *Film Quarterly*, 55(3), pp. 16–28.
Bordwell, D. (2002b). Film futures. *SubStance*, 31(1, 97), pp. 88–104.
Bordwell, D. (2005). *The cinema of Eisenstein*. New York: Routledge.
Bordwell, D. (2006). *The way Hollywood tells it*. Berkeley and Los Angeles: University of California Press.
Butler, M. J. (2003). *Unlocking the groove: Rhythm, meter, and musical design in electronic dance music*. PhD thesis, Indiana University.
Butler, M. J. (2006). *Unlocking the groove: Rhythm, meter, and musical design in electronic dance music*. Bloomington and Indianapolis: Indiana University Press.
Cage, J. (1961). The future of music: Credo. In: *Silence*. Hanover, NH: Wesleyan University Press.
Calavita, M. (2007). "MTV Aesthetics" at the movies: Interrogating a film criticism fallacy. *Journal of Film and Video*, 59(3), pp. 15–31.
Caracciolo, M. (2014). Tell-tale rhythms: Embodiment and narrative discourse. *Storyworlds: A Journal of Narrative Studies*, 6(2), pp. 49–73.
Chion, M. (1994). *Audio-vision: Sound on screen*. Edited and Translated by C. Gorbman. New York: Columbia University Press.
Cook, N. (1998). *Analysing musical multimedia*. Oxford: Clarendon Press.
Donnelly, K. (2013). Extending film aesthetics: Audio beyond visuals. In: J. Richardson, C. Gorbman and C. Vernallis, eds., *The Oxford handbook of new audiovisual aesthetics*. New York: Oxford University Press, pp. 357–371.
The Editors. (2003). Panel discussion on film sound/film music: Jim Buhler, Anahid Kassabian, David Neumeyer, and Robynn Stilwell. *The Velvet Light Trap*, 51(Spring), pp. 73–91.

Flinn, C. (2004). The music that Lola ran to. In: N. Alter and L. Koepnick, eds., *Sound matters: Essays on the acoustics of modern German culture*. New York: Berghahn Books, pp. 197–213.

Gorbman, C. (2007). Auteur music. In: D. Goldmark, L. Kramer and R. Leppert, eds., *Beyond the soundtrack: Representing music in cinema*. Berkeley and Los Angeles: University of California Press, pp. 149–162.

Iyer, V. (2008). Book review: Unlocking the groove: Rhythm, meter, and musical design in electronic dance music, by Butler Mark. *Journal of the Society for American Music*, 2, pp. 269–276.

Kulezic-Wilson, D. (2015). *The musicality of narrative film*. Basingstoke: Palgrave Macmillan.

Kulezic-Wilson, D. (2017). Sound design and its interactions with music: Changing historical perspectives. In: M. Mera, R. Sadoff and B. Winters, eds., *The Routledge companion to screen music and sound*. New York and London: Routledge, pp. 127–138.

Laine, T. (2017). *Bodies in pain: Emotion and the cinema of Darren Aronofsky*. New York and Oxford: Berghahn Books.

Lipscomb, S. (1999). Cross-modal integration: Synchronization of auditory and visual components in simple and complex media. *The Journal of the Acoustical Society of America*, 105(2).

Lipscomb, S. (2013). Cross-modal alignment of accent structures in multimedia. In: S. Tan, A. Cohen, S. Lipscomb and R. Kendall, eds., *The psychology of music in multimedia*. Oxford: Oxford University Press, pp. 192–213.

MacMillan, M. (2009). It's all in the soundtrack: Music as co-constructor of postmodern identity in Tykwer's end-of-millennium text, "Run Lola Run." In: *Proceedings of the Joint Conference of the 31st ANZARME Annual Conference and the 1st Conference of the Music Educators Research Center (MERC)*. Melbourne: ANZARME and MERC, pp. 107–115.

Marshall, S. And Cohen, A. (1988). Effects of musical soundtracks on attitudes toward animated geometric figures. *Music Perception*, 6(1), pp. 95–112.

McAllister, P. (2007). Romantic imagery in Tykwer's "Lola Rennt". *German Studies Review*, 30(2), pp. 331–348.

McLeod, K. (2001). Genres, subgenres, sub-subgenres, and more: Musical and social differentiation within electronic/dance music communities. *Journal of Popular Music Studies*, 13, pp. 59–75.

Minett, M. (2013). Beyond the badass: Electronic dance music meets film music practice. *New Review of Film and Television Studies*, 11(2), pp. 191–210.

Mitry, J. (1997). *The aesthetics and psychology of the cinema*. Translated by C. King. Bloomington: Indiana University Press.

Nye, S. (2010). Review essay: Run Lola Run and Berlin Calling. *Dancecult: Journal of Electronic Dance Music Culture*, 1(2), pp. 121–127. Available at: https://dj.dance cult.net/index.php/dancecult/article/view/296/282 [Accessed June 2018].

O'Sickey, I. M. (2002). Whatever Lola wants, Lola gets (or does she?): Time and desire in Tom Tykwer's Run Lola Run. *Quarterly Review of Film and Video*, 19, pp. 123–131.

Rubenstein, K. (2010). Run, Lola, Run's unique approach to gender. *ReelRundown*. Available at: https://reelrundown.com/movies/Run-Lola-Runs-Unique-Approach-to-Gender [Accessed June 2018].

Rudnev, V. (2003). Run, matrix, run: Event and intertext in modern post-mass cinema. *Third Text*, 17(4), pp. 389–394.

Spring, K. (2010). Chance encounters of the musical kind: Electronica and audiovisual synchronization in three films directed by Tom Tykwer. *Music and the Moving Image*, 3(3), pp. 1–14.

Vuust, P. and Witek, M. (2014). Rhythmic complexity and predictive coding: A novel approach to modeling rhythm and meter perception in music. *Frontiers in Psychology*, 5, pp. 1–14.

Wedel, M. (2009). Backbeat and overlap: Time, place, and character subjectivity in *Run Lola Run*. In: W. Buckland, ed., *Puzzle films: Complex storytelling in contemporary cinema*. Malden, MA: Wiley-Blackwell, pp. 129–150.

Whalen, T. (2000). The rules of the game: Tom Tykwer's *Lola Rennt*. *Film Quarterly*, 53(3), pp. 33–40

Widgery, C. (1990). *The kinetic and temporal interaction of music and film: Three documentaries of 1930s America*. PhD thesis, University of Maryland, College Park, UMI order no. 9121449.

Wooller, R. and Brown, A. (2008). A framework for discussing tonality in electronic dance music. In: S. Wilkie and A. Hood, eds., *Proceedings of the sound: Space—the Australasian computer music conference*, Sydney, pp. 91–95.

Filmography

Amores Perros. (2000). Directed by Alejandro González Iñárritu. Mexico.

Assault on Precinct 13. (1976). Directed by John Carpenter. USA.

Battleship Potemkin. (1925). Directed by Sergei Eisenstein. Soviet Union.

Berlin Calling. (2008). Directed by Hannes Stöhr. Germany.

Blade Runner. (1982). Directed by Ridley Scott. USA.

Fight Club. (1999). Directed by David Fincher. USA and Germany.

Go. (1999). Directed by Doug Liman. USA.

Hanna. (2011). Directed by Joe Wright. Germany, UK, USA and Finland.

Human Traffic. (1999). Directed by Justin Kerrigan. UK and Ireland.

Lara Croft: Tomb Raider. (2001). Directed by Simon West. Germany, Japan, UK, USA.

La Roue. (1923). Directed by Abel Gance. France.

Layer Cake. (2004). Directed by Matthew Vaughn. UK.

Man on Fire. (2004). Directed by Tony Scott. USA and UK.

Midnight Express. (1978). Directed by Alan Parker. USA, UK and Turkey.

Pi. (1998). Directed by Darren Aronofsky. USA.

Run Lola Run. (1998). Directed by Tom Tykwer. Germany.

Sexy Beast. (2000). Directed by Jonathan Glazer. UK and Spain.

Spy Game. (2001). Directed by Tony Scott. USA.

Strike. (1925). Directed by Sergei Eisenstein. Soviet Union.

The Beach. (2000). Directed by Danny Boyle. USA and UK.

The Birth of a Nation. (1915). Directed by D. W. Griffith. USA.

The Matrix. (1999). Directed by The Wachowski Brothers. USA.

The Saint. (1997). Directed by Phillip Noyce. USA.

The Sentinel. (2006). Directed by Clark Johnson. USA.

Trainspotting. (1996). Directed by Danny Boyle. UK.

Soundscape Composition
Listening to Context and Contingency

John L. Drever

14.1. Introduction

Soundscape composition does not easily qualify as a delineated genre with clear aesthetic or procedural criteria; rather it represents a common set of attitudes and values that emerge out of the themes, methods and strategies associated with the study of the soundscape, itself, a circa 50-year-old subject area that has perpetually undergone development and subsequently revivification by allied disciplines (such as acoustics, biology, geography, ethnomusicology, sociology, etc.). In the same vein, any meaningful exploration on *soundscape composition* should nurture and prompt action and experimentation across disciplines, not nostalgically cling to the preservation of ossified art genre specifications. In this regard, the most comprehensive and I think enthusing definition of soundscape composition that primes practice has been the articulation by Hildegard Westerkamp, a soundscape artist and activist, who also resists pinning it down too tightly:

> [I]ts essence is the artistic, sonic transmission of meanings about place, time, environment and listening perception.
> (Westerkamp, 2002, p. 52)

It goes without saying that such all-encompassing themes have great resonance with commercial sound design; however, soundscape composers are permitted to dedicate all their efforts to their exploration, not subservient to an external narrative or the strictures of film sound clichés etc. Westerkamp leads by example: soundscape compositions such as the stereo acousmatic works, *Talking Rain* (1997) and *Kits Beach Soundwalk* (1989) and the eight-channel *Into the Labyrinth* (2000), each address and activate these themes in a continuously creative and context responsive manner. These works don't just present edited, juxtaposed and superimposed field recordings, but they enquire into the methods and modes of their construction, and most importantly they are not exclusively framed through Westerkamp's highly attuned listening. They involve, include and inform the listener in their practices of listening and sonic ways of knowing.

This chapter will explore the amalgams of orthodoxies and orthopraxis of those salient soundscape concepts that are exercised in manifold configurations in soundscape compositions and related practices. Naming, attending to and more deeply examining these aural vicissitudes, it is hoped, will inform future expressions of sound design practice in a deeper manner, as habitually experienced by our everyday listening and encountered in our everyday soundscape, a discussion that should provoke the sound designer's ever-increasing dependency on tried and tested stock sound effects.

14.2. Ambient/Immersive/Atmospheric Soundscape

Within its comparatively short life, the term *soundscape* has become a somewhat nebulous descriptor; customarily used in conjunction with *ambient*, *immersive*, or *atmospheric*, it loosely refers to a ubiquitous presence of an unceasing, ateleological sonic dimension (i.e. not fatalistically heading toward an inevitable end like a Beethoven symphony or a rock ballad) of a suggested environment(s), be that enclosed or boundless, vague and indeterminate or clearly articulated, identifiable and/or particular. It is not contained within a focused zone within the surrounding sphere of audible space; rather, a soundscape is spatially permeable or diffuse, or the combination of copresences of multiple conflicting (perhaps contrasting or expanding or blending over time) dimensions of spaces. A soundscape is never a void; rather it tends to imply presence and activity, even if it teeters on the thresholds of awareness. A singular fixed implied pulse, a fluctuating polyrhythmic palimpsest, a randomly yet controlled texture of background/ foreground peppering or ebb and flow or interweaving of layers, it often resides at the margins of auditory perception. Quietly affecting and informing but not drawing attention to itself due to its predictable nature, this *keynote sound* becomes the setting for *figure* activities and behaviors to be situated and enacted (Truax, 2001, pp. 24–25). And inescapably, this sonic texture plays a crucial yet intrinsically subliminal role in the perception of acoustic atmospheres: "the character of a space is responsible for the way one feels in a space" (Bohme, 2017, p. 128).

When we talk of *soundscape composition*, in happenstance we may be putting much of the preceding into play; however, there are notions, strategies and tendencies (some explicit, others tacit) that inform the makeup of what constitutes a *soundscape*, much of which originates from a transdisciplinary field, acoustic ecology from the late 1960s–early 70s and subsequent developments.

It is axiomatic that the process of listening,[1] a sense of place[2] and this slippery term "soundscape" are deeply and richly intertwined. So, in place of a fanfare to herald our case study on soundscape composition, there is no better place to start than to practice and reflect on our individual listening wherever we may be located. As it happens, we are all soundscape experts!

14.2.1. Everydayness Listening Exercise

> Everydayness is more or less exclusively associated with what is boring, habitual, mundane, uneventful, trivial, humdrum, repetitive, inauthentic, and unrewarding. At everyday level, life is at its least interesting, in opposition to the ideal, the imaginary, the momentous.
>
> (Sheringham, 2009, p. 23)

With these underwhelming adjectives in that epigraph in mind, have a listen to your everydayness. Wherever, whenever, whatever, whomever—temporally step out of that habitual or assigned mode of activity you find yourself in. This environment is now your designated field of study.

The goal is to reconsider your *background listening* behavior: "It occurs when we are not listening for a particular sound, and when its occurrence has no special or immediate significance to us" (Truax, 2001, p. 24).

You could be sharing the experience, somewhat akin to an audience attending a classical concert, or a take a hermetic approach, whichever, it is important to find a means of discouraging verbal social interaction for the duration of this exercise. It may be beneficial to set your smart device to the do-not-disturb option.

You may be mobile, traveling through the acoustic environment or in a fixed location allowing the prevailing acoustic environment to swell and unfold around you. You may prefer an open time period or a specific duration. You can decide.

Nota bene: please ensure you are not putting yourself or others into a greater state of risk or hazard than you otherwise would be.

The instruction or, better still, the invitation is the humble and physically and cognitively undemanding task: Listen![3]

So how was your listening experience?

For what it is worth, this is my commentary, written following the exercise, not during, so as to not interrupt the flow of listening experience during the exercise.

I chose the fixed location option, my office at work in South London, and gave myself four minutes and thirty-three seconds duration, unashamedly borrowed from John Cage's titular, three movement composition, *4'33"* (1953). The exercise commenced at 13:10 on a Wednesday lunchtime in mid-July.

I'm seated in smallish office on the first floor of a converted terraced house, perpendicular to a sash window to my left, level with my head and torso. I endeavour to give my 100% concentration to the task, which is always difficult when I am at my place of work. I attend to and imbibe the prevailing acoustic environment, allowing my erstwhile perception of background and foreground to intermingle. What I mean is, I consciously attempt to shift my *everyday listening* behavior, a highly filtered personalized listening predicated on "listening to things going on around us, with hearing which things are important to avoid and which might offer possibilities for action" (Gaver, 1993a, p. 2).

It is remarkably quiet, almost all I can hear beyond my breathing is what an acoustician may refer to as noise ingress—the environment outside is buzzing. Beyond the barely audible muffled voice of a colleague talking with a student in the office below, the predominant ingress arrives from the window, through the air gaps or transferred through the glass. I open the window, and a rich spatial profile extends from my left ear. I'm placed in two contradictory acoustic architectures, to my left, expansive and open, to my right, enclosed and dry.

I find myself mentally mapping out the expansive external environment from this auditory information: the nearby primary school playground populated by the joyful (albeit piercing) cries of kids, the main road whose full length is articulated and demarcated by an assortment of different sirens from individual emergency services that are taking some time to pass in and out of earshot. The train line is on the horizontal perimeter of the *acoustic horizon* (i.e. how far I can hear) (Truax, 2001, p. 67), which makes itself present a couple of times, and there is almost continual aerial presence of aircraft overhead, whose incessancy makes it hard to differentiate between individual airplanes, as the subtly fluctuating drone with a pitch sweeping flange effect intermittently cuts through the strata of white noise. I hear the planes *high-up* in my mind, and of course I know that the planes are elevated in the sky, but what are the acoustic cues that are telling me that, none?

I am reaffirming sociospatial, topographical knowledge that I have acquired due to physically traversing the terrain for more than ten years, and beyond that my auditory life to date of acquired spatial environmental knowledge.

The handheld school bell is rung a couple of times, an unbroken *sound signal* from my own childhood and way before. I wonder if it is an adult or child who has been given the power to sound it, probably an adult due to the assuredness of the gesture. [. . .] [I]n response, the children's voices steadily diminish. I have a strong memory of my playground jollities being interrupted by the One O'Clock Gun fired from Edinburgh Castle, jolting my auditory perception, never failing to get the adrenaline going. Returning more than two decades later to the same playground, I failed to register the gun. It transpires that in 2001 the caliber of the gun was updated from a 25 pounder to a 105 mm Light Gun, the gun no longer penetrates the south of the city any more, rendering it a *disappeared sound* and remaining only as a *sound romance* to be retold by an *aural witness* such as myself.

A lorry reverses, signaled by an automated warning beep. These intermittent sound of bells, beep and sirens (automated and manual) are examples of consciously designed sound, not haphazard by-products but sounded with the clear intention of drawing attention to themselves and their concomitant message. I'm conscious that this tonal bleep beep reversing *sound signal* is under threat, with the preference for more acoustically sophisticated directional multifrequency/white noise–based signal. Will anyone regard its demise nostalgically as a *sound romance*?

As my commentary implies, I find myself psychospatially arranging and compartmentalizing "remote physical events" (Gaver, 1993b, p. 285), I'm putting them in boxes—playground, aircraft, trains, cars, voices—and that helps me to disentangle them from one another. This is independent of their spectral similarity or my localization of their attributed cause, i.e. my perception of the physical location of "sound-producing events" (Gaver, 1993b, p. 288), but is rather due to what J. G. Gibson calls their *affordances*:

> The *affordances* of the environment are what it *offers* the animal, what it *provides* or *furnishes*, either for good or ill.
> (Gibson, 1986, p. 127, emphasis as in original)

I need to get beyond my predilection for hearing *affordances*, which are hijacking my listening. I need to refresh my listening. The advice of Georges Perec, who was more preoccupied with the visual rather than the audible on this matter, is to "exhaust the subject":

> Make an effort to exhaust the subject, even if it seems grotesque, or pointless, or stupid. You still haven't looked at anything, you've merely picked out what you've long ago picked out.
>
> (Perec, 1997, p. 50)

In this endeavor, in his series of Practical Exercises from his *Espèces d'espaces* [Species of Spaces] (first published in 1974), he encourages us to: "decipher," "distinguish," "describe," "apply yourself," "note the absence of [. . .]," "force yourself to write down what is of no interest," "detect a rhythm," "deduce the obvious facts," "set about it more slowly, almost stupidly," "strive to picture yourself, with the greatest possible position" (Perec, 1997, pp. 50–54).

This rigor of first-person sensory analysis of what Perec calls the *infraordinary* (as opposed to the extraordinary) is in stark contrast to Cage's dictum on *New Music: New Listening* in his article on "Experimental Music" (first presented in 1957 and published in his collection of writings *Silence*):

> Not an attempt to understand something that is being said, for, if something were being said, the sounds would be given the shapes of words. Just an attention to the activity of sounds.
>
> (Cage, 1987, p. 10)

I resume my listening. I endeavour to recalibrate, from attribution of sources to the physical features of the space that those sounds have encountered on their journey to my ears. An adult voice ricochets off the corrugated façade, the direct and reflected voices are blended, obscuring speech intelligibility. This building was erected several years after I moved into the office: I imagine how this voice may have carried when this was an open space. I'm reminded of Murray Schafer's voice calling out into the desolate winterscape of Manitoba, in the soundscape composition, *Winter Diary* (WDR, Schafer, 1997) in collaboration with sound recordist Claude Schryer. We hear the close-by crunching of snow from his footsteps as he passes across the spatial image of the stereo microphone. There are no planes, trains or cars within earshot. He calls out, but there is no ricocheting or echoing, the voice is unimpeded by surfaces bar the absorbent flat snowy ground.

I consider the effect of masking: what sounds are blocking other sounds, rendering them inaudible from my position, and how does this change over time. I know trains also pass by regularly, but why is their audibility so infrequent—is that due to masking? It is a fine day; wind and rain would quickly add a blanket of broadband white noise, curtailing my acoustic horizon and masking the subtle details of acoustic information I am currently perusing.

I am suddenly aware of the presence of a bird call, a sparrow, close-up and then another; had they been calling throughout the duration of my listening, that only now had come into my consciousness as my attention shifted? I was recently reading a study done by ornithologists on the challenges of age-related hearing loss and bird surveying, with a particular regard to bird calls in the upper frequency range such as Goldcrest and Treecreeper (Tucker, Musgrove and Reese, 2014). I think about how my age and sex are irrevocably acting as an EQ. I imagine what a springtime dawn chorus would be like here, having never been in my office at 4:30, before the daily aircraft activity.

I now hear what I think is the recycling of glass bottles being dropped into a container, although I am not certain, I don't remember hearing this from my office window before, so I question my assumed attribution of source to cause. I know there is building work going on, maybe it is pieces of scaffolding being dropped into a skip. Oh and now someone is whistling and coming close to my office window, and then stops. And now I hear the front door of my building opening and closing.

What I have failed to mention is a constant high-pitched pure tone hitting my right ear. I know this sound well; this is the tinnitus that accompanies the right side hemisphere of my listening. Sure, it is a phantom sound, but is it a bona fide feature of my soundscape?

And then I hear another emergency service vehicle making its steady passage along the main road. Visitors are often distressed by the regularity of the audible presence of emergency services here. Their sheer omnipresence throughout the day has dulled my response, it has become necessary to place them into the *ground* of my (un)conscious, but I'm reminded how after the London bombing of July 2005, they had moved back into the *figure*—they had got louder! When I went to record this augmentation with the sociologist Les Back, equipped with a directional shotgun mic, for the duration of our recording my microphone did not register any sirens; in fact it picked up the clip-clop of hooves from mounted police, which listening back to, suggested a pastoral scene, incongruous with the view. Back remarked: "You never hear what you are listening for" (Back, 2007, p. 118).

Out of the blue, I now hear a continuous texture of a combination of bubbly and whirring sounds, looking out the window, I assume it is the air extractor fan from an adjacent building. Now I can't stop hearing it. This task has drawn my attention to features that are clearly present that I was erstwhile unaware of.

On some level, the setting is predictable; the architecture and infrastructure are not going to suddenly mutate, and I can reliably predict what I will hear, and yet what I do hear is wonderfully surprising and capricious and constantly shifting. Noise as unwanted sound has no meaning in this exercise (if I was trying to lecture or study, this would be a different matter); everything I hear is information rich, it has a story to tell, and is sonically fascinating. Cage's oft repeated quote comes to mind:

Wherever we are, what we hear is mostly noise. When we ignore it, it disturbs us. When we listen to it, we find it fascinating. The sound of a truck at fifty miles per hour. Static between the stations. Rain.

(Cage, 1987, p. 3)

I could go on . . . What I am describing is the soundscape, as heard, felt and made sense of from a first-person perspective, within a specific time bracket, and as such it is unrepeatable. Contrary to my introduction, the exercise is certainly not mundane or minimal; it is expansive and infinitely transdisciplinary. This complexity coupled with the paucity of the instruction "Listen!" presents a conundrum: how am I listening, and can I actively choose how to listen? And pivotally, despite borrowing Cage's randomly generated time brackets, the soundscape attitude is far from satisfied with, "Just an attention to the activity of sounds" (Cage, 1987, p. 10).

As I am reading back through the text, I become acutely aware how unreliable and partial a listener I am. There is so much to hear, and it appears that not only are sounds coming and going, but my hearing is perpetually fluctuating, notwithstanding my digressing mind. The homogeneity or heterogeneity of the soundscape seems to be contingent on my attention. And I am extrapolating discrete information from the overabundance of incoming sound from multiple sources and events. We have the remarkable capacity to selectively attend to certain aspects of the acoustic environment, while electively deafening our ears to specific signals. This is particular effective with regard to attending to competing human voices, while our attention can be hijacked on hearing our name mentioned in a neighboring conversion. This capacity is called the *cocktail party effect* (Cherry, 1953).

Vigilant not to be auraltypical with my pontifications on aural perception, with the onset of my own age-related hearing loss, specifically limited high-frequency sound perception, this is a skill that will be increasingly diminished (Drever, 2017a). In contrast to the context of my Everydayness Listening Exercise, a good sound designer knows that the cocktail party effect within the fixed stereo image is limited. With a pragmatic concern for the limits of perceptual and cognitive load on the filmgoer, the great sound designer Walter Murch judiciously selects the amount of auditory information within the audiovisual mix he presents to the listener.

> There is a rule of thumb I use which is never to give the audience more than two-and-a-half things to think about aurally at any one moment. Now, those moments can shift very quickly, but if you take a five-second section of sound and feed the audience more than two-and-a-half conceptual lines at the same time, they can't really separate them out. There's just no way to do it, and everything becomes self-cancelling. As a result, they become annoyed with the sound and it appears "loud" even at lower levels.
>
> (Murch, 2018)

The everyday soundscape doesn't function like this. When is there only two-and-a-half things being sounded? The soundscape is not rarefied or meticulously controlled, ultimately it is haphazard and inherently complex. All soundscapes are in a sense are diegetic[4] but there is no master guiding narrative, and like the running shower in Hitchcock's *Psycho*, the soundscape is generally *anempathetic*—"it seems to exhibit conspicuous indifference to what is going on" (Chion, 1994, p. 221). Their presence in the soundscape is not due to some figure of irony, metaphor, metonymy or synecdoche.

Despite the wonders of audition such as the cocktail party effect, on the other hand, cognitive psychology tells us of our slothful predilection to *auditory change blindness*:

> The latest research on the phenomenon of auditory change blindness unequivocally shows that in the absence of attention, people simply have no conscious awareness of the majority of the auditory stimuli around them.
>
> (Spence and Santangelo, 2010, p. 266)

And the more familiar the *ground*, the more hidden from your perception it may be. As well as unmediated listening exercises such as the preceding ones, the use of sound recording can be helpful in reconfiguring our habitual auditory change blindness. In *City as Classroom*, McLuhan, Hutchon and McLuhan (1977) developed an exercise of recording everyday sounds from familiar locations, editing them down into short compositions and exchanging them with colleagues:

> Try to transmit the feeling of being there, not the impression of a "trip through": avoid story-lines sequence (*figure*) in order to concentrate on the *ground*.
> (McLuhan, Hutchon and McLuhan, 1977, p. 12)

Finally, and yet most troubling, my ostensibly benign listening experience has highlighted a somewhat troubling issue: what is the ethical dividing line between overhearing and eavesdropping?

14.3. Discovery of the Social

When Luc Ferrari—a composer fired up by Cage's radical adoption of chance procedures yet schooled in Pierre Schaeffer's *musique concrète*, a compositional approach founded on the manipulation of recorded sound and phenomenologically oriented auditory perception—stepped into the Paris streets with a Stefan Kudelski Nagra III Tape Recorder (one of the first commercially available high-quality mobile tape recorders, launched in 1958), he discovered a new dimension to his compositional palette:

> As soon as I walked out of the studio with the microphone and the tape recorder, the sounds I would capture came from another reality. That led to the unexpected discovery of the social.
> (Caux, 2012, p. 129)

While Cage espoused the disappearance of the "social, political, poetic, and ecological" (Kahn, 1997, p. 557) from his listening, like the scales that fell from Paul's eyes on the road to Damascus (apologies for the negative blind metaphor; see Hull), the social, which had always been there, had become audible through this subversive *musique concrète* composer's tools of sound recording, taken out of the sequestered recording studio and onto the streets. And now, with it out into *the everyday*, Ferrari was opened to an array of arts practice fixated with reframing the objects they encountered, such as the Duchampian *readymade*, the Bretonian *objet trouvé* (found object) and the *collages, décollage* and *assemblages* of the *Nouveau réalisme* typified by Daniel Spoerri's *Tableau Piège* (Snare-picture), a series of works that he commenced in 1960:

> [O]bjects, which are found in randomly orderly or disorderly situations, are mounted on whatever they are found on (table, box,

drawer, etc.) in the exact constellation they are found in [. . .]. By declaring the result to be a tableau, the horizontal becomes vertical. For example: the leftovers of a meal are mounted on the table and the table is then hung on a wall.

(Spoerri, 2018)

Ferrari went on to create a wide range of sonic works, most radically within the community of *musique concrète* composers at the Groupe de recherches musicales (GRM); his *Presque Rein* series of works embraced often raw field recordings as a central pursuit engaging with the social, often on a playful and idiosyncratic level. In his *Presque rien n° 1 ou le lever du jour au bord de la mer* (1967–1970), the listener is presented with what at first glimpse appears to be unedited field recordings, which he tells us depicts seamlessly edited highlights of the daybreak recorded from a fixed position by a fishing harbor on the island of Korcula in Croatia. At the time, he was unaware of the term "soundscape" and had adopted the term *anecdotal music* to his practice but retrospectively recognized his affinity for soundscape. His colleague at the GRM, François-Bernard Mâche, named this approach *phonography*.

The idea of a possible sound equivalent with the prestige that photography had been able to acquire for itself.

(Mache, 1992, p. 191)

Mâche cites Walter Ruttmann's *Weekend* (*Wochenende*, 1930), a film work with a blank screen accompanied by a sound collage of fragments resembling weekend-type activities of Berliners, as the pioneering work. Of course, we can find many early examples of recorded location sound/field recording that could correspond with soundscape themes.

Before the invention of the mobile tape recording, field recording was very limited and cumbersome. An early example of urban field recordings is *London Street Noises* (1928), produced by Columbia Records in collaboration with the tabloid newspaper the *Daily Mail* to spearhead a discussion on the increasing noise in the streets of London. The producers felt it necessary to include a running commentary, so a Commander Daniel introduces the date, time and location and comments on the activities; fittingly, his voice is momentarily masked by passing traffic.

The *Mass Observation* infused use of everyday location sound in Humphrey Jennings's GPO documentary film work such as *Spare Time* (1939) and his Crown Film Unit wartime documentary of everyday life in *Listen to Britain*, in collaboration with Stewart McAllister, are fascinating examples where the image is cut to the location sound. Or the African savannah is brought to life through the juxtaposition of field recordings made in

London Zoo by Ludwig Koch (who become synonymous with birdsong recording) in his sound book, *Animal Language* (1938), made with biologist Julian Huxley.

It was the New York radio producer, Tony Schwartz, who liberated field recording from the shackles of a recording van. With his own specifications for a battery-powered portable tape recorder, he was out recording in the street and into people's homes. One of his most powerful yet beautifully simplistic concepts is *History of a Voice* included on his *Folkways* LP of 1962, *You're Stepping on My Shadow* (Nine Sound Stories Conceived and Recorded by Tony Schwartz). We hear the story of Nancy's voice from a neonate to a strongly opinionated nine-year old. Schwartz also introduces us to "her voice's family tree" with the introduction of her parents' and grandparents' voices.

With the availability of mass-produced tape recorders such as the Grundig, Minivox and Telefunken in the 1960s, there was a huge blossoming of amateur tape recording, evidenced by magazines such as *The Tape Recorder* and *Amateur Tape Recorder* and recording clubs.

14.4. Definitions of Soundscape

Functioning as a qualitative counter to a wholly pejorative concept of environmental noise, the term *soundscape* was resolutely presented to the world in a series of pamphlets by the composer R. Murray Schafer, first directed toward music teachers drawing on his radical expansive approach to music education, such as *Ear Cleaning* (1967) and *The New Soundscape* (1969), and then coalescing in the summation of his soundscape thinking, *The Tuning of the World* (first published in 1977 and republished as *The Soundscape: Our Sonic Environment and the Tuning of the World* 1994). In his *Book of Noise* (1998, first published in 1968), moving from the classroom to the citizen concerned with noise pollution, he describes the soundscape as "the entire acoustic environment of our lives, wherever we may be, at home, at work, indoors" (1998, p. 4). Music as metaphor and an attitude to sound is always close at hand in Schafer's world; here in very much an enabling manner, the idea of soundscape is proposed as a symphony:

> And we are simultaneously the audience, the performers and the composers.
>
> (Schafer, 1998, p. 4)

Musical analogies are unsurprising, Schafer is first and foremost a composer, and, conspicuously, his research group, based in Simon Fraser

University in Vancouver, grandly titled the World Soundscape Project (launched in 1971), was overwhelmingly comprised of composers, most notably Barry Truax and Hildegard Westerkamp. The group coined a number of helpful terms to help analyze, communicate and educate aspects of the soundscape: soundmark, keynote sound, lo-fi/hi-fi soundscape, sound romance, sound event, ear witness, disappearing sound etc. And they honed a number of fieldwork methods including soundwalking (Drever, 2009), sound mapping, sound diary, the recording of ear witness accounts, the making and archiving of field recordings.

This new term "soundscape" etymologically feeds on a more ancient term, "landscape." Landscape is primarily concerned with the surfaces of the earth, but more than that, its Germanic etymological roots *Landschaft* suggests a set of active relationships: the suffix *-schaft* denoting a "state or condition of being" (OED, 2018). The suffix from related terms such as kinship and friendship, "-ship" also deriving from *-shaft*. Consequently, landscape concatenates an experiential tie with the notion of land. Crucially, this everyday expression comes with baggage, as each pictorial representation and subsequent reading is culturally and historically contingent, bound up in territorial pulls of power and resistance (i.e. exemplified in the practice of cartography). The cultural geographer Denis Cosgrove has unpicked these issues, addressing landscape as an "ideological concept" (Cosgrove, 1998, p. 15):

> Landscape is not merely the world we see, it is a construction, a composition of that world. Landscape is a way of seeing the world.
>
> (Cosgrove, 1998, p. 13)

He goes on:

> It represents a way in which certain classes of people have signified themselves and their world through their imagined relationships with nature, and through which they have underlined and communicated their own social role and that of others with respect to external nature.
>
> (Cosgrove, 1998, p. 15)

Landscape tends to arrive at us via a singular fixed point of view or latterly extended into a panorama, delineated by a quadrilateral frame. The canvases of Friedrich, Turner and Constable come to mind, reinforced by the writings of Goethe, Shelley, Coleridge, Wordsworth . . . the pervading specter of Romanticism. We observe a landscape as a spatially detached observer, yet stimulating aesthetic values predicated on latent notions of the picturesque and the sublime.

The analogy with landscape is somewhat tenuous, however, for example according to the pioneering Finnish landscape geographer, Johannes Gabriel Granö:

> In order for a landscape to be perceived, a given minimum amount of lighting is necessary. [. . .] Moreover, we must be a given minimum distance away from the landscape if we want to see it.
> (1997, p. 49, first published in 1929)

As we all know, the soundscape doesn't stop in darkness, and we tend not to listen on to a soundscape, we tend to listen or speak from within the soundscape, a concept taken up by Westerkamp (Westerkamp, 1998). And, of course, despite disciplinary prejudices, as we all know, our perception of the environment is always multimodal; we perceive via the sensorium, so we need to be careful not to consider the senses in some kind of independent isolation, with synesthesia—the involuntary cognitive linkage of senses such as seeing a sound or tasting a words—as a profound example.

In his book *Acoustic Communication*, Barry Truax refined many of these concepts. Here, soundscape is predicated on the act of auditory perception and the making sense of that perception·

> An environment of sound [. . .] with emphasis on the way it is perceived and understood by an individual, or by a society. It thus depends on the relationship between the individual and any such environment.
> (Truax, 1999)

The soundscape is a work-in-progress, as the physical environment is always in flux, and our perception of our surroundings is constantly being formed, rendering it ephemeral. When one's perception ceases, so does the soundscape for that person.

In the recent codification by an acoustics subcommittee for the International Organization for Standardization (BS ISO, 2014), acoustic environment is clearly demarcated from soundscape. Soundscape corresponds with Truax's articulation, where the related term "acoustic environment" refers to its physical manifestation:

> [T]he sound from all sound sources as modified by the environment. Modification by the environment includes effects on sound propagation, resulting for example from meteorological conditions, absorption, diffraction, reverberation and reflection.
> (BS ISO, 2014, p. 2)

Consequently, unlike the soundscape, it is not contingent on the vagaries of the perception of a listener.

In her introduction to *The Soundscape of Modernity: Architectural Acoustics and the Culture of Listening in America, 1900–1933* (2002), Emily Thompson provides an elaboration on the definition of soundscape that more closely resonates with Cosgrove's notion of landscape:

> Like a landscape, a soundscape is simultaneously a physical environment and a way of perceiving that environment; it is both a world and a culture constructed to make sense of that world. [. . .] A soundscape's cultural aspects incorporate scientific and aesthetic ways of listening, a listener's relationship to their environment, and the social circumstances that dictate who gets to hear what. A soundscape, like a landscape, ultimately has more to do with civilization than with nature, and as such, it is constantly under construction and always undergoing change.
>
> (Thompson, 2002, pp. 1–2)

This definition reminds us of the inherent snowballing of challenges of grasping, making sense of or attempting to present some kind of definitive soundscape analysis, if there was such a thing.

Advancing from the somewhat anthropocentric orientation of acoustic ecology, the evolving concept of soundscape has given fresh impetus to pressing new fields of scientific research such as bioacoustics or soundscape ecology:

> [T]he collection of biological, geophysical and anthropogenic sounds that emanate from a landscape and which vary over space and time reflecting important ecosystem processes and human activities.
>
> (Pijanowski et al., 2011)

14.5. Soundscape Composition

As already mentioned, field recording was an important tool for the World Soundscape Project (WSP). It allowed the team to document and analyze the acoustic environment in the controlled and reproducible environment of the lab and to compare and contrast with other recordings from different locations in the developing WSP archive. Communication with the public had always been central to Schafer's work; field recordings could play an unparalleled role in getting the message out in the era of mass radio and gramophone listening, exemplified in their annotated double LP, *The Vancouver Soundscape* (1973). Falling between public art, social science and education, this publication heralded a new form of sonic composition:

soundscape composition. Schafer inserted the following guidance notes to prime the general listener:

> To record sounds is to put a frame around them. Just as a photograph frames a visual environment, which may be inspected at leisure and in detail, so a recording isolates an acoustic environment and makes it a repeatable event for study purposes. The recording of acoustic environments is not new, but it often takes considerable listening experience to begin to perceive their details accurately. A complex sensation may seem bland or boring if listened to carelessly. We hope, therefore, that listeners will discover new sounds with each replay of the records in this set—particularly the first record, which consists of some quite intricate environments. It may be useful to turn off the room lights or to use headphones, if available. Each of the sequences on these recordings has its own direction and tempo. They are part of the World Symphony. The rest is outside your front door.
>
> (Schafer, 1973, pp. 1–2)

Each track took a unique approach to editing, exploring themes, such as *The Music of Horns and Whistles*, specific events such as *New Year's Eve*, aural history or geographical or sociological zones. The copious editing of *sound events* from diverse locations and times is seamlessly brought together in what would be most easily classed as composition. The program notes and accompany booklet ensured that the context is adequately disclosed. In track 3, *Entrance to the Harbour*, in just over 7 minutes, the listener journeys above water through a careful editing of multiple recordings, coming close and closer into *Vancouver Harbour*, ending on land and indoors of a waiting room.

14.6. Sound Events/Effects

I'm careful to name the individual segments of recordings, *sound events*, rather than *sound effects*; a soundscape recording is not by default a sound effect, although it may turn out to be a highly effective sounds effect. A sound effect has little to do with the everyday. Sound effects correspond to "the ideal, the imaginary, the momentous" (Sheringham, 2009, p. 23). A successful sound effect is highly efficient, unambiguous/stereotypical, sonic shorthand representing or evoking an activity, an event, an environment or location and its desired mood, and like a meme it is extremely promiscuous and has a high propensity for deterritorialization. Dating back to the earliest days of theater and radio drama (see Frank Napier's guide to *Noises Off: A Handbook of Sound Effects* from 1936, for example), the semiotic power of the sound effect is so successful, it almost functions as a lingua franca across radio, theater, cinema, TV and computer games, augmented reality, VR, its power never waning.

The perception of a *sound event*, in contrast, is not necessarily immediately audible or readable. It is dependent on knowledge of its territory as was abstracted, including all "social and environmental" (Truax, 1999) aspects of its original "spatial and temporal" context:

> A nonabstractable point of reference, related to a whole of greater magnitude than itself.
>
> (Schafer, 1994, p. 274)

The sound event is a context-specific phenomenon that:

> whether foreground or background in perception, only acquires meaning through its context, that is, its complete relationship to the environment.
>
> (Truax, 2001, pp. 52–53)

And it is fundamental that the remediated material is a document of an *actual* spatial-temporal occurrence. Or, in the language of Peircean semiotics (Peirce, 1955), it pertains to an indexical relationship where sound or a moment of relative silence does not refer to itself; rather it functions as an initial sign in the service of another sign in terms of causation. There is an umbilical-like contiguity between what is re-presented (that is, the selective sensuous translation of vibrating air molecules) and what gave rise to that sound: its cause, for example a school hand bell. The quality of the recording does not compromise its status as sound event. The audio technology and its routing and my damaged hearing, for that matter, are all part of the context.

For the Academy of Urbanism's *2015 Great Place Award*, I composed short soundscape compositions of three shortlisted places (Academy of Urbanism, 2018). My task was to examine and present the role that the soundscape played in the nature of great place making. One of those locations was St. Pancras Station in London, which is marketed as both "Europe's Destination Station" and a shopping and cultural venue in its own right. Once permission was granted by the station's management, and I acquiesced to various constraints related to safety, such as the restriction of booms, I traversed the station from very much a human scale perspective of commuter and traveler. I did not go with the aim of capturing sounds but rather to record what I encountered throughout a day, contingent on all the possible factors that come into play in such a location, which I presented back in an edited-highlights fashion. Despite my efforts aimed at faithful representation, if you were making a radio drama that has a departure scene set in St. Pancras, unless you ascribe to Dogme 95, you may well not want to use my recordings, as on the whole they don't exude train station-ness; even my recordings of the electric Eurostar trains don't particularly sound like trains—none of that muscularity or romance is evoked. The recordings have the authenticity of documentary, and they tell you about that specific place at that specific time, not an idealized or completely fictional notion of it. Although the worlds tend to seep into each other, as the ingrained sound effect takes hold of your imagination, dictating what things should sound like.

Mark Vernon played with this ambiguity between the sound event and sound effect in his LP record *Sonograph Sound Effects Series vol. 1: Sounds of the Modern Hospital*. The LP mimics the presentation and form of a 1970s sound effects record, with the tongue-in-cheek guidance, "The Recordings are issued for amateur purposes only"; however, all 33 tracks recordings are actual location recordings, including informative titles, such as Side A, track 10 *Anaesthetic Machine*—two one-way valves rattle in tandem as the patient inhales and exhales. There is no attempt to be evocative or compositional with the editing. The LP includes the following note:

> Whilst every effort has been made to record the subject in as great a degree of isolation as possible, the sound recordings you will hear on this LP record were made in a busy working hospital and not under controlled conditions. Therefore, on occasion, you may hear some unavoidable background noise, conversations and other extraneous sounds.
>
> (Vernon, 2013)

As sound can't always speak for itself and veers toward ambiguity, some soundscape artists feel the need to provide voice-over commentary or detailed annotations accompanying the work. In his *Sounds from Dangerous Places* (2012) project, Peter Cusack included a book with pictures and factual details in prose. For Cusack, it is important that we have an understanding of the context of his recordings, as it is fundamental that he has a thorough and sensitive understanding of the context he is re-presenting. For example, on discussing the field recordings from inside the Chernobyl Exclusion Zone, he states that it is necessary:

> to inform myself as far as possible, but also to listen to the small voices, to the environment itself, to those whose personal knowledge of the area goes back generations, to those on the front line and to those whose lives have been changed forever by events over which they have no control.
>
> (Cusack, 2012, p. 18)

14.7. Conclusion

It is not just about having the right kit and knowing how to use it, meaningfully rich soundscape composition is the result of persevering and diligent work. For Jana Winderen, an artist who spends many arduous hours
recording ice and underwater creatures in difficult-to-access environments,
such as the North Pole or coral reefs in the Caribbean, she has no means
of seeing what she is recording through the hydrophones, and yet it is
imperative for her to concentrate fully on the act of listening while making
the recordings. Significantly, she works closely with biologists on these
projects in a mutually informative manner. She marvels in these complex
sonic textures and helps communicate these sound worlds to a larger public, but there is also a serious biological research dimension to this work.

There is the field recordist who rushes through environments in hunt
of fugitive sounds, never before heard or recorded by humans. As they
already somehow know what they are looking for, this Indiana Jones–like
character will miss those ground sounds and will not build up a meaningful relationship with that environment.

Perhaps it is necessary to collaborate or cocompose soundscape composition in order to generate that meaningful connection and open up own-
ership of the recordings and the process. Sonic Postcards, an education
project led by Sonic Arts Network, facilitated soundscape activates in
primary schools in the UK, resulting in the making of hundreds of short
soundscape compositions of the environments that the pupils inhabit. This
was followed up by the exchanging of the sonic postcards with classes
throughout the country who live in radically different sonic environments.
This was an invaluable experience that one could imagine being part of the
permanent curriculum for all schoolchildren.

Soundscape composition is a complex and challenging field. There are
no disciplinary givens, and the potential research implications could be
overwhelming, but it is through opening up to the context and contingency
of the environment that makes this kind of sonic work so enriching and an
activity of perpetual learning.

Notes

1. Psychologists regard the act of listening as a five-dimensional process: (a) cognitive,
    (b) affective, (c) behavioral/verbal, (d) behavioral/nonverbal and (e) behavioral/inter-
    active dimensions provide a conceptually meaningful framework for explicating the
    listening process (see Halone et al., 1998).
2. Concepts of *place* should not be treated as a given. For key theories on place, see dis-
    cussion on Cresswell (2004).

3. From 1966, Max Neuhaus presented a range of artworks using the work *LISTEN* in uppercase as a prompt to the audience (see Drever, 2009). There are a number of collections of Listening exercises; for example, see Oliveros (2005) and Schafer (1992).
4. "Narratively implied spatiotemporal world of the actions and characters" (Gorbman, 1980).

References

Academy of Urbanism. Available at: www.academyofurbanism.org.uk/acoustic-ecology-of-great-places/ [Accessed July 2018].

Back, L. (2007). *The art of listening*. Oxford: Berg.

Böhme, G. (2017). *Atmospheric architectures: The aesthetics of felt spaces*. Edited and Translated by T. Engels-Schwarzpaul. London: Bloomsbury.

BS ISO 12913–1, (2014). Acoustics—Soundscape—Part 1: Definition and conceptual framework.

Cage, J. (1987). *Silence*. London: Marion Boyars.

Cage, J. (1953) 4'33". Second Tacet Edition. Leipzig/London/New York: Edition Peters.

Caux, J. (2012). *Almost nothing with Luc Ferrari*. Translated by J. Hansen. Berlin: Errant Bodies Press.

Cherry, E. C. (1953). Some experiments upon the recognition of speech with one and two ears. *Journal of Acoustical Society of America*, 25, pp. 975–979.

Chion, M. (1994). *Audio-vision: Sound on screen*. New York: Columbia University Press.

Cresswell, T. (2004). *Place: A short introduction*. Oxford: Blackwell Publishing.

Cosgrove, D. E. (1998). *Social formation and symbolic landscape*. Madison: University of Wisconsin Press.

Cusack, P. (2012). *Sounds from dangerous places*. ReR MEGACORP & Berliner Künstlerprogramm des DAAD.

Drever, J. L. (2007). Nostophonics: Approaches to grasping everyday sounds from a British perspective. In: R. Bandt, M. Duffy and D. MacKinnon, eds., *Hearing places*. Newcastle upon Tyne: Cambridge Scholars Press, pp. 173–185.

Drever, J. L. (2009). Soundwalking: Creative listening beyond the concert hall. In: J. Saunders, ed., *The Ashgate research companion to experimental music*. Farnham, UK: Ashgate, pp. 163–192.

Drever, J. L. (2017a). The case for aural diversity in acoustic regulations and practice: The hand dryer noise story. In: *Conference proceedings: ICSV24, The 24th international congress on sound and vibration*. London: The International Institute of Acoustics and Vibration and the Institute of Acoustics.

Drever, J. L. (2017b). Field recording centered composition practices: Negotiating the "out-there" with the "in-here". In: M. Cobussen, V. Meelberg, B. Truax, eds. *The Routledge companion to sounding art*. New York: Routledge.

Gaver, B. (1993a). What in the world do we hear? An ecological approach to auditory event perception. *Ecological Psychology*, 5(1), Lawrence Erlbaum, pp. 1–29.

Gaver, B. (1993b). How do we hear in the world? Explorations in ecological acoustics. *Ecological Psychology*, 5(4), pp. 285–313.

Gibson, J. J. (1986). *The ecological approach to visual perception*. Hillsdale, NJ: Lawrence Erlbaum.

Gorbman, C. (1980). Narrative film music. In: Altman, R., ed., *Yale French Studies, Cinema/Sound*, 60, pp. 183–203.

Granö, J. G. (1997). *Pure geography*. Baltimore: Johns Hopkins University Press.

Halone, K. K., Cunconan, T. M., Coakley, C. G. and Wolvin, A. D. (1998). Toward the establishment of general dimensions underlying the listening process. *International Journal of Listening*, 12(1), pp. 12–28.

Kahn, D. (1997). John Cage: Silence and silencing. *The Musical Quarterly*, 81(4) (Winter), pp. 556–598.

Mâche, F.-B. (1992). *Music, myth, and nature, or, the dolphins of Arion*. Translated by S. Delaney. Chur, Switzerland: Harwood Academic Publishers.

Murch, W. Available at: http://filmsound.org/murch/waltermurch.htm [Accessed July 2018].

McLuhan, M., Hutchon, K. and McLuhan, E. (1977). *The city as classroom: Understanding language and media*. Agincourt and Ontario: Book Society of Canada.

Napier, F. (1936). *Noises off*. London: Frederick Muller.

Oliveros, P. (2005). *Deep listening: A composer's sound practice*. New York: iUniverse.

OED Online. (2018). Oxford University Press. Available at http://0-www.oed.com.catalogue.libraries.london.ac.uk/view/Entry/105515?rskey=YwajBL&result=1 [Accessed July 2018].

Perec, G. (1997). *Species of spaces and other pieces*. London: Penguin Books.

Peirce, C. (1955). *Logic as semiotic: The theory of signs*. New York: Dover Publications.

Pijanowski, B. C., Farina, A., Gage, S. H., Dumyahn, S. L. and Krause, B. L. (2011). What is soundscape ecology? An introduction and overview of an emerging new science. *Landscape Ecology*, 26, p. 1213. https://doi.org/10.1007/s10980-011-9600-8.

Schafer, R. M. (1967). *Ear cleaning*. Ontario: BMI Canada.

Schafer, R. M. (1969). *The new soundscape*. Ontario: BMI Canada.

Schafer, R. M. (1973). The Vancouver Soundscape LP, republished on CD, Vancouver: Cambridge Street Records, CSR-2CD 9701, 1997.

Schafer, R. M. (1992). *A sound education: 100 exercises in listening and sound-making*. India River and Ontario: Arcana Editions.

Schafer, R. M. (1994). *The soundscape: Our sonic environment and the tuning of the world*. Rochester, VT: Destiny Book.

Schafer, R. M. (1996). *The Vancouver soundscape 1973*. Vancouver: Cambridge Street Records.

Schafer, R. M. (1998). *The book of noise*. India River and Ontario: Arcana Editions.

Schafer, R.M. (1997). *Winter diary*. WDR.

Sheringham, M. (2009). *Everyday life: Theories and practices from surrealism to the present*. Oxford: Oxford University Press.

Spence, C. and Santangelo, V. (2010). Auditory attention. In: C. J. Plack, ed., *The Oxford handbook of auditory science, hearing*. Oxford: Oxford University Press, pp. 249–270.

Spoerri, D. Available at: www.danielspoerri.org/web_daniel/englisch_ds/werk_einzel/05_fallenbild.htm [Accessed July 2018].

Street Publishing. Available at: https://www.sfu.ca/sonic-studio-webdav/handbook/ [Accessed July 2018].

Thompson, E. (2002). *The soundscape of modernity: Architectural acoustics and the culture of listening in America, 1900–1933*. Cambridge, MA: MIT Press.

Tucker, J., Musgrove, A. and Reese, A. (2014). The ability to hear Goldcrest song and the implications for bird surveys. *British Birds*, 107(April), pp. 232–233.

Truax, B. (1999). *Handbook of acoustic ecology*, 2nd ed. Vancouver: Cambridge Street Pub.
Truax, B. (2001). *Acoustic communication*, 2nd ed. Westport, CT: Ablex Publishing.
Vernon, M. (2013). *Sonograph sound effects series, Vol. 1: Sounds of the modern hospital*. Glasgow, UK: Meagre Resource Records.
Westerkamp, H. (1998). Speaking from inside the Soundscape. Available at: https://www.sfu.ca/~westerka/writings%20page/articles%20pages/speakingsound.html [Accessed July 2018].

15

From Feeling Vibrations to Building Audiovisual Scenes

The Perceptual Practice of Storytelling With Sound

Isabelle Delmotte

15.1. Introduction

The auditory system does not favor certain sounds over others, unless individual physical limitations demand so. Nevertheless, humans can be inattentive to aspects of their acoustic environment, be willingly ignorant of some sounds and forgetful of others. Life in sonic vibrations encompasses audibility and inaudibility: it thus requests a grasp of our own and others' perceptual abilities. Technically, sounds are measurable vibrations passing through solids, liquids and gases. These vibrations travel on our skin, reach and permeate our flesh and follow our bone structure. Being aware of sonic activities while exposed to sound and vibrations is a way of knowing the world and the self.

When we bite into an apple, is it the traveling airborne "sound" of the flesh of the apple that our ears respond to? Or are the vibrations that resonate through our jaws conveying experiences of "sound"? Can we describe the experience as other than the sounds of "biting into an apple"? Adding that the apple was green or red, or old, and was eaten by the sea in autumn or by a river in summer will add variations that affect sensory experiences and creative vision. How we feel and communicate sonic worlds to ourselves and others influences how we bring moving images into the mind's eye or onto the screen. Although wordy descriptions facilitate the indexing of sound, the skills that influence creative collaborations are not always informational but often affective. Max van Manen defines pathic knowledge not as a form of passive comprehension but rather as a process of sensed and visceral understanding that is "relational, situational, corporeal, temporal, actional" (van Manen, 2007, p. 20). A pathic knowledge allows individuals to decipher bodies and gestures, feel and absorb atmospheres and explore and assess the moods of others.

15.2. Flesh

Acoustics and psychoacoustics provide measurable perspectives on the materiality of sound, its perceptual qualities and various environmental impacts. Pressure and intensity can be objectively measured, but loudness, for example, has different values and meaning, depending on an individual's perceptual range, environment and state of mind. Being aware of these variables leads to the description of a phenomenon from the first-person point of sensing. We sometimes tend to ignore the variety and variations of sonic signals that reach us and often dismiss the importance of the vibrations born within us. Individual subjectivity and embodied experiences are often at the core of the discourse on sound. Philosophers, acousticians, musicologists, environmentalists and filmmakers have provided encompassing methodologies and terminologies leading to sonic cultural embodiments. Collectively, they aim to link sonic manifestations with the performing body, cognition, cultural frameworks, creative outputs and ecologies of contexts. In that respect, the construction of audiovisual narratives digs into the richness of individual perceptions and their communication to others. Individual variations enhance the ways we listen to and create sonic environments. However, many of us are not always attuned to the value and individuality of our perceptual abilities.

Phenomenology involves an experiential approach to the world and focuses on the manner in which lived experiences are perceived and interpreted. German philosopher Edmund Husserl aimed to understand what an individual perceives, how this perceptual experience manifests to consciousness and what it means for that individual. "Doing" a phenomenology is a complex and transparent process that rejects assumptions and prejudices and gives a disciplined approach to the human experience. Husserl aimed to interpret this holistic experience: the practice of his philosophy entails a precise methodology that starts with empirical observation followed by introspection and reflection. However, philosopher Don Ihde warns that introspection is not "*the* method of phenomenology" (Ihde, 2012, p. 10, italics in original). This chapter doesn't pretend to lead a detailed exploration of sonic experiences in phenomenological terms but rather to develop procedures of sensory awareness that can be practically shared without too many words. Nevertheless, some key sentences from phenomenologists are trigger points for our sensory exploration of the properties of vibrations, the role of the holistic body in testifying to their occurrence and the different approaches to communicate their presence.

Sounds are elusive and manifest in space and time. In this realm of sensations, perceptual memories and acts of awareness "are continuously constituted (and reconstituted) as past, present and future, respectively, so that it looks to the experiencing subject as if time were permanently flowing off" (Beyer, 2016). Momentary perceptions, or primal impressions, are engaged in "continuous alteration" (Husserl, 1964, p. 50). They lead to a consciousness of the sonic object that just passed, a retention, followed by an anticipation of the future of that same perceived object, a protention. This imperceptible chain of events triggers a reflection that embraces the notion of evidence. Husserl defined evidence as an acquisition of conceptual insight, a process that also requires cognitive activity (Ströker, 1997, p. 203). All elements of the natural and cultural worlds can be made evident; Husserl outlined phenomenology as a specialized study of experiences and consciousness, as well as a quest for "the essence of things," what lies behind existence.

Vibrations are continuous and often imperceptible. Bodies absorb and listen to the audible and nonaudible. Composer and sound theoretician Pauline Oliveros formulated the concept of "sonosphere" and focused on the bandwidth of audio frequencies and the resonant sensations they generate. She described it as "the sonorous or sonic envelope of the earth," in which:

> [h]umans sense the Sonosphere according to the bandwidth and resonant frequencies and mechanics of the ear, skin, bones, meridians, fluids and other organs and tissues of the body as coupled to the earth and its layers from the core to the magnetic fields as transmitted and perceived by the audio cortex and nervous system. (All of this with great variation of course). All cells of the earth and body vibrate.

Oliveros adds that the *sonosphere* can be molecularly, auditorily and nervously sensed "by human beings, as well as by all living creatures, plants, trees and machines" (Oliveros, 2010, pp. 22–23, italics in original).

The sonosphere is unruly and ever present: its renewed reciprocity integrates the human experience with the world. In that same spirit, French philosopher Maurice Merleau-Ponty acknowledged the primacy of the body as actor, transmitter and receptor and brought to the fore the permeability of the being in an immersive multidimensional world: "the world is wholly inside and I am wholly outside myself" (Merleau-Ponty, 2008, p. 474).

At the other end of the spectrum, composer and acoustic ecologist R. Murray Schafer discriminates among sonic activity; he emphasizes changes in environments and the dominance of low frequencies due to the increase of mechanistic and anthrophonic sonic activities (Schafer, 1994). Schafer coined the expression "soundscape" and declared that a soundscape consists of "events *heard* not objects *seen*" (1994, p. 8, italics in original). Bryan Pijanowski and other practitioners of soundscape ecology propose that a soundscape is "the collection of biological, geophysical and anthropogenic sounds that emanate from a landscape and which vary over space and time reflecting important ecosystem processes and human activities" (Pijanowski et al., 2011, p. 1214). The concept of "sound space" brings to the fore a sense of physical, emotional and social engagement that is specific to individual experiences (Fluegge, 2011). This notion of sound space also highlights the territory being listened to and the modes of performing the act of listening.

Building sound spaces necessitates a critical awareness to sound in order to crystallize perceptions. This process weaves holistic interactions of humans with the sonosphere at the personal and communal levels. Sound grounds us; its fleeting presence makes us part of the world, but an awareness of *a* sound and the perception of its particularity are far from "pure." *This* sound, toward which our ear tends, is encompassed in the maze of a soundscape: we hear it in "the mist of hum, rustling, rumble, static, clatter, or racket" (Lingis, 1996, p. 54). Listening involves attentiveness to both the presence and absence of sound, but consciousness registers slowly the impulses absorbed by the holistic body, and our ears perceive and remember "in short syntheses" the sound as it evolves (Chion, 1994, pp. 12–13).

The human auditory apparatus performs rapid processing, followed by perception and comprehension. This perceptual system is our "mere focal organ of hearing" (Ihde, 2007, p. 44); it collects frequencies rather than deciphers them, and its performance often relies on the position of our body in relation to sources of sonic activities. Composer, filmmaker and theoretician Michel Chion isolates four modes of listening that outline forms and functions of acoustic activities (1994, pp. 25–34). Causal listening focuses on the source of the manifestation, and reduced listening concentrates on the structures of the sound regardless of its source. Semantic listening allows for coded interpretations of the sound object. Acousmatic listening is integral to our everyday life: it entails hearing sounds with no visible cause and being aware of their invisibility (Chion, 1983, p. 11). In an acousmatic situation, we hear sounds without being able to see their cause and can ask ourselves, using reduced listening, what they are and what they look like. An acousmatic listening would generate a "mental visual representation" if the cause of the unseen sound had been previously known (Chion, 2009, p. 465). This practice is also embedded in our listening to all recorded material and exemplified, for example, when interacting with telephones.

Composer and researcher Pierre Schaeffer noted, "For years, we have been performing phenomenology without knowing it [. . .] and it is better, all things considered, to practice phenomenology rather than just talk about it" (Pierre Schaeffer cited in Solomos, 1999, p. 57, my translation).

Schaeffer outlined four modes of listening: listening, perceiving, hearing and comprehending. The relationship between these four perceptual processes is summed up as follows: "I perceived [*ouïr*] what you said despite myself, although I did not listen [*écouter*] at the door, but I didn't comprehend [*comprendre*] what I heard [*entendre*]" (cited in Chion, 1983, p. 20, italics in original). This example maps the temporal and spatial paths of sounds reaching cells, infiltrating the auditory apparatus, triggering impulses and easing, or not, comprehension. It also suggests different states of engagement with sonic activities that are yet to be heard: passive immersion (hearing), focused attention (listening) and appropriation (giving meaning).

15.3. Exercise One

All modes of listening play essential roles in the feeling, visualizing and imagining of the audible and inaudible signals permeating our environments. This exercise aims to enhance an individual's appreciation of the abilities to reconcile physical vibrations with the audible information that is primarily noticed. There are many places of vibrational experiences that we sometimes cannot escape from but that we can often choose to be in. One such place is an elevator: an area that vibrates, from which we can escape intermittently, but which also offers ways to control the duration of sensory experiments.

Aims

1.   To notice the role of the body's center of gravity
2.   To experience the processes of cognitive anticipation and causal listening

Technical Requirement

1.   Access to elevators/lifts

Tasks

1.   Position yourself in different parts of the elevator during a few ascents and descents of different durations.
2.   Repeat the experience with opened and with closed eyes.
3.   Repeat the experience standing up, crouching and kneeling on the floor of the elevator.
4.   Describe to yourself the resulting physical sensations and auditory perceptions.

5. Take notes between each ascent and descent.
6. Reflect on the experiences that you describe in your notes.
7. Talk to others about your experiences in the elevator and ensuing reflections, and listen to their impression of their same experience.
8. Discuss.

15.4. Mediating

The previous exercise highlights the possibilities for feeling and reflecting on a lived experience in a contained space that moves vertically. This seemingly one-way experience of a soundscape and the repercussions on our senses prepares our bodies for "being a means of communication with the world" (Merleau-Ponty, 2008, p. 106). The listening practices outlined by Schaeffer and their relationships to other realms of perception are intrinsic to our daily life. Diverse sensing modes activate different essences of sonic manifestations from abstract to concrete and from objective to subjective. Soundwalks offer other discoveries linked to our body cutting the space, our breath meeting the air and our feet attacking or brushing over the earth. During this dynamic symbiosis, the skin of our head and ears expands our personal sound space. We weave connections between different sounds, appraise their clarity or muddiness, estimate their durations and bring them "home" to our bodies. Over time, this sensory automation will modify our behavior to sound and "requesting sound quality becomes a natural activity" (Westerkamp, 2001).

The use of audio and visual recording equipment increases our ways of experiencing the sonosphere in which we live. The mechanical interface expands the possibilities of being at the core of an audible space regardless of our physical location. Magnetic and digital recording technologies have transformed the bodily absorption of multilayered soundscapes: we can isolate specific sounds after having been exposed to them. Oliveros appreciated technology as a tool that can expand the limited range of human audible abilities and can explore "the sonorous body that we inhabit" (Oliveros, 2011, p. 163). Spanning from the cosmos to the ocean floor and our viscera, distant and intimate places of sonic activity magnetize our ears. Mechanical modes of sound transmission and reception have recalibrated human cognitive, haptic and social relationships to sonic activities.

Starting with the carbon microphone at the end of the 1800s, the experiences of grasping audible and inaudible vibrations have provided us with unique and repetitive reinterpretations of the life lived. Pauline Oliveros reminisced about her 21st birthday gift from her mother: the reproducible capabilities of a tape recorder changed her perceptual relationship to sound (Oliveros, 2010, pp. 75–76). Each recording generates a unique approach to acquisition, and each listening to that same recording is a new process of impression and self-reflection leading to new knowledge. Sounds captured but unheard at the time of recording can surprise and provoke wonder, thus giving us food for thought on both the emission and reception of vibrations. With this technology, the ear can be exposed to sound by choice rather than by imposition: unwanted vibrations can be disposed of or silenced before, during or after recording.

Recorded sounds operate as anthropological tools of precision in the quest to feel, interpret and comprehend the world. When he was a schoolboy, at his insistence, the family of esteemed sound and image editor Walter Murch acquired a tape recorder. Murch captured the sounds of New York City by holding a microphone outside his window and also discovered tape editing. Around that time, he heard a radio program that sounded familiar to him and in line with his acousmatic experiments: it was Pierre Schaeffer performing his *musique concrète* made from recorded sounds (Ondaatje, 2002, pp. 6–7). For musician and acoustic ecologist R. Murray Schafer, the resulting split between a sound source and its electroacoustic reproduction amounts to a state of "schizophonia" (1994, p. 273). For Schafer, remoteness between spaces of creation and listening environments builds an abstract representation of sound events. However, this process of abstraction can also morph into an accurate and powerful instrument for listening critically to live, rather than recorded sonic signals. The following exercise takes advantage of the impossibility of always ascertaining the specific sources of mediatized sounds. Here we experiment with listening as a discovery of forms and the invisible, as outlined by Don Ihde when he wrote that "*it is to the invisible that listening may attend*" (Ihde, 2007, p. 14, italics in original).

It is a mental survival trait to give invisible auditory signals a form and a presence: the virtual modeling triggered by sounds brings to the fore visualizations of potential sources of danger or pleasure. Nevertheless, perceptions of sound don't automatically demand an interpretation of the signals or a sense of its cause and purpose. Don Ihde notes that we primarily recognize sounds as "sounds of things" and that their relative and abstract shapes reveal themselves to us (Ihde, 2007, p. 60). This mental process of spatial representation doesn't need perceptual knowledge to imagine unseen experiences (Schwartz and Heiser, 2006, p. 293). Still, listening to this unseen acoustic spatiality while wearing headphones emphasizes the construction of a personal sound space and generates a critical appraisal of the different sonic activities brought to the ears. We perform an active dissociation of audible and haptic perception rather than an awareness of specific sonic activities. One difficulty lies in the relentless constancy of auditory signals, in contrast to the ease of fragmenting visual information.

A second difficulty is to morph these sensory audio signals into indexical and sensory signs.

Human auditory and sensory propensities should make us perceive the world around us in similar ways. However, our individual subjectivity and perceptual intentions can make us feel and interpret vibrations in unique manners. Percussionist Evelyn Glennie explains that it is not her deafness that makes her feel vibrations differently but rather her refined sense of listening to vibrations with her whole body. Glennie learned to sense the ways high- and low-pitched sounds were mapping her body and has thus developed a new awareness for the frequencies she used to perceive aurally. She now feels low sounds in her legs and feet while high pitches reach her upper body parts (Glennie, 2015).

We could expect both similar and dissimilar characteristics and descriptions of sonic perceptions: low or high pitch, dull or strident, quiet or loud, harsh or soft, fat or thin, fast or slow, close or far and so on. Individuals might also note "absent" sounds from their sound space. Describing lived phenomena can be a fraught process: "creation tries to break through to the world itself [. . .] Nothing is quite as easy as naming, describing, conceiving" (Serres, 2008, p. 133). Aware of these difficulties, Pauline Oliveros proposed an expandable vocabulary to "express the meaning of sound and sounding" that includes nouns and adjectives ranging from "knell, sonorous, resonant" to "silence, noise, resound, racket" (Oliveros, 2011, p. 166). The nomenclature that she suggests does not position the human body at the center of sonic activities: the texture and substance of sounds take priority. It is our call to expand the list of expressions through our individual auditory and haptic abilities rather than privilege our images of sonic activities, causes and purposes.

In the context of this chapter, though, we should explore ways to unveil the sounds of the everyday and make them visible to our imaginations and those of others. All immediate senses and memories are thrown into the act of deconstructing, describing, naming and defining the essence of entwining sensations. This practical experiment can be performed in succession by multiple participants, but the microphone should be placed at a single location.

15.5. Exercise Two

Aims

1. To dissociate visual certitude and sonic perception
2. To experience semantic and reduced listening

Technical Requirement

1. Portable recorder with a long XLR cord (Tascam, Zoom and the like)
2. Headphones
3. Condenser microphone
4. Access to a building

Tasks

1. Chose a place to listen (all participants will use the same location).
2. Plug in the microphone into the recorder using a long XLR cable.
3. The microphone will be positioned behind a closed door or hung from a window, preferably by a nonparticipant. The microphone will be out of sight and outside the space occupied by the listeners. The microphone will not be moved during the exercise.
4. Do not record; simply use the recorder as an extension of your ear.
5. Participants will listen with headphones while sitting and while standing.
6. Participants will listen with eyes open and also eyes closed.
7. Participants will take notes of their sensations and thoughts for each period of listening.
8. Participants will meet afterward and share their experiences by describing sounds as causal and semantic entities. It is a challenge to express the mental modeling of shapes that come to mind, and participants should try to perform this process.

15.6. Sensory Sharing

Listening is an activity that demands self-reflection, corporal and spatial awareness, as well as sensory imagination and trust in your intuition. We might instinctively remember a sound heard but equally realize that we don't know what that sound could be. However, signals don't address one sense exclusively; no sense works in isolation, and perceptions are multilayered. Merleau-Ponty also examined sound as a facilitator for the enrichment of other senses. He wrote at length on synesthesia, a neurological condition that implies a joining or blending of sensations such as the sight of sounds, the taste of shapes, the color of numbers or the hearing of colors (Sacks, 2005, p. 37). Affect is a phenomenon that is not linked to emotions but is the body's first response to stimuli "at a precognitive and pre-linguistic level" (Labanyi, 2010, p. 224). Affect doesn't live by opposition to cognition or cultural paradigms but rather stimulates them. True to his beliefs that all senses and objects communicate with one another, Merleau-Ponty noted:

> The sensory experience has only a narrow margin at its disposal: either the sound and the colour, through their own arrangement, throw an object into relief, such as an ashtray or a violin, and this object speaks directly to all the senses; or else, at the opposite end of experience, the sound and the colour are received into my body, and it becomes difficult to limit my experience to a single sensory department: it spontaneously overflows toward all the rest.
>
> (Merleau-Ponty, 2008, p. 264)

The previous exercises brought primary senses to the forefront of our contact with the audible and inaudible. They also, hopefully, generated feelings of wonder and an instinct to present pictures to our self, based on sonic activity and scenes generated by unseen sources. We might be able to imagine the purpose of their incidence, the source of their provenance, and shape a representation of their presence. Attention to these parameters and awareness of cultural assumptions influence individual and collaborative practices of storytelling.

In this respect, a sound walk can provide shared reflections of our lives in sound as it is "an exploration, and an attempt to understand, the sociopolitical and resonances of a particular location via the act of listening" (McCartney, 2004, p. 179). We can stop from time to time to feel the story of the outside, eyes closed and immobile, and fill our individual sonosphere to connect the molecular to the auditory and the synaptic. If phenomenology, as a theory and philosophy, "'thinks' the world, practice 'grasps' the world—it grasps the world pathically" (Van Manen, 2007, p. 20). This sophisticated mode of comprehension seems almost innocuous, but our reliance on its practice is at the core of all collaborations.

Finding words to describe the feelings transmitted by subtle layers of auditory sensations can be challenging. We should keep in mind Merleau-Ponty's words on the revealing of the world: "Perception does not give me truth like geometry but presences" (Merleau-Ponty, 1964, p. 14). Yet questioning one another on axial body sensations paired to characteristics of sensory signals can bring subtlety to the experience of "presences." We can segment our body cartographically, with our right ear being our upper right, our forehead our upper front, our left hip becoming our middle left and so on. Two visual aids, one consisting of an outline of a body shape and the other of a circle divided into eighths, can help us to position the trajectory of sonic activities toward the self and around our vertical posture and center of gravity. We can use these simple visual aids in the next two exercises.

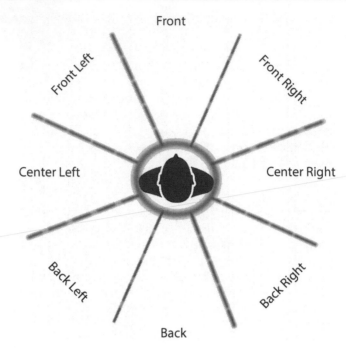

Figure 15.1 Segmented World.

We can feel the length of our legs as the low band; our torso is our middle band. Our high band goes from our shoulders to our temporal lobe. The top of our scalp is the upper and under our feet is the under. A sweeping sound could, for example, go from low center left to the middle back, taking what seemed like a second. It felt large and diffused. What about that high pitch? It traveled fast from my middle front left to high back right. It felt close and narrow. The loud wind brushed from my high front right to top back right. Adding visuals and forms to this mode of expression can diminish language barriers. In addition, sonic experiences don't need an extensive vocabulary to be discussed: a hand floating in the air or brushing a garment, a face expressing surprise, a head turning, or fingers pressing the ears could complement words and drawings.

The proposed practice exposes attentiveness to sonic environments and range of perceptions among participants, as well as highlighting the importance of imagination to the construction of a spatial narrative. Indian film director Mani Kaul believed that sounds alone could build a script and unveil the corporality of a story. He commented on his desire to find a producer who would agree to finance a movie in which a story would

Upper

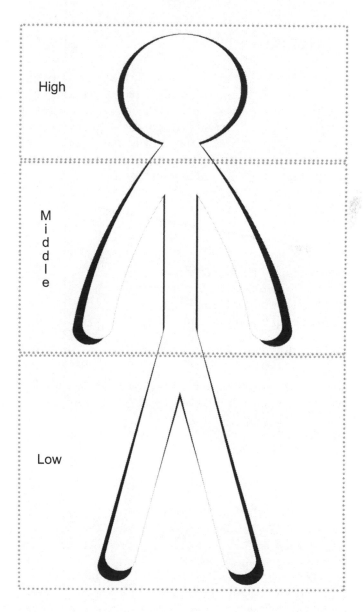

Figure 15.2 Segmented Silhouette.

be conceived from sounds recorded at particular locations (Kaul, 2003, pp. 219–220).

15.7.   Exercise Three

This practical exercise aims to fulfill aspects of Kaul's ideas on the physicality of sound as the genesis of audiovisual storytelling. The process of using locative recordings requires an increase in sensory awareness and methods to communicate personal sensations to others. This exercise invites reflection and affective collaboration by connecting an individual's sensory and spatial awareness with that of others. Here, we aim to build a narrative from shared experiences of being in the same place at the same time, in which we trust one another's perceptions, as different as they may be, of that event.

Aims

1. To develop environmental sonic awareness
2. To become a microphone before using one
3. To engage with visual imagination
4. To practice group communication

Technical Requirements

1. A portable recorder (use only the integrated microphone)
2. Access to different settings at a walkable distance from a specific location
3. A space for a group meeting (preferably with a whiteboard)
4. A group of four to six members

Tasks

1. You will walk to different places of your choice to first listen to the environment and afterward record isolated sounds. Each location should be called a station and be numbered. At these stations, you should be able to see the causal sources of audible sounds.
2. Capture audio-scenes of 10–20 seconds' duration.
3. Write a short note about the sonic activities at each station. Use the two visual aids to map the origin of sounds reaching and passing you. Number the notes, and match them with the recording station.
4. Without explaining the nature of your recordings, play them to a group of people who have not listened to them before.

5. While listening to each audio-scene, each member of the group will use the two visual aids (the outlined body shape and segmented circle) to position their perceptions.
6. After listening to each audio-scene, each member of the group will write a short comment about:
   6.1. How does the recording make them feel?
   6.2. What are the sounds (or types of sounds) that are most noticeable to them?
   6.3. Can they associate a recording station with a previous locative experience?
   6.4. Where do they visually situate the scene?
   6.5. Where do they visualize the sounds coming from within the scene?
   6.6. Can they visualize the trajectory of the sounds within the scene?
   6.7. What unseen activities are occurring during the listening?
   6.8. Can they feel the presence, or not, of people or animals?
   6.9. Can they guess the time of day?
   6.10. What does the weather sound like?
7. Ask all members of the group to collaboratively discuss a story based on some of the sounds that have been played.
8. Discuss the choice of sounds.
9. Classify the selection of sounds based on sensations and colors.
10. Together, write a story, and outline its visual treatment in correspondence with the chosen sounds.

15.8. Mediating the Story: An Affective Practice

As discussed and experienced in the previous three exercises, sensory awareness and attention to sound implies that places of audition, perception, physiology, behavior, memories and subjectivities are integral parts of individual and shared sonic experiences. The listening practices outlined by Pierre Schaeffer and their relationships to other realms of perception are intrinsic to our daily life. This engagement might not always be worded as "phenomenology"; it is a lived encounter with the knowing, an incarnation of the practice of living. In this context, professional occupations of all kinds depend on the sense of the body and individual "relational perceptiveness, [. . .] and other aspects of knowledge that are in part pre-reflective, pre-theoretic, pre-linguistic" (Van Manen, 2007, p. 20). The phenomenon of affect relates to visceral sensations, often unconsciously, that can stem and travel from some individuals to others and generate collaborative actions. It is thus pertinent to associate the notion of "affect" to aspects of individual and collaborative creative practices related to audiovisual storytelling.

In film productions, a typical creative chain of events usually starts with a discussion about a story between producers and directors. This step can lead to a script and be followed by a visual concept that will culminate in a film shoot and subsequent postproduction stages. Audio production was often, and still can be the last element to be considered in the segmented production process. However, according to sound designer Randy Thom, sound designers are people "in a position to actually shape the film to use sound effectively and efficiently; powerfully and subtly" (Brophy, 2000, p. 6). The concept of "sound design" itself is a recent and sometimes unclear addition to the list of filmmaking practices. Film editor and sound designer Walter Murch coined the expression "sound designer," and stated that:

> The origin of the term "sound designer" goes back to *Apocalypse Now* when I was trying to come up with what I had actually done on the film [. . .]. I thought, "Well, if an interior designer can go into an architectural space and decorate it interestingly, that's sort of what I am doing in the theater. I'm taking the three-dimensional space of the theater and decorating it with sound." I had to come up with an approach, specifically for *Apocalypse Now* that would make that work coherently. In my case, that was where "sound designer", the word, came from.
>
> (Jarrett, n.d.)

The designing of a soundtrack is a spatial modeling process that is performed with the ears and bodies of other filmmakers and the perceptual experience of audience members in mind. Film writer, director and independent producer Rolf de Heer attends to sound right from his devising of a story and its scriptwriting, all the way to the postproduction stages. During an informal conversation with me, the writer-director made a gesture with his two hands, positioning them near his head, explaining that while writing, sound sits "next to him, just there" (Delmotte and De Heer, 2012). De Heer has worked consistently with sound designer James Currie since *Incident at Raven's Gate* (De Heer, 1988). Their experimental and adventurous collaboration has covered many films in which De Heer gives a central narrative role to sound(s).

In 1993, the director used a binaural recording to disclose to the audience the real-world atmosphere discovered through the ears of the main protagonist of his film *Bad Boy Bubby* (De Heer, 1993). Bubby, who has been trapped in a small room for 30 years, explores the complex and intense soundscapes: each of his motions resonates with spectators' bodies and auditory apparatus. The experimental recording device had to be very small to be hidden behind actor Nicholas Hope's ears and was constructed by a former employee of the Australian Security Intelligence Organization

(Zielinski, 2010, p. 149). De Heer conceived this unique sound-catching equipment with the intention of moving the affective being of audience members and initiated this transmission from a human body in motion.

It is pertinent to associate the notion of "affect" with individual and collaborative creative practices that involve physical sensations essential to creation and perceptions of audiovisual storytelling. Sound design played an essential part in the writing process for the film *The Hurt Locker* (Bigelow, 2008). Director Kathryn Bigelow, writer Mark Boal and sound designer Paul N. J. Ottosson worked together to achieve the sophisticated simplicity of immersing audiences in a space between the skin of the two military bomb disarmer protagonists and the texture of their 50-kilogram protective suits. Their collaboration involved, at times, working together in the same space to perfect the preproduction stage and script (Coleman, 2010). An outcome of close collaborations and physical engagement between individuals can translate in a transmission of affect, a flow across bodies and subjectivities also described as "a process that is social in origin but biological and physical in effect" (Brennan, 2004, p. 3).

15.9. Exercise Four

This last exercise combines the iterative experiences of the preceding ones and aims to develop an affective practice. Bodies of makers and audiences alike are engaging with sound to morph the audible and inaudible into sensations and emotions. The exercise entails working in pairs and without headphones in the same space. Breaking the traditional cycle of production and summoning at the same time, a sense of knowing and perceptual imagination can lift sensory storytelling. Designing sound for the moving image but without watching the footage can reconcile at the same time lived sensations and memories of past experiences, as well as spatiotemporal imagination, perceptual discovery and cognitive expectations.

Aims

1. To dissociate visual and sonic perceptions
2. To activate the audiovisual imagination
3. To build an audio story based on an existing script
4. To forge a collaborative pathic practice

Technical Requirements

1. A visual extract of an existing film and its scripted scene
    1.1. The scene should be one to two minutes' duration and chosen by someone other than the audio practitioners.

1.2. Only the script of the scene should be made available to the audio practitioners.
1.3. The person who chose the scene should give its exact duration, as well as time code references for some locative parameters, if applicable (inside or outside location, night or day, city or countryside, for example).
1.4. The scene shouldn't involve dialogue so that the ubiquitous sounds of the everyday or sounds specific to particular activities can take center stage.
2. A DAW station with audio and visual editing suite
3. Access to a sound bank or personal sound library

Tasks

1. Read the script of the scene carefully.
2. Each participant should read it aloud to the person they are paired with. Repeat this step between each of the following steps.
3. Divide the written text into plot sequences. Imagine each sequence as a self-contained story.
4. Together, imagine visual snippets for each sequence.
5. Together, imagine audio elements for each sequence.
6. Together, imagine audio-scenes for each sequence.
    6.1. Read aloud the first sequence.
    6.2. Use your physical and emotional reactions to perform this task: ask each other how you would situate yourself in the scene.
    6.3. Use the two visual aids (the outlined body shape and segmented circle) to express your expectation of the sensations that sounds should create on your body.
    6.4. Take notes of the different sonic activities that you think could be effective in making you feel the sequence.
    6.5. Repeat these procedures for each scene.
7. Using your notes, start building each audio sequence. Listen with your eyes closed. Ask each other if the sequence "feels" right. What physical sensation, even minimal, does the assemblage of sounds trigger?
8. Mix the soundtrack of each sequence.
9. Combine the sequences to bring the audio story of the script to life. Listen with your eyes closed.
10. Keep asking each other how this audio-scene makes you feel; make handwritten notes.
11. Lay your soundtrack (without any musical score) onto the original visuals.
12. Show your work to others, have them listen to the piece with closed eyes at first.
13. Get feedback.

14. Afterward, show the original audiovisual scene to the same audience.
15. Discuss your perceptions of the different audiovisual versions as a group.
16. Discuss publicly the expectations you both had, as makers, of the script prior to starting this experiment.

15.10. Conclusion

In this chapter, we have performed an iterative approach to sensory engagements with our sonic environments. We have made delicate incursions into the soundscapes that we inhabit and that fuel our intimate sound spaces. The sense of place of sound and sense of bodies in sound take us alongside acoustemologist and musician Steven Feld's notion of reciprocity with our lived environment "that as place is sensed, senses are placed; as places make sense, senses make place" (Feld, 2005, p. 179).

Vibrations permeate our surface and viscera. We have explored the body as a medium that can be stimulated and expanded but not supplanted by technological means. The displacement of vision has highlighted the need to attend more carefully to sound as a holistic and haptic entity, as well as to sharpen the use of our auditory apparatus. These various experiential and relational phenomena of sensations and perception continually change and adapt to facilitate moments of shared emotions, meaning-making and visual storytelling.

Our experimental creative collaborations have been driven by affective and pathic predispositions. These approaches have considered the wording of sensations and gestures as means of communication. We haven't denied the richness of known terminologies but have acknowledged our responsibility to shape new expressions to better share and unveil sonic sensations. This process has hopefully reconciled individuals' impressions and knowledge of the world with the sensory experiences of others. The accumulation of lived and felt experiences and their sharing may have allowed unexpected exchanges to surface between cocreators. We cannot control ever expanding sonic frameworks, but we can attempt to channel their richness, as they can give life to audiovisual stories that can be felt by our skin contours.

References

Beyer, C. (2016). Winter edition. Edmund Husserl. In: E. N. Zalta, ed. *The Stanford encyclopedia of philosophy*. Available at: http://plato.stanford.edu/archives/win2016/entries/husserl [Accessed August 2017].

Bad Boy Bubby. (1993). [Film]. Directed by Rolf de Heer. Australia: Blue Underground.

Brennan, T. (2004). *The transmission of affect*. Ithaca, NY: Cornell University Press, pp. 1–23.

Brophy, P., ed. (2000). *Cinesonic, cinema and the sound of music*. Sydney: Australian Film and Radio School, pp. 1–28.

Chion, M. (1983). *Guide to sound objects, Pierre Schaeffer and musical research*. Paris: Institut National de l' Audiovisuel & Éditions Buchet, Chastel, pp. 11–20.

Chion, M. (1994). *Audio-vision, sound on screen*. New York: Columbia University Press, pp. 3–34.

Chion, M. (2009). *Film, a sound art*. New York: Columbia University Press, pp. 465–500.

Coleman, M. (2010). 2010 CAS awards 'Meet the Winner' panels: The Hurt Locker. *Sound-Works collection*. Available at: https://vimeo.com/10546973 [Accessed December 2017].

Delmotte, I. and De Heer, R. (2012). Informal conversation with film director/writer Rolf de Heer.

Encounter at Raven's Gate. (1988). [Film]. Directed by Rolf de Heer. Australia: FGH.

Feld, S. (2005). Places sensed, senses placed: Towards a sensuous epistemology of environments. In: D. Howes, ed., *Empire of the senses, the sensual culture reader*. Oxford: Berg, pp. 179–191.

Fluegge, E. (2011). The consideration of personal sound space: Toward a practical perspective on individualized auditory experience. *Journal of Sonic Studies*, 1(1), Available at: http://journal.sonicstudies.org/vol01/nr01/a09 [Accessed July 2016].

Glennie, E. (2015). *Hearing essay*. Available at: www.evelyn.co.uk/hearing-essay [Accessed December 2017].

The Hurt Locker. (2008). [Film]. Directed by Kathryn Bigelow. USA.

Husserl, E. (1964). *The phenomenology of internal-time consciousness*. Bloomington: Indiana University Press, pp. 40–97.

Ihde, D. (2007). *Listening and voice, a phenomenology of sound*. 2nd ed. Albany: State University of New York Press, pp. 3–72.

Ihde, D. (2012). *Experimental phenomenology: Multistabilities*. New York: State University of New York Press, pp. 3–13.

Jarrett, M. (n.d.) *Sound doctrine, an interview with Walter Murch*. Available at: http://www2.yk.psu.edu/~jmj3/murchfq.htm [Accessed December 2017].

Kaul, M. (2003). The rambling figure. In: J. Sider., D. Freeman and L. Sider, eds., *The school of sounds lectures 1998–2001*. London: Wallflower Press, pp. 209–220.

Labanyi, J. (2010). Doing things: Emotion, affect, and materiality. *Journal of Spanish Cultural Studies*, 11, pp. 223–233.

Lingis, A. (1996). *Sensation: Intelligibility in sensibility*. Amherst, NY: Humanity Books, pp. 53–66.

McCartney, A. (2004). Soundscape works, listening, and the touch of sound. In: J. Drobnick, ed., *Aural cultures*. Toronto and Ontario: YYZ Books, WPG Editions, pp. 179–185.

Merleau-Ponty, M. (1964). The primacy of perception and its philosophical consequences. In: J. M. Edie, ed., *The primacy of perception and its philosophical consequences*. Evanston, IL: Northwestern University Press, pp. 12–42.

Merleau-Ponty, M. (2008). *Phenomenology of perception*. Oxford: Routledge.

Oliveros, P. (2010). *Sounding the margins: Collected writings 1992–2009*, Kingston, NY: Deep Listening Publications, pp. 22–31.

Oliveros, P. (2011). Auralizing in the sonosphere: A vocabulary for inner sound and sounding. *Journal of Visual Culture*, 10, pp. 162–168.

Ondaatje, M. (2002). *The conversations, Walter Murch and the art of editing film*. London: Bloomsbury, pp. 3–7.

Pijanowski, B. C., Farina, A., Gage, S. H., Dumyahn, S. L. and Krause, B. L. (2011). What is soundscape ecology? An introduction and overview of an emerging new science. *Landscape Ecology*, 26, pp. 1213–1232.

Sacks, O. (2005). The mind's eye: What the blind see. In: D. Howes, ed., *Empire of the senses, the sensual culture reader*. Oxford and New York: Berg, pp. 25–32.

Schafer, R. M. (1994). *The soundscape: Our sonic environment and the tuning of the World*. 2nd ed. Rochester, VT: Destiny Books, pp. 3–12.

Schwartz, D. L. and Heiser, J. (2006). Spatial representations and imagery in learning. In: K. R. Sawyer, ed., *Handbook of the learning sciences*. Cambridge: Cambridge University Press, pp. 283–298.

Serres, M. (2008). *The five senses, a philosophy of mingled bodies*. London: Continuum, pp. 85–151.

Solomos, M. (1999). Schaeffer phénoménologue. In: *Ouïr, entendre, écouter, comprendre après Schaeffer*. Paris: Buchet, Chastel-INA, GRM, pp. 53–67.

Ströker, E. (1997). Evidence. In: E. A. Behnke, D. Carr, J. C. Evans, J. Huertas-Jourda, J. J. Kockelmans, W. Mckenna, A. Mickunas, J. N. Mohanty, T. Nenon, T. M. Seebohm and R. M. Zaner, eds., *Encyclopedia of phenomenology*. Dordrecht: Kluwer Academic Publishers, pp. 202–204.

Van Manen, M. (2007). Phenomenology of practice. *Phenomenology & Practice*, 1, pp 11–30.

Westerkamp, H. (2001). Soundwalking. *eContact!* 4(3). Available at: https://econtact.ca/4_3/Soundwalking.html [Accessed October 2017].

Zielinski, A. M. (2010). *Conversations with a sound man*. Norwood, South Australia: Peacock Publications, pp. 147–164.

Index

Note: Numbers in **bold** indicate tables and numbers in *italics* indicate figures on the corresponding pages.

2nd-order 3-D system 161
3D films: acoustic space 199–203; contemporary soundtrack and 206–8; 3D immersion, layering sounds for 203–6; industry 194–8; language and describing space 198–9; sound and vision 191–4
3rd-order 3-D system 161
4-K resolution 150n1
7.1 Dolby Surround format 195
8dB Sound 132
9th Life of Louis Drax, The (2016) 78
12-tone equal temperament 230
16-bit digital system 180
2015 Great Place Award 374

Aadahl, Erik 194
Aardman Animation 132
Ableton Live 270
absentmindedness 128
absorption coefficients 6–7, 125
Academy Awards 151n11, 197
Acord, David 204
acousmatic listening 383
acoustical warmth 105–6, 109
Acoustic Communication 371
acoustic guitar sounds 179
acoustic horizon 361, 364
acoustic instrument EQ 179–80
acoustic phonetics 287
acoustic pianos 179
acoustic signal 296–7

acoustic space 199–203
acting 61, 63, 75, 364
actor's performance 61–2, 66, 79
actor's visual presence 71
acuteness, sound color 288–9
additive broadband noise 84
ADR *see* automatic dialogue replacement (ADR)
ADSR (attack/decay/sustain/release) envelope 36
advanced authoring format (AAF) 151n6
aesthetic integrity 148–9
air band 175
algorithmic reverb 120
Allen, Ioan 206
allure, sound object 284
Amateur Tape Recorder 369
ambiences and atmospheres 203
ambient/immersive/atmospheric soundscape 359–60
ambisonics 161, 203
Amen break 344, 349, 350, 355n11; pattern kicks 346
Amores Perros (2000) 329
amorphism 286
amplification factor 116
amplitude 11–12; distortion 124; level, loudness 31; panning 165–6; variation 293
AMS Neve DFC 141
Anaesthetic Machine 375
analog metering 161–2

analog signal processing 84
analog–to-digital conversion (ADC) entry points 84
analogue dynamic range compression 48
analog VU (volume unit) meters 161
anamorphism 286
anecdotal qualifier 129
angle of incidence 10
angle of reflection 10
Animal Language (1938) 369
annotation propagation mechanisms 83
Anthology of Noise and Electronic Music, An 1–7 323
anthrophony 294
anthropological tools of precision 386
antiphase 18
API *see* application programming interfaces (APIs)
application programming interfaces (APIs) 249, 265–6, 278
Aronofsky, Darren 328, 343, 344
Arrival (2016) 17–18
articulatory phonetics 287
artificial panning 115
artificial reverberation 120
Art of Noises 304
Assault on Precinct 13 (1976) 327
Association of Motion Picture Sound (AMPS) 197
ateleological sonic dimensions 359
atmos 102
Atmosphères (1961) 242
atmosphere sounds editing 99–102, *101*
atoms (basic sound snippets) 269
attack 17, *18*, 284
attack-decay 284
attack dispersion 292
attack time 22, 92, 118, 181–2, 184
attention span 269
attribute sounds from freesound *260*
audible frequencies 121–2
audio aggregation process 130
Audio Commons 276; APIs 268, 278; content 248–9, 270; content, automatic analyses 253; Ecosystem (ACE) 264–5, *265*, 265–6, 268, 270, 278; Retrieval Interface *270*

audio effects: balance 114; digital storage and processing equipment 114; distortion 124; dynamic range processing 117–19; equalization (EQ) and filtering 116–17; hyperrealism 125–6; imagination 127–8; modulation effects 121–3; panning 114–16; perception and interaction 126–7; pitch shifting 123–4; realism 125; reverberation 119–20; vocoder 123
Audio Engineering Society (AES) 158, 197
Audiofinder 298
audio material types 85, *86*
audio mix supports 153
audio postproduction industry 154
audio resolution 84
Audio Set 298
audio technology developer DTS 196
AudioTexture 270–1, 275
audiovisual mix 366
audiovisual productions and sound elements 104
audiovisual rhythm 327, 335, 340, 347, 352–3
audiovisual scenes 380; exercises 384–5, 387–8, 392–3, 395–7; flesh 381–4; mediating 385–7, 393–5; sensory sharing 388–92
audiovisual storytelling 393
audiovisual strategy 329, 336, 344, 348
audiovisual structural congruency 339, 352–3
auditory change blindness 366
auditory information 361
auditory system 380
augmentation 218
augmented reality (AR) application 297
aural witness 362
Auro 11.1 206
AURO 3D recommendation 159–60
Auro Technologies 192, 196
Australasian Computer Music Association 323
automated onboard mixing 91
automatic dialogue replacement (ADR) 47; creative and technical challenges

in 70–1; mixer and recordist roles 68; performance in 65–7; recording 63–4; role 67–70; techniques to record 67
automation systems 186
auxiliary sends 164
auxiliary tracks 165
aux-sends 164, 169
Avatar (2009) 191, 195, 197
Avid Downmixer 145

Back, Les 364
background effects 102
Bad Boy Bubby (1993) 394–5
balance 114
ballistics of compressor 118
Bandcamp 250
band pass 176; filter 117, 123
band stop 117
bandwidth (Bw) 176; richness 84
Barthet, Mathieu 255–6, 262, 264, 272–3
Basehead 298
bass-drum pattern 331
bass instrument EQ 178
bass managed systems 158
Battleship Potemkin (1925) 333
Bayle, François *309*, 310
Beach, The (2000) 329
beats 238–9, 250; machine-accurate alignment of 330; oscillating beat pattern 214
beats per minute (BPM) 250, 334
Beethoven symphony 359
Berlin Calling (2008) 328
Best Motion Picture of the Year 197
bidirectional (or figure-of-eight) microphone 41, 43, *45*, 46
Bigelow, Kathryn 395
biophony 294
Birth of a Nation, The 333
bit depth 53–5, *54*, **55**, 58, 59n17, 272
bit rate 53–4, **55**
Blade Runner (1982) 327
Blue Bloods (2018) 72
Boal, Mark 395
body movements 71
Bones, O. 297

Book of Noise (1998) 369
boosting peaking filter 123
bootlegging 273
Bordwell, David 335
Boyle, Danny 328–9
Brave (2012) 70, 206
breathing 118
breath pass 66
brightness 177, 271, 292, 294
broadband sounds 64
broadcast-compliant R128 mix 145
broadcasters 86, 131–2, 134, 144, 150n3, 189
broadcast sound and film mixing studios 156
broadcast wave (.bwav) 151n7
Brown, Andrew 332
Burge, Brent 207
Butler, Mark J. 327, 330–1, 354n1, 354n4
buzzing 121, 344, 347, 361

Cage, John 211, 225, 352, 361, 363, 365, 367
Calavita, M. 354n6
camera movement 334–5, 338, 340, 350
Cameron, James 191, 195, 197, 237
Canadian Electroacoustic Community (CEC) 323
capacitor microphones 38
car audio system 180
carbon microphone 385
cardioid microphones 42, 57
Carol (2015) 72
Carpenter, John 327
"catch-all" digital audio workstation (DAW) 131
categorizing and classifying sounds 283
causal listening 310
CCMixter 263
center frequency (Fc) 176
central top speakers 115
channel-based audio distribution systems 192
channel faders 164
Chernobyl Exclusion Zone 375
Cherry Orchard, The 220

Chion, Michel 310, 348, 383
chorus effect 122
choruses 173
Chowning, John 316
chromatic alteration 233
chromatic scale 231
Chuckie Egg 235, 245
Cinema Audio Society (CAS) 197
cinematic performances, design of *see*
    automatic dialogue replacement (ADR)
cinematic voices: types 62
City as Classroom 366
clangs 217, 218
clarity 48, 99
classical music and opera 212
claustrophobia 204
client-server architecture 261
clip gain 165
clipping 31–2, *33*
close-miked bass drums recordings 178
close-miked instruments 175, 184
cloud-based applications 260–1
cloud-based live coding 264
cochlea 23
cocktail party effect 365
Cohen, A. 337
Cohen, Max 343–4
Cohen, Rabbi 344
coherent wavefront 10
Columbia Records 368
combined panning techniques 166–7, *167*
communication 73
compactness 291
composer, compositional techniques
    211–12
compositional ideas and techniques 303
compositional techniques: classical
    music and opera 212; composer
    211–12; conceptual sound designer
    213; enthusiasm 212; graphical scores
    223–5; modes and methods 221–2;
    motifs and melodies, creation 216–17;
    music composition 213; "note-for-
    note" notation 213; rhythm 213–16;
    sonic sculptor 212; soundscapes 211,
    212–13; timbre and texture 222–3;

tonic, designing from 219–21; variation
    217–19
comprehending *[comprendre]* 310
compression 2; and expansion 50; ratio 92;
    threshold 92
compressors 48–50, 92, 118–19, 126–7,
    139–41, 178, 180–4, 186
computational auditory scene analysis
    (CASA) 296
computerized environment arrangement *83*
computer music in North America 316–18
Computer Music Tutorial 297
concatenative synthesis 270
conceptual sound designer 213
condenser microphones 38–40, *39*, 48,
    74–5
Conformalizer 147
conformance- complementation-contest 340
congruence-associationist model 337
console 99, 105, *108*; mixing in live
    performances *109*
consonants 116, 143, 287, *287*, 311
constant long-term characteristics 269
contact microphones 40–1
contemporary audio software 52–3
contemporary cinema sound 191
contemporary soundtrack and 3D films
    206–8
content-based queries 249, 278
content documentation and management
    86, *87*
content-repository and management
    infrastructure 84
context-based composition 318
contextual [C] sound 251
convolution reverb 69–70, 120
copyright 257–8
Cosgrove, Denis 370, 372
Creative Commons (CC) 248, **258**,
    259, 272
creative commons licenses 249
critical band 240
critical bandwidth 27–8
cross-classifications 295
cross-modal alignment of accent
    structures 352

Crowdsourced CC audio content 273
Crown Film Unit wartime documentary 368
CSOUND 316
Cuarón, A. 197
Cuckos Reaper 199
cutting rhythm *see* editing rhythm

Daily Mail 368
dampening 4
DAWs *see* digital audio workstations
   (DAWs)
dBFS and loudness standards 189
dB FS target levels 163
decay 17
decibels 31, *32*, 33
de-esser 119, 139, 175
degree of membership 296
degree of polyphony 332
degrees of freedom 258
de Heer, Rolf 394–5
delay effect 92, *95*, 120, 128
delay panning 115, 166
democratization of music technologies 303
d'Escrivan, Julio 307
design objective and sounding object
   motif 217
desirable sound 45
dialogue(s) 62, 63; in animated pictures
   62; compression 183; dynamic range
   139–40; editor 68; lines of 64; stem
   133; target level 156
diaphragm microphone 39, 69
Diatonic scales 231–2
diegetic and non-diegetic sound
   effects, 343
diffuser acoustic panels *11*
Digidesign 91
digital audio: bit depth 53–4; continuous
   analogue and discrete digital waves
   51; digital compression 54; digital-
   to-analogue conversion (DAC) 54–5;
   rendering and reality 56–7; sample rate
   51–3; working practices in 55–6
digital audio workstations (DAWs) 6, 82,
   88, 89–91, *90*, 164, 215, 230, 249, 261,
   263, 272; mix stems 131; systems 84
digital cinema 195

Digital Cinema Package (DCP) 198;
   specification 196
digital clicks 160
digital compression 54
digital 3D films 191, 193–5, 197, 206
digital film exhibition 198
digital limiters 181–2
digital metering 162, *162*
digital musical interfaces 261
digital NLE software and plugins 88
digital noise-removal tools 64
digital recording 48; noise levels of 48
digital rights management (DRM)
   systems 86
Digital Sound Cinema (DSC) 192
Digital Surround (5.1) format 195
digital systems management 130
digital techniques 317
digital-to-analogue conversion (DAC)
   54–5
digitization parameters 84–5
diminution 218–19
directional sonic propagation *5*
direct sound 8–10, *9*, 109, 120, 125,
   160, 170
disappeared sound 362
"displaced" voices 63
distortion 31, 34, 50, 55, 58, 59n12, 69, 84,
   114, 118, 124, 127, 131, 138, 162, 178,
   182; and clipping 48, 59n13
dithering 55
diverse sensing modes 385
Doc Martin (2004–present) 133
Doctor Who 316
documentary film mix template session
   154
Dolby Atmos 142, 195–6, 197, 198, 206;
   speaker configuration 160; streams 130
Dolby E 150n4
Dolby Surround 7.1 206
doubling effect 170–1
downmixing 142
downward compression 117–18
downward expansion 118
drama mix 133
drum-machine-based constructions 344
drums EQ 178

DSC *see* Digital Sound Cinema (DSC)
dual-mono stereo signal 165
dubbing 120; mixers 129
Dublin Core 86
ducking effect 119
Dunkirk (2017) 25
duration 35
dynamic, sound object 284
dynamic EQ 119
dynamic microphones 37–8, *38*
dynamic noise reduction 182
dynamic processors: compressors 181; dialogue compression 183; dynamic noise reduction 182; dynamic range 180; expanders and gates 182; leveling amplifiers 180–1; limiters 181–2; multiband compression 183; musical instrument dynamic processing 184; vocal and soloist compression 183
dynamic range 92, 180; compression 49–50, 117–18; management 139; processing 117–19
dynamics processing 92–6

ear *21*; hearing range 23; inner ear 23; middle ear 22; outer ear 21–2
Ear Cleaning (1967) 369
early reflections 8
Ebert, Roger 193
echo 9, 120, 169, 192, 194, 200, 204, 207, 244
edit decision lists (EDLs) 86–7
editing: automated onboard mixing 91; criticism 88; DAWs 88, 89–91, *90*; digital NLE software and plugins 88; GUIs 88; magnetic tape recorder editing procedure 89, *89*; multitrack tape recorders 88; NLE 88; overdubbing 88; punch-in-and-out technique 88; reference tone 87; shuttling 89; stand-alone units 89; transfer medium 87; "undo" function 91
editing rhythm 334
EDLs *see* edit decision lists (EDLs)
EDM *see* electronic dance music (EDM)
effects (FX) stem 133

eight-channel 358
electric and acoustic guitar timbres 179
electroacoustic music 303–4; computer music in North America 316–18; Daphne Oram and radiophonic workshop 314–16; defined 303; Edgard Varèse 305–6; experiments in film 306–7; futurists 304; gesture/texture 319–22, *321*; Pierre Schaeffer and GRM 307–10; practices 322–3; recording and musical possibilities 304; soundscape 317–19; Stockhausen and WDR 311–14
Electroacoustic Music Studies Network 323
electronica 327
electronic dance music (EDM) 327–30; audio and visuals 333–5; breakbeat, noise and glitch 343–52; features 330–3; noise-based sound of 343–52; synthetic and 353; techno and kinesis 335–43
electronic synthesis 331
elektronische Musik 311
emotional and cultural landscapes 221
emotional involvement 193
emotional variations 66
energy-motion trajectories 322
energy-motion trajectory 320
Enthusiasm 307
Entrance to the Harbour 373
envelope: attack 17; decay 17; diagram *18*; follower 121, 123; sustain 17
environmental noise 296, 369
environmental sounds 294–5
equalization (EQ) 67, 69, 168; acoustic instrument EQ 179–80; bass instrument EQ 178; drums EQ 178; equalizers 176–7; and filtering 116–17; filters 175–6; frequency band approach to mixing 177; frequency bands 173–5; guitar EQ 179; keyboard instrument EQ 179; mix stems 132, 143–4; panning 166; parameters 262; voice EQ 177–8
equalizers (EQ) 15, 27, 91–2, 176–7; in DAW *93*; layout of 116
error 48

ethnic-sounding drums 339
eumorphism 286
Euphonix console 150
Europeana 265–6, 273
European Broadcast Union (EBU) 189, 197
everydayness listening exercise 360–7
excitation (*arousal*) 256
expanders 50, 92; and gates 182
experienced mixers 132, 149
'external' DSP mix technology 141

facial expression 71
far-field monitors 109–10
Farmer, David 192, 200, 205
Ferrari, Luc 367, 368
Fields, K. 297
Fight Club (1999) 329
figure movement 334
Files, Will 205, 207
film dialogue 171
Film Journal International, 2012 194
film soundtrack 331
film technology companies 193
filters 91–2, 175–6
final mix creation 187
Fincher, David 329
Finding Fatimah (2017) 132
first-person sensory analysis 363
fixed reference point 29
flangers 122, 173
flat 229
Fletcher-Munson Curve 30, *30*
Flickr 248
Flight (2012) 25
Flinn, Caryl 333, 354n9
Foley: art of 71–2; performing 75–9;
    recording 72–5; sound effects 71–2;
    see also automatic dialogue replacement
    (ADR)
Foley, Jack 71–2, 102
Foley effects 71–2, 102
folksonomies 249, 253, 278, 297
foreign language versions, film 64
formants 35–6, 173
formant space 288
foster intense attention 102

*Foundations in Sound Design for
    Embedded Media* 295
Fourier, Joseph 311
four-on-the-floor bass drum pattern 331
free culture 258
free-form tags 253
Freesound 248, 250, 265
Freesound Explorer 255, *255*
frequencies 11–12, 306; differences, pitch
    28; response 109–10; sweeping 123;
    value 27
frequency bands 173–5;
    characteristic,approach to mixing 177
frequency equalization (EQ) processing
    91–2
Frère Jacques 216, 236, 239, 244
full-range speakers 173
fully filled M&E mix 144
fundamental frequency 173
futurist movement 304

Gabrieli, Giovanni 313
gain management 139
gain structure, mix stems 131, 138–41
Gance, Abel 333
gates 182
gating 119
geophony 294
gesture 223, 319–22
gesture-texture 284
Gibson, J. G. 362
Glazer, Jonathan 329
Glennie, Evelyn 387
glitch aesthetics 344
Go (1999) 328
Goldcrest 364
graduated continuant 284
grain, sound object 284
granular synthesis 317
graphical notation 223–4
graphical scores, compositional techniques
    223–5
graphical user interfaces (GUIs) 88, 262,
    277
Gravity (2013) 25, 41, 197, 205
Great Gatsby, The (2013) 207

groove-based musics 330–1, 333, 352
Groupe de recherches musicales (GRM)
    307–10, 368
grouping, mix stems 132, 142–3
GUI *see* graphical user interfaces (GUIs)
guitar EQ 179
Guyot, F. 295
Gygi, B. 295

hairs-on-the-back-of-your-neck-
standing-up moments 149
half-note intervals 347
half-step *see* minor second
Hamlet 220
Hanna (2011) 329
hard disk recorders 48–9
hard knee, compression 50
hard-panned sound sources 171
harmonic (s) 173; content 221–2;
    distortion 124; oscillator 13; partials *13*,
    14; progressions 227; series 230;
    timbre 284
headphone (s) 109–10, 160; listener 160;
    monitoring 160, 173–4
headroom 48–9
head tracking system 161
hearing: damage 34; frequency ranges **25**;
    loss 127; range 23
hearing and listening: ear 21–3;
    psychoacoustics 23–37
hearing *[entendre]* 309, *310*
Hedges, Michael 149, 150, 151n11
height channels 159–60, 170
Heil, Reinhold 353
Here and Now (2014) 132
hertz (Hz) 27, 176
higher-order ambisonic systems (HOAs)
    161
high frame rate (HFR) 195
high pass filter (HPF) 117, 176
high-Q filters 176
high-resolution digital systems 174
high shelf filter 117
hip hop sequences 348
Hitchcock, Alfred 366
Hitchhiker's Guide to the Galaxy 316

Hobbit, The: An Unexpected Journey
    (2012) 195
Hollywood Reporter 195
Hope, Nicholas 394
Houix, O. 295
Hugo (2011) 197
human dialogue 62–3
"human limiter and compressor" 69
Human Traffic (1999) 328
Hurt Locker, The (2008) 395
Husserl, Edmund 381, 382
Hutchon, K. 366–7
hybrid "object-based" panning
    model 196
Hyde, James 68
hydrophones 41, 72, 376
hypercardioid microphone 43
hyperrealism 125–6

Ihde, Don 381, 383, 386
imagination 127–8
immersion 193
immersive 3D sound 198
immersive film sound technologies 192
immersive formats, 3D 205
Immersive Sound Conference (2014) 206
impulse 69; response 69; response
    measurement 120
Iñárritu, Alejandro González 329
incoherent reflection pattern 10–11
independent pitched streams 332
individual instruments 171–2, *172*
individual sound attributes 296
inflection 218
information: architecture, mix stems 131,
    133–5, **135–8**; function 104
infraordinary 363
inharmonic partials *13*, 14
inharmonious noises 127
in-line channel dynamic range
    management 140
inner ear 23, *24*
inner orientation 104
in phase 18–19, *19*, 35, 47
input gain 181
inserts 164

integrating audio commons content using web APIs 265–9
intellectual property issues 86
intelligibility 64, 116; of dialogue 126
intensities 228, 306, 334
interactive (weighted) tag cloud *253*
intermodulation distortion (IMD) 124
internal monologues 62, 63
internal pressure, mixer 148
International Computer Music Association 323
International Confederation of Electroacoustic Music 323
international music clearance 147
International Phonetic Alphabet (IPA) 287; consonants *287*; vowels *288*
International Phonetic Association (IPA) 287
internet-connected DAWs 261–3, *263*
interval, pitch 29
interval ratios **28**
in-the-box mixing 149, 151n10
intimacy 109
intonarumori 304–5, *305*
Into the Labyrinth (2000) 358
Inverse Square Rule 6, *7*, 46
inversion 218
Is-a hierarchy 296
Ives, Charles 341
Iyer, Vijay 330

Jackson, Peter 149–50, 195
Jamendo 250, 252, 265
Japanese Broadcast Network 204
Jaws (1975) 216, 230
Jazz recordings 171
Jekyll and Hyde (1932) 307
Jennings, Humphrey 368
Johnson, Clark 329
Jones, Gareth Rhys 72–8
JSON file 298

Kaempfert, Bert 230
Kane, Doc 62, 67, 69
Kaul, Mani 390–2
Kaye, David 66
Kellough, Bethan 319
Kerrigan, Justin 328
keyboard instrument EQ 179

key input 181–3
keynote sounds 294, 359
King Kong (1933) 307
Kits Beach Soundwalk (1989) 318, 358
Klimek, Johnny 353
knee 50, 118, 181, 183
Koch, Ludwig 369
Kontakte 313
Kray, Nick 62, 64, 65, 69–70
Krumhansl, C. L. 231
Kulezic-Wilson, Danijela 328, 343, 348

Laban method 66
landscape 294, 370–2, 383
Lang, Mary Jo 74
Lara Croft: Tomb Raider (2001) 329
large diaphragm condenser microphones 39, *40*, 69
La Roue (1923) 333
late reflections 8–9
Latin music productions 185
laxness, sound color 288
Layer Cake (2004) 329
LCR panning 169
LCR stereo pan-pot 169
lead expectation 102
leading tone 220
Lee, A. 197
Lefebvre, Henri 237, 338
left surround and right surround channels 169–70
legal-level mix 131
Legend of the Guardians: The Owls of Ga'Hoole (2010) 201
Lego Movie, The (2014) 207
Le Sound 270, *270*, *271*
Lessig, L. 248, 257–8
leveling amplifiers 180–1
Library Monkey 298
license restrictions when creating remix **260**
licensing and attribution 257–9
Life of Pi (2012) 197
Ligeti, György 242
Liman, Doug 328
limiters 50, 92, 181–2
limit mode 181
linear audio production 259–61
linear pitch progressions 235

lines of dialogue 64
linguistic rhythmical patterns 214
Lipscomb, Scott D. 337
lip synchronization 63; lack of 63
listen and repeat in ADR 67
listening *[écouter]* 308, 388; act of 376n1; cognitive aspects 298
Listen to Britain 368
live coding 249, 261, 264, 269; with MIRLC 275–6, *275*
live collaborative music making with playsound 276–8
live performances and mixing 105
Lobanova, A. 297
lo-fi 344, 370
Logic Pro X 270
lo-/hi-fi categorization, soundscape 294
Lola 337–9, 341
London Street Noises (1928) 368
long release, compression 50
Looperman 250
looping 64
Lord, P. 207
Lord of the Rings (2002) 25, 149
LoRo 141, 145, 150n2
loudness 127; amplitude level 31; boosting amplitude 33; clipping 31, 33, *33*; decibels 31, 33; exposure and risk chart *34*; fixed reference point 29; Fletcher-Munson Curve 30, *30*; hearing damage 34; measurement *32*; metering 163; pink noise 30, *31*; scales of 29; SPL meters 30–1; threshold of hearing 31; white noise 30, *31*
loudness unit (LU) 163
loudness unit full scale (LUFS) 163, 189
loudspeaker-based cinema sound system 192
loudspeakers placement 109–10
low cut filter 117
"low-fidelity" effect 175
low-frequency effect (LFE) 156, *157*, 169; channel 170; rotary pot or slider 170; signal 156–7
low-frequency oscillator (LFO) 121
low pass filter (LPF) 117, 123, 176
low-Q filters 176
low shelf filter 117
LUFS *see* loudness unit full scale (LUFS)

Luhrmann, B. 207
Lux Aeterna (1966) 242

Mâche, François-Bernard 368
machine learning 255
Macmillan, Maree 336, 341
Mad Max: Fury Road (2015) 197
magnetic tape 82, 88, 307, 311; recorder editing procedure 89, *89*
magnum opus 139
major second *see* minor second
major third 229
makeup gain 118, 163, 181
Mamoulian, Rouben 307
Mancini, Henry 230
Man on Fire (2004) 329
Man with a Movie Camera (1929) 307
Marcell, M. M. 251, 295
Marshall, S. 337
Martian, The (2015) 197
mass profile, sound object 284
master faders 164
mastering: dBFS and loudness standards 189; music for CD 188; music streaming, loudness standards for 189
Material Exchange Format (MXF) 86
materializing indices 57
mathematical stereo panning laws 166
Matrix, The (1999) 329
Matthews, Max 316
maximizer plugins 181–2
McAllister, Grant 342
McAllister, Stewart 368
McLuhan, E. 366–7
McLuhan, Marshall 345, 366–7
mechanical sound effects machines 71
media assets 83
media management for audio production: additive broadband noise 84; analog–to–digital conversion (ADC) entry points 84; annotation propagation mechanisms 83; audio material types 85, *86*; computerized environment arrangement *83*; content documentation and management 86, *87*; content-repository and management infrastructure 84; digital audio workstation (DAW) systems 84; digitization parameters

84–5; intellectual property issues 86; media assets 83; multichannel matrix encoding 85; quality preservation status 84; spatial enhancement 85
melodic contour 234; and motion 235
melodic lines 330
melodic profile, sound object 284
Merkley, R. 248, 258
Merleau-Ponty, Maurice 382, 388–9
metadata 249
meter 331
metering: analog metering 161–2; dB FS target levels 163; digital metering 162; loudness metering 163
microphones and recording technology: condenser microphones 38–40, *39*; contact microphones 40–1; digital audio 51–7; dynamic microphones 37–8, *38*; hydrophone 41; microphone positioning 44–7; polar patterns 41–4, *42*; signal chain/gain staging 47–50
middle ear 22
Midnight Express (1978) 327
midrange frequency band mix balance 187
mid-side-based split technique 169
mid/side (M/S) pairing 43
mid-side (MS) processing 168
mid-side (MS) signal 168
Miller, C. 207
Miller, G. 197
minor second 229
minor third 229
Mitry, Jean 334–6
mixable recording 75
mix automation 186
mix balance 184–5; checklist 186–7; final mix creation 187; midrange frequency band mix balance 187; mix automation 186; music and effects levels 185–6; priority 184–5; voice levels 185
mixdown 105
mixer community 131
mixers 99, 129–32, 148–50, 170, 199, 207
mixing: aesthetic qualities 152–3; dynamic processors 180–4; equalization 173–80; mastering 188–9; metering 161–3; mix balance 184–7; mix session preparation 153–4, *155*; monitoring

155–61; panning 165–73; signal flow 163–5
mix session preparation: OMF 153; session organization 153–4, *155*
mix stems: anecdotal qualifier 129; audio aggregation process 130; digital audio workstation (DAW) 131; digital systems management 130; dubbing mixers 129; EQ process 132, 143–4; gain structure 131, 138–41; grouping 132, 142–3; information architecture 131, 133–5, **135–8**; monitoring paths 131, 141–2; output formats 132, 144–5; output formats impact 132, 145–8; outside broadcast (OB) facilities 130; picture-mixes 131
mobile phone 304
modern compressors 181
modulation effects: delay 122; filters 123; level 121–2
monitoring: ambisonic systems 161; bass managed systems 158; headphones 160; height channels 159–60; object-based mixing 160–1; paths, mix stems 131, 141–2; speaker placement and calibration 155–6; subwoofers 156–8; surround sound systems 158–9
mono and phantom sources 165
mono and stereo stems 153–4
monophonic reproduction 191
monotonous robot voices 123
mono to split stereo 169
Moodplay web-based social music player 256, *256*, *257*
motifs and melodies, creation 216–17
Motion Picture Sound Editors (MPSE) 197
Moussinac, Léon 333
Moving Picture Expert Group (MPEG) 197
moving sound source 47
MS encoding matrix 168
"MTV-style" filmmaking 337
multiband compression 119, 183
multiband dynamics processing 96–8, *98*, 178, 183
multichannel audio 103, 126; signal processing 103
multichannel matrix encoding 85

multichannel microphones 43
multichannel sound installation 306
multi-dimensional audio (MDA) 196
multidimensional panning effect 166
multimedia content 248
multimono modulating effects 173
multitrack tape recorders 88
Murch, Walter 193, 366, 386, 394
"musical" element 346
musical instrument dynamic processing
    184
musicalization 215–16, 251
musical motion: stability and metaphors of
    motion and force 235–7
musical pitch: intervals and tonal attraction
    230–4; intervals and Western scale
    structures 228–30; and sound design
    228
musical-sonic discourse 336
musical sound 177, 225, 265, 270,
    304, 312
musical tone 228
Music and Effects (M&E): levels 185–6;
    stems 134–5
music and FX stems 135, 145
Music and Sound Search Tool (MuSST)
    252; query results 254
music-as-moving-force (2007) 236
music beds in surround 171
music for CD 188
MUSIC I 316
MUSIC IV 316
Music of Horns and Whistles, The 373
music production with SampleSurfer 272
music sessions 154
music stem 133
music streaming, loudness standards for
    189
music theory for sound designers 227;
    counterpoint and contrast 244–5;
    intervals and tonal attraction 230–4;
    intervals and Western scale structures
    228–30; melodic contour 234–5;
    musical motion, stability and metaphors
    of motion and force 235–7; musical
    pitch and sound design 228; rhythm
    and pulse, rhythm and pitch 237–40;
    sensory consonance and dissonance,

notes and noise 240–2; tonal
    consonance and harmonic function
    242–3
musique concrète 308, 310, 316, 367–8,
    386

"nails on a chalkboard" effect 175
Napier, Frank 373–4
narrative films 327
naturalness 65–6, 70
near-field monitors 109–10
negative parallax 202–3, 205
Neill, John 206
Nelson, Andy 206
nesting 132, 144
Neuhaus, Max 377n3
Neumann U87 69
neutral parallax 202
New Music: New Listening 363
New Soundscape, The (1969) 369
Neyrinck SoundCode 145
NLE see nonlinear editing (NLE)
noise 14, 15–16, 305, 312, 365; floor 163;
    ingress 361; machines 304; pitch and
    15–16
noisemes 296–7
Noises Off: A Handbook of Sound Effects
    373–4
noisy frequency bands 182
nondestructive editing 82
nonlinear editing (NLE) 82, 88
nonverbal communication, ADR 68
normalization 165
notch filter 117
"note-for-note" notation 213
note-noise 284
Noyce, Phillip 329
Nugen Downmixer plugins 145
Nyquist frequency 55

object-based audio system 192
object-based mixing 160–1
object-based panning, MDA 196
objects-interactions-sounds associations 76
objet trouvé 367
octaves 173, 219–20, 229; equivalence
    229; spectrum 116
oeuvre of electroacoustic music 303

off-beats 239
Oliveros, Pauline 377n3, 382, 385, 387
omnidirectional microphones 41, 46, 57
Ondes Martenot 306
online application 260
online audio content 248–9; audio
    commons ecosystem 264–5; cloud-
    based live coding 264; integrating
    audio commons content using web
    APIs 265–9; internet-connected DAWs
    261–3; licensing and attribution 257–9;
    live coding with MIRLC 275–6; live
    collaborative music making with
    playsound 276–8; music production
    with SampleSurfer 272; repurposing
    audio commons content 269; retrieving
    251–7; soundscape composition
    272–5; sound texture generation with
    audiotexture 269–72; transforming
    linear audio production 259–61;
    uploading 249–51; web-based music
    production 263–4
On Sonic Art (1996) 320
Ontology for Media Resources: OWL Web
    Ontology Language 297
Open Media Framework (OMF) 150n5,
    153
openness, sound color 288
optical mode 181
optical sound film 307
Oram, Daphne 314–16, *315*
ORAMICS machine 314–15
oscillating beat pattern 214
Ottosson, Paul N. J. 395
outer ear 21–2, *21*, 160
outer orientation 104
out of phase 18–19, *19*, 43, 123, 330, 337,
    340, 349
output formats, mix stems 132, 144–5,
    145–8
outside broadcast (OB) facilities 130
overdubbing 88
Oxford Dictionary of Music 216

panning 105, 114–16, 115; amplitude
    panning 165–6; combined panning
    techniques 166–7; delay panning 166;
    equalization panning 166; height channels
170; individual instruments 171–2, *172*;
    LCR panning 169; left surround and right
    surround channels 169–70; low-frequency
    effects (LFE) channel 170; mid-side
    (MS) processing 168; mono and phantom
    sources 165; mono to split stereo 169;
    music beds in surround 171; spaciousness
    and envelopment, creation 172–3; stereo
    phase-correlation meter 168; stereo
    sources 167–8; vocals and dialogue 170–1
pan-pots 165–6
parallax 202, *202*
parametric equalizer 116–17
Paramount 132
Parker, Alan 327
Park Road Post Production 206
Parmegiani, Bernard *309*
partials and composite waveforms:
    EQ (equalizer) 15; fundamental 13;
    harmonic oscillator 13; harmonic partials
    13, 14; inharmonic partials *13*, 14; noise
    14, *15–16*; pitch 14, *15–16*; timbre 14;
    transients 14; vibration patterns 12
Part-of hierarchy 296
Pashley, Wayne 207
passing tone 217
pass listening 188
pathic knowledge 380
peaking filter 117
perceiving [Ouïr] 308
perception and interaction 126–7
perceptual [P] sound 250
Perec, Georges 363
perfect cadence 243
performance 71
permissions culture 258
personal radio microphone 68–9
Phantom Power 39
phantom source 115
phase: cancellation 19, *20*, 22, 43, 47;
    panning 115; sound wave 18–19;
    vocoder 124
phasers 173; effect 123
phenomenology 381, 389, 393
Philips Pavilion 306, *306*
phonography 368
physical sound source 1–6; compression
    2; dampening 4; rarefaction *2*; shape and

structure of 5; string, vibration patterns of *3*; textures *4*
Pi (1998) 328, 329, 343–52
picture-mixes 131
Pigrem, J. 264, 272
Pijanowski, Bryan 383
Pinewood 132
pink noise 30, *31*; reference level 156, 158
Pink Panther (1963) 230, *230*
Pirrie, Julien 62, 72–4, 75–6, 78
pitch 14, *15–16*, **29**, 228, *312*; chroma 25, 26; critical bandwidth 27–8; frequency differences **28**; frequency value 27; interval 29; interval ratios **28**; radically divergent frequency differences **27**; rate of oscillation 27; resolution 27; Shepard tone 27; shifting 123–4
pitch category: and tonal hierarchy relationships *233*; and tonal proximity relationships *232*
pitch divisions of piano keyboard 228
pitch/noise ratio 332
pitch stability 228
plagiarism, protection against 257
playback speed 36, 311
play free music improvizations 264
Playsound.space 264, 276
Poème électronique 306
polar patterns *42*; bidirectional 43; cardioid 42; hypercardioid 43; multichannel 43; omnidirectional 41; shotgun microphone 43, *44*
polyrhythmic palimpsest 359
Poly WAV files 132, 144, 151n8
pop mix 154
positional data 199
positional rendering 199, 200
positiveness *(valence)* 256
positive parallax 202
postproduction 82
postsound recordings: sound effect 102–3; voiceovers and overdubs 99–102
preamplifiers 47–8
preliminary synchronization 64
PRESENCES Électronique Festival *324*
primary sound source 44–6
principal component analysis (PCA) 299
processing log files 86–7

production shoot 62, 65
professional tape machines 180
profile 96, 285–7, 316, 361
propagated sound: absorption coefficient 6–7; angle of incidence 10; angle of reflection 10; coherent wavefront 10; diffuser acoustic panels *11*; direct sound 8, *9*; early reflections 8; flat and uneven surfaces *11*; incoherent reflection pattern 10–11; inverse square rule 6, *7*; late reflections 8–9; reverberation 7–10, *10*; specular reflection 10; vibration energy 6
Prores video file 145
Pro Tools 64, 75, 91, 133, 141; 12.8 199; side-chain compressor 139
proximity effect 46
Psycho (1960) 366
psychoacoustics 23; audio tools and technologies development 36–7; duration 35; hearing frequency ranges **25**; loudness 29–34; pitch 27–9; sonic tricks 25–7; timbre 35–6
public address (PA) system 142
pulse *312*
pulse beats 218
pulsing mechanical drone 214
pumping 118
punch-in-and-out technique 88
pupitre d'espace 313

quality of sound 75
quality preservation status 84
quantization bit length 84

Radiophonic Workshop 314–16
random number-based generator 121
rarefaction 2, *2*
rate of oscillation 27
Rate of Tonal Change 332
ratio 181
RDF-based vocabularies 297
readymade 367
realism 102, 110, 113, 125–7, 193
Reaper 270
recorded performance, naturalness 65–6
recorded sounds 12
recording and musical possibilities 304

rectification of recordings 82–3
Red Tails (2012) 206
reduced listening 310
reel-to-reel audio tape recorders 84
reference mix 185–6
reference tone 87
relative sensory consonance/dissonance
    240, *241*
release time 92, 118, 181–2, 184, 218
remixes 259, 264
Renaissance SFX 298
renderer 160–1
repurposing audio commons content 269
re-recording: dialogue 120; mixer
    148, 149
resonance frequencies 125
REST API 265
restoration and noise reduction 96; spectral
    de-noise procedure 96, *97*
retrieving online audio content 251–7
retrograde 218
reverberation 7–10, *10*, 119–20, 167, 173;
    defining 204–5; space and sense 204;
    types 120
reverse reverb 120
re-versioning 145
rhythm 213–16, 227, 306, 330; digital
    audio workstation (DAW) 215;
    linguistic rhythmical patterns 214;
    musicalization 215–16; oscillating beat
    pattern 214; and pulse, rhythm and pitch
    237–40; of pulses 312; quarter notes
    214; 7/8, counting patterns for *215*; slow
    pulsing mechanical drone 214; in sound
    designer's toolkit 213; text to *215*
rhythmical displacement 218
rich sound description 250–1, *251*
Rimsky-Korsakov, N. 237
ring modulation 122
rock ballad 359
Roesch, John 74, 77
Rosam, Ian 133; Calrec Apollo
    Microphone EQ Settings **144**; Calrec
    Apollo Mixing Console Channel **135**–**8**,
    140; HD and UHD outputs **146**
rotating speaker table *314*
Run Lola Run (1998) 328, 329, 335, 336,
    337, 338, 353, 354n2

Russell's arousal/valence (A/V) model 255
Russolo, Luigi 304, *305*
Ruttmann, Walter 307, 368

Saint, The (1997) 329
Sallabank, Alan 132–3
sample-based synthesis 270
sample rate 51–3; difference 51–2, *52*;
    and frequency *53*
SampleSurfer 272
scales 229
Schaeffer, Pierre 306, *309*, 367, 383, 386,
    393; and GRM 307–10
Schafer, R. Murray 317, 363, 369, 373,
    374, 382, 386
schizophonia 386
Schönberg's 12-tone technique 303
Schryer, Claude 363
Schwart, Tony 369
Scorsese, M. 197
Scott, Ridley 197, 327
Scott, Tony 329
segmented silhouette *391*
segmented world *390*
self-contained symbolic unit 348
semantic audio 249, 253, 256, 278
semantic interoperability 297
semantic listening 310
semantic [S] sound 250
semantic technologies 248
semicircular canals 23
semitone *see* minor second
sensitivity and linearity 109–10
sensory consonance and dissonance, notes
    and noise 240–2
sensory dissonance, notes and noise 240–2
sensory imagination 388
sensory sharing 388–92
Sentinel, The (2006) 329
Sexy Beast (2000) 329
sharp 229
sharpness 291
Shaun the Sheep (2007) 72
Shepard tone 25, 27
Shepperton Studios group 132
shifting planes 306
short attack, compression 49–50
short release, compression 50

shotgun microphones 19, 43, *44*, 68–9
shot-to-shot relations 333
shuttling 89
side-chain functionality 119
signal chain/gain staging: digital recording
48; dynamic range compression 49–50;
error 48; headroom 48–9; preamplifiers
47; variables 48
signal flow: auxiliary sends 164; auxiliary
tracks 165; channel faders 164; clip
gain and normalization 165; inserts 164;
master faders 164; trim pots 163
signal processing and mixing 103
signal sensing path 119
signals sent postfader 164
signal-to-noise ratio 46, 50, 53, 64, 163
silences 188
sinc wave oscillation 12
singular value decomposition (SVD) 299
Skywalker Sound 204
Slawson sound color 288
slow attack, compression 49–50
small diaphragm condenser microphones
39–40, *40*
Smalley, Denis 284, 319–20, 322
smallness, sound color 288
snare drum sound 184
Snyder, Z. 201
social media revolution 249
Society for Electroacoustic Music in the
United States (SEAMUS) 323
Society of Motion Picture and Television
Engineers (SMPTE) 196–7
socio-cultural context 103
soft knee, compression 50
sonic diversity 305
sonic mind maps *224*, 224–5
sonic objects 228
sonic relationships **222**
sonic sculptor 212
sonic signals 381
sonic textures 223
sonic tricks 25–7
sonic vibrations 380
sonograms 287
Sony Pictures 132; Imageworks 207
sound 381; classes 296; classification 154;
color 288–91, *290*; design elements
207; design manipulation 143; editors
104; effects and ambiences 102–3,
203; events/effects 373–5; files 82;
object motif 217, 218; objects 217, 284;
physical sound source 1–6; propagated
sound 6–11; range 84; recording 304;
romance 362, 370; signals 294, 362;
source, primary 1, 44–6; space 383;
spectra, annotation 287, *289*; texture
generation with audiotexture 269–72;
textures *4*; tools and techniques to
capture 20; walk 225, 389
SoundCloud 248, 249, 275
Sound Ideas Series (6000, 7000 and
10000) 298
sound-making gesture 319
soundmarks 294, 370
sound-masses 306
SoundMiner 298
sound mixing 103–10, *106–7*
sound ontologies: analytical approaches
294–7; ecological approaches 294; Is-a
hierarchy 296; noisemes 296–7; Part-of
hierarchy 296; phenomenological
approaches 283–6; psychoacoustic
approaches 291–4; single tone 296;
sound source 296; sound source group
296; technical tools and applications
297–9; voice-based approaches
286–91
sound pressure level (SPL) 30–1, 37–8;
meters 30–1, 157, 158
soundscape 102–3, 211, 212–13,
221, 294, 317–19; ecology 383;
etymologically 370
soundscape composition 272–5, 358–9,
372–3; ambient/immersive/atmospheric
soundscape 359–60; defined
369–72; discovery of social 367–9;
everydayness listening exercise 360–7;
sound events/effects 373–5; using
online audio content 272–5
*Soundscape of Modernity, The:
Architectural Acoustics and the Culture
of Listening in America, 1900–1933*
(2002) 372
Sounds from Dangerous Places (2012) 375
Soundsnap 250

soundtracks 329; designing 394; presentation 194–5
sound wave 50; amplitude 11–12; envelope 16–18; frequency 11–12; harmonic and inharmonic partials 13, 13–14; partials and composite waveforms 12–15; phase 18–19; phase cancellation 19, 20; properties 11–12; recorded sounds 12; sine wave oscillation 12, 12; tools and techniques development 20
source-event 297
spaciousness and envelopment, creation 172–3
Spare Time (1939) 368
spatial enhancement 85
spatial sound language 198
speaker equalization settings 156
speaker formats 192, 199–200, 203
speaker layouts: 5.1 and 7.1 formats 195
speaker placements 159; and calibration 155–6
speaker simulation on headphone processors 160
spectral centroid 293
spectral de-noise procedure 96, 97
spectral density 285
spectral energy distribution 292, 292
spectral flux 293
spectral motion 284
spectrograms 6, 276
spectromorphology 322
specular reflection 10
speech: forms 104; intelligibility 363
Spielberg, Steven 150, 230
SPL see sound pressure level (SPL)
Spoerri, Daniel 367
Spring, Katherine 337
Spy Game (2001) 329
stand-alone units 89
Stapleton, Oliver 193
Star Trek: The Motion Picture (1979) 244
steady rhythm 213–14
Stefan Kudelski Nagra III Tape Recorder 367
Steiner, Max 307
stereo acousmatic works 358

stereocilia 23
stereo image processors 172
stereo microphone pair 115
stereo monitoring 155–6
stereo phase-correlation meter 168
Stilwell, Robynn 332
Stockhausen, Karlheinz 311–14; rotating speaker table 314; WDR 311–14
Stöhr, Hannes 328
Stolfi, A. 264, 276, 278
storytelling 62; postproduction voice replacement 62; with sound (see audiovisual scenes)
Strangers in the Night (1966) 230
"stretching" of sound 99–102
Strike (1925) 333
structural function 104
structural tone 216, 216–17
structural tones 217
structure-borne vibrations 41
studio-grade subwoofer 157
sub-bass 173
subharmonic frequencies 191
subjective-state, attributes 297
subwoofers 156–8, 173
surround multichannel reproduction 105
surround sound systems 158–9
sustain 17, 286; sound object 285
suture 193
sweet spot 156, 205
Sync 75
synchronicity 292
synchronization 67; of sound 71
synth-electronica 327
synthesizers and modern tools 306
system noise 160

Tableau Piège 367
tag clouds 252
tags describing a sound 252
"talk box" 123
Talking Rain (1997) 358
Tape Recorder, The 88, 367, 369
target spectrum 119
technical [T] sound 250
tempo 213–14; and intensity 214
Tenney, James 211, 213, 217–18, 220

Terminator, The (1984) 237
text-based queries 249
textures 223, 319–22; sound *4*
Theremin 306
Thom, Randy 194, 206, 394
Thompson, Emily 372
three-layered 22.2 immersive sound format 204
threshold 181; of hearing 31
THX Certified with Dolby Atmos monitoring 150
THX "Deep Note" 228
timbral difference 36
timbral fusion 36
timbral space 291, *293*
timbre 14–15, 35–6
timbre and texture: compositional techniques 222–3
time (reverb and delay) processing 92, *95*
"tinnitus" tone 127
Titanic (2012) (TV series) 70
tonal and noisy aspects of sound 291–2
tonal consonance and harmonic function 242–3
tonal dissonance 240
tonality 219–21
tonal stability 332
tone and space matching, ADR 64
tone quality 228
tonic, designing from 219–21
tonicity 332
tonic note 219–20, 231
touch sensitivity and flying faders 186
Toy Story 3 (2010) 195, 197, 206
track fader on DAWs 164
track management 141
track organization *155*
traditional left–right leveling technique 105
Trainspotting (1996) 328
transfer medium 87
transients 14, *17*
transverse waves 6
Treatise on Harmony (1772) 219
Treecreeper 364
tremolo effect 121
trim pots 163

Truax, Barry 221, 272, 316, *317*, 370, 371, 374
Tuning of the World, The 369
tunnel chase (Fiedel) Reduction *237*
two-channel stereo format 203
Tykwer, Tom 328, 335, 353
typology, sound object 284, 285, *286*

UHDTV sound layers **204**
ultra-high definition (UHD) television 150n1
Unanswered Question 341
"undo" function 91
unidirectional microphones 42
unified time structuring 311
uniform noises 96
unity 118, 163–4
universal loudness standards 188
Unkrich, L. 195, 197
unprocessed mono signal 169
unwanted sound 45, 87, 96
uploading online audio content 249–51
upper midrange frequency band 175
upward compression 118
upward expansion 118

Vancouver Harbour 373
Vancouver Soundscape, The (1973) 372
van Manen, Max 380
Varèse, Edgard 305–6
variation, compositional techniques 217–19
Vaughn, Matthew 329
vector-based amplitude panning 192
verbal categories and listeners 291
Vertov, Dziga 306–7
vibrations 382; energy 6; patterns 12
vinyl records 84
visualization techniques 255
vocalization 99
vocals 332; abstraction 289–90; and dialogue 170–1; performances 63, 71; and soloist compression 183; tone 69
vocoder 123
voice: coder 123; editing 99–102, *101*; EQ 177–8; levels 185; tonal match 68–9
voice-overs 62, 63; and overdubs 99–102

voice postproduction recordings *see*
  automatic dialogue replacement (ADR)
voltage controlled amplifier (VCA) 186
vowel color 291
vowels 287
VU meters 105

Wachowski Siblings 329
Washington, Dinah 341
waveform 82, 308
waveforms: and spectrograms 105
wave patterns 14
Waves Audio LTD 272
web application 260
Web Audio 263
web-based music production 261, 263–4
Wedel, M. 338
Weekend (1930) 368
West, Simon 329
Westerkamp, Hildegard 316, 370
Western cinema 197
West German Radio (WDR) 311
Weta Studios 149–50
Whalen, T 354n10

white noise 30
white noise–based signal 362
whole-step *see* major second
Williams, John 230
Winter Diary 363
Wise, R 244
Wishart, Trevor 320; gesture archetypes
  321
Wochenende 307
Wooller, Rene 332
word-by-word clip gain 185
working balance and panning plan 154
World Series of Sound 298
World Soundscape Project (WSP)
  318, 370
World Wide Web Consortium (W3C) 297
World Wide Web (WWW) ecosystem 248
Wright, Joe 329
WSP Group 318–19, *318*

YouTube 248

Zero Parallax System (ZPS) 202
Zettl, H. 104

Printed and bound by CPI Group (UK) Ltd, Croydon, CR0 4YY

18/10/2024

01776242-0006